Electroactive Polymer Electrochemistry

Part 2: Methods and Applications

Electroactive Polymer Electrochemistry

Part 2: Methods and Applications

Edited by

Michael E. G. Lyons

Trinity College
University of Dublin
Dublin, Ireland

Springer Science+Business Media, LLC

Library of Congress Cataloging-in-Publication Data

On file

DOI 10.1007/978-1-4899-1715-7

Originally published by Plenum Press, New York in 1996
MyCopy version of the original edition 1996

10 9 8 7 6 5 4 3 2 1

Contributors

Philip N. Bartlett • Department of Chemistry, University of Southampton, Highfield, Southampton SO9 5NH, United Kingdom

John F. Cassidy • Chemistry Department, Dublin Institute of Technology, Dublin 8, Ireland

Paul A. Christensen • Chemistry Department, Bedson Building, University of Newcastle upon Tyne, Newcastle upon Tyne NE1 7RU, United Kingdom

John Cooper • Department of Electronics and Electrical Engineering, University of Glasgow, Glasgow G12 8LT, United Kingdom

J. Corish • Department of Chemistry, Trinity College, University of Dublin, Dublin 2, Ireland

Andrew Hamnett • Chemistry Department, Bedson Building, University of Newcastle upon Tyne, Newcastle upon Tyne NE1 7RU, United Kingdom

Simon J. Higgins • Department of Chemistry, Donnan Robert Robinson Laboratories, University of Liverpool, Liverpool L69 3BX, United Kingdom

Emmanuel I. Iwuoha • School of Chemical Sciences, Dublin City University, Dublin 9, Ireland

David Kelly • School of Chemical Sciences, Dublin City University, Dublin 9, Ireland

Dónal Leech • Département de Chimie, Université de Montréal, Montréal, Québec, Canada

D. A. Morton-Blake • Department of Chemistry, Trinity College, University of Dublin, Dublin 2, Ireland

Malcolm R. Smyth • School of Chemical Sciences, Dublin City University, Dublin 9, Ireland

Johannes G. Vos • School of Chemical Sciences, Dublin City University, Dublin 9, Ireland

Preface

The development of "tailormade" electrode surfaces using electroactive polymer films has been one of the most active and exciting areas of electrochemistry over the last 15 years. The properties of these materials have been examined by a wide range of scientists from a variety of perspectives, and now electroactive polymer research is considered to be a reasonably mature area of research endeavor. Much is now understood about the fundamental mechanism of conduction in these materials. A wide range of electrochemical techniques may be used to probe the conductivity processes in these materials, and more recently, a number of *in situ* spectroscopic techniques have been used to further elucidate the structure of these materials. The *in situ* spectroscopies and allied techniques have also been used to obtain correlations between structure and redox activity. The applications found for electroactive polymers are many and varied, and range from thin film amperometric chemical and biological sensors, electrocatalytic systems, drug delivery devices, and advanced battery systems through to molecular electronic devices.

The research literature on electroactive polymers is truly enormous and can daunt even the most hardened researcher. The vast quantity of material reported in the literature can also intimidate beginning graduate students. Hence the present book. The original idea for this book arose as a result of a series of lectures on chemically modified electrodes and electroactive polymers given by the writer to final-year undergraduates at Trinity College Dublin. At the time (1991) the only course material available was the original literature and a few review articles. When faced with such a situation one has to bite the bullet and do something about it. It is hoped that the present book will prove useful for advanced undergraduate/first-year graduate courses in electrochemistry/materials science. The material is written with the demands of graduate students very much in mind, and the individual chapters are written at a level that graduate students will appreciate.

Active research scientists are very busy people, and the editor has been fortunate to have been able to persuade a number of the leading contributors to the development and understanding of electroactive polymer materials to contribute to this project. I thank them for their labors and for the first-rate contributions produced.

A single volume work was originally envisaged. However, as the project proceeded it was realized that the area of electroactive polymer electrochemistry was too wide-ranging to be encompassed within the covers of a single volume. It was therefore decided to produce a three-part work. The first part (Chapters 1 to 4) presents a survey of the fundamental principles underlying the electrochemical behavior of electroactive polymer materials. In that volume the fundamentals of charge percolation and conductivity behavior allied with the membrane properties of electroactive polymer-modified electrodes is discussed. In Part 2 (Chapters 5 to 11) the topics of methodology and applications of electroactive polymer films is addressed. Finally, in Part 3, the important topic of polymer ionics is discussed. This topic is currently of much interest and has led to the invigoration of the classical subject of ionics. At the time of this writing, Part 3 is still in the planning stage.

Part 1 dealing with fundamentals begins with a long chapter by Lyons, outlining both the fundamental mechanisms of charge percolation in electroactive polymer films, and the various techniques which may be used to quantify the kinetics of charge percolation. The various theoretical models used to describe redox conduction are presented in detail. Furthermore, the quantitative aspects of the various techniques used to examine charge percolation are also addressed. In Chapter 2 (Lyons) a number of theoretical models describing electrocatalysis by using electroactive polymer films is presented. The approach here is to show in some detail how one can set up simple models that capture the essence of the electrocatalytic process and subsequently to solve the diffusion/reaction equations to produce approximate analytic expressions for the current response in terms of system specific parameters. In Chapter 3 (Doblhofer and Vorotyntsev) the membrane properties of electroactive polymer films are discussed in careful detail. This is a very important topic, and the authors are leading exponents of the methods discussed. The conduction behavior of redox polymers is well understood in terms of intersite electron hopping. In contrast, the situation for electronically conducting organic polymer materials is not as well resolved. In the final chapter of Part 1, Albery and Mount (Chapter 4) present a lucid discussion of their dual rail transmission line approach that can be used to model the mixed conduction (electronic/ionic) behavior of electroactive polymers. Much of the material presented in this chapter has been only recently developed by the authors, and the material contains many novel insights and ideas.

The emphasis changes in *Electroactive Polymer Electrochemistry, Part 2: Methods and Applications,* where methodology and applications is addressed. The volume begins (Chapter 5) with a contribution by Morton-Blake and Corish. These contributors have been very active in the area of atomistic simulation of matter transport phenomena in solid materials. In the present contribution they describe in a lucid manner the results of their recent work in applying the methodology of atomistic simulation to quantify dopant transport processes in electroactive polymers. The idea of simulation is continued in Chapter 6, where Cassidy carefully describes the application of digital simulation protocols to charge transport in electroactive polymer films. A novel feature of Chapter 6 is the inclusion of detailed FORTRAN computer codes for a number of specific problems often encountered in practice.

The contribution by Higgins, Christensen, and Hamnett (Chapter 7) is very timely and discusses *in situ* ellipsometry and FTIR spectroscopy applied to electroactive polymer-modified electrodes. These coauthors have been particularly active in the latter area, and their overview, although brief, conveys many original insights.

In Chapter 8, coauthored by Kelly and Vos, the electrochemical behavior of osmium and ruthenium poly(pyridyl) redox polymers is discussed in some detail. Vos has made significant contributions in this area. This chapter ties in well with the more general discussion presented by Lyons in Chapters 1 and 2, in that many of theoretical concepts addressed in the latter chapters are again discussed by Kelly and Vos with specific reference to redox-active metallopolymer materials.

In Chapter 9, Bartlett and Cooper discuss the applications of electroactive polymers in bioelectrochemistry and bioelectronics. This is a very exciting and rapidly developing field, and it is proper that the volume includes this topic. Electroactive polymer materials will feature strongly in future developments in this area. Again Bartlett and Cooper have made major contributions in this field.

In Chapter 10, Leech discusses the analytical applications of polymer-modified electrodes. In Chapter 11, the analytical theme is continued and again, Iwuoha and Smyth survey the applications of electroactive polymers in electroanalytical chemistry. In this case the important area of biosensors is examined. The areas described in these chapters have attracted significant research activity in recent years, and the material presented in these chapters is timely, and is written by well known and experienced practitioners in the field.

One is aware that one cannot be all things to all men, and the choice of material presented in this book has been largely dictated by the interests and expertise of the editor. The major emphasis has been on polymer-modified electrodes. The related topic of polymer ionics has not been addressed at all. This is a rapidly developing and important research area. One is constrained by length limitations when producing a book of this type, and a hard decision had to be made to omit some topics. This is unfortunate, but such is life!

The editor is grateful to the senior editor of Plenum Press, Ms. Amelia McNamara, for her patience in waiting for the book when deadlines were long overdue. One begins a project such as this with high hopes and over-optimistic expectations with respect to the length of time required to complete the project. One quickly learns that the task of an editor is onerous and time consuming, and that events conspire to thwart smooth progress with increasing frequency as the deadline approaches. Perhaps there is a new physical law to be discovered here!

I finally would like to acknowledge the help and support of my wife, Catherine, and my children, Michelle, Edward, and Rebecca. Many precious hours were stolen from them during this project. Daddy has now stopped working at the computer screen, for the time being at least. I now hope to resume those long walks along Bray Head with the family, now that this work has finally come to fruition.

Michael Lyons

Trinity College
Dublin

Contents

Chapter 5

Atomistic Simulation Investigations of Electroactive Polymers

D. A. Morton-Blake and J. Corish

Chapter 6

Numerical Simulation Methods Applied to Electroactive Polymer Films

John F. Cassidy

Chapter 7

In Situ Ellipsometry and FTIR Spectroscopy Applied to Electroactive Polymer-Modified Electrodes

Simon J. Higgins, Paul A. Christensen, and Andrew Hamnett

Chapter 8

Osmium and Ruthenium Poly(pyridyl) Redox Polymers as Electrode Coatings

David Kelly and Johannes G. Vos

Chapter 9

Applications of Electroactive Polymers in Bioelectrochemistry and Bioelectronics

Philip N. Bartlett and John Cooper

Chapter 10

Analytical Applications of Polymer-Modified Electrodes

Dónal Leech

Chapter 11

Polymer-Based Amperometric Biosensors

Emmanuel I. Iwuoha and Malcolm R. Smyth

Electroactive Polymer Electrochemistry

Part 2: Methods and Applications

5

Atomistic Simulation Investigations of Electroactive Polymers

D. A. Morton-Blake and J. Corish

1. INTRODUCTION

1.1. The Theoretical Approach

A theoretical treatment of a chemical system, i.e., an atom or a molecule, ideally implies an investigation that relies on established physical principles, and calculation rather than experiment, to furnish the properties of interest. Such a procedure is valuable in (1) Computing a property inaccessible to measurement, since the ensuing provision of a structural description of the system on an atomic level may then be used to interpret quantities measured on the bulk sample (2) Testing underlying theory, to provide an understanding of the behavior of the system and an ability to predict other properties.

The best nonempirical calculations, which ideally involve no assumptions nor approximations, and are thus described as *ab initio*, should therefore be capable of achieving good accuracy provided their underlying theory is complete. But although over the past decade, the advent of powerful computing hardware has greatly extended the size of molecules to which good quality *ab initio* quantum chemical methods can be successfully applied, demands made by many kinds of real chemical systems on computing time and machine capacity have not as yet been met. However when applied to small and medium-sized isolated (gas-phase) molecules, these methods are now predicting energies, geometries, and many molecular properties with rapidly improving accuracies.[1,2]

D. A. Morton-Blake and J. Corish • Department of Chemistry, Trinity College, University of Dublin, Dublin 2, Ireland.

Electroactive Polymer Electrochemistry, Part 2: Methods and Applications, edited by Michael E. G. Lyons. Plenum Press, New York, 1996.

In other quantum chemical methods, the rigidity of the *ab initio* philosophy is relaxed in a trade-off for the ability to treat larger chemical systems, consisting of tens even hundreds of atoms. This is achieved by replacing the very large number of electronic integrals, which are calculated explicitly in *ab initio* methods and responsible for the strain imposed on computing resources, by parameters derived from experimental quantities. These procedures constitute the semiempirical methods, which when used with care, may predict good molecular geometries and relative energies.[3–5]

The limiting factor of the application of quantum chemistry is not necessarily decided by the size of the system. *Ab initio* and other quantum chemical calculations have been performed even on polymers[6,7] and crystal lattices[8] by exploiting their one- and three-dimensional translational periodicities. However the requirement of space-group symmetry by these methods is obviously restrictive. For example, the absence of long-range order in an amorphous solid or liquid at present precludes the application of rigorous methods in investigating such systems or in studying the special characteristics of a polymer arising from its chain-conformational freedom. The treatment of these systems would require the introduction of severe approximations.

1.2. Atomistic Simulation

When a chemical system defined by a set of atoms in space is considered in a simulation method, no explicit account is taken of the electrons and nuclei, and the description of the system is classical, not quantum mechanical. Each atom is represented without an internal structure by its position, and it is specified by its charge and mass. Associated with the atoms are potential energy functions, which describe their interactions with all other atoms in the system; these are generally referred to as potentials. The complete set of potentials constitutes the force field of the system, and knowledge of it enables energy associated with the atoms to be calculated. The subsequent modification of the geometry and particularly for polymers, whose intramolecular torsions are often facile, the conformation of the system to minimize its calculated energy is, in terms of computing time, the lengthiest part of a simulation calculation.

Atomistic simulation of ionic lattices has been widely conducted to investigate matter transport in systems where charged atoms are associated with sites that are effectively stabilized by the strong electric fields of neighboring ions. Parameters associated with transport in such systems have been calculated, and these agree with experimental kinetics or diffusion studies.[9] For polymer systems, the quest for the same accord between theory and experiment is hampered by the heterogeneous nature of the samples. Despite crystallinity levels of 70–80% in some conductive polymers, e.g., polyacetylene[10] diffractograms have been the subject of conflicting interpretations, and there is some evidence that this may be caused by a distribution of various crystalline regions with different lattice structures throughout the sample.[11,12] This and the presence of large amorphous regions in currently available polymer samples result in their description as microcrystalline or paracrystalline. Since the quantities provided by experiment are usually associated with the sample as a whole, including contributions from the (sometimes many) crystalline and amorphous regions, the data may be difficult to interpret by describing events on an atomic scale. This is where atomistic simulation, for which methods are available for application to both the crystalline and amorphous states, can provide descriptions that may hitherto be unavailable from experiment.

Simulations may be classified as static and dynamic. In a static simulation no explicit account is taken of thermal motions in the system, which is therefore treated as if it were at a temperature of absolute zero. A molecular dynamics simulation, on the other hand, requires the specification of a temperature, which defines the kinetic energy to be distributed between the available degrees of freedom. By solving a set of classical equations of motion, such thermally induced reorientation phenomena as the vibrations, rotations, and translations of the system may be described.[13] The two approaches have their advantages, which become clear in the following sections.

The small size of the crystalline regions in typical polymer samples means that the ordered arrangement of atoms is unlikely to extend very far on a macroscopic scale. But although three-dimensional periodicity is not a prerequisite for electronic conductivity,[14] nevertheless in many polymer systems, positive correlations have been found between sample crystallinity and conductivity.[15] Moreover if as is widely accepted, the electronic conductivity of doped polymers involves electron transport along the chains by an energy band mechanism,[6,7] an interruption of the translational order would constitute traps hindering the transport of electronic charge carriers. We would therefore expect the sample to contain regions in which polymer chains and dopant species exhibit translational order. The (micro)crystalline regions may therefore be assumed to play an important role in transporting dopant ions through the polymer material. Although the crystalline regions are clearly not the only ones involved in the ion transport (the amorphous phase must also be discussed), the inaccessibility of detailed structural knowledge about these regions means that simulations cannot yield comparable degrees of quantitative information.

Solid-state systems of particular interest in this book are conductive polymers with the ability to occlude dopants entering the bulk of the polymer sample, thus conferring on it its special electrical properties.[16] Some of these properties are a consequence of the mobility of the dopant ions in the host polymer material, and these properties are responsible for such technological applications as battery electrodes,[17,18] ion gates,[19] and electrochromic devices,[20,21] which depend on a field-induced oxidation of the polymer specified by its doping level. Various diffraction methods[22–26] and tunnelling electron microscopy[27–35] reveal that these materials usually consist of a mixture of microcrytalline and amorphous phases, diffraction of the former producing peaks, while the latter result in diffuse scattering. In the crystalline regions diffraction studies reveal a regularity defined by parallel chains, a short chain repeat segment and for polymers with planar backbones, a well-defined setting angle, i.e., the angle between the plane and a chosen crystallographic axis. In the amorphous phases the absence of regularity may be due to a distribution of various chain conformations and/or chain entanglements.

1.3. Calculating Energy

Both static and dynamic methods rely on a proposed atomistic structure and a force field in which interactions are classified as follows.

1.3.1. Electrostatic Energy

Each distinct pair of atoms bearing electrostatic charges q_i and q_j with separation r contributes an energy term given by Coulomb's law

$$\mathrm{V}(r) = \frac{q_i q_j}{4\pi\varepsilon_0 r} \tag{1}$$

When the atomic charges $\{q_i\}$ required in Eq. (1) are not decided by symmetry, they are usually obtained by a quantum chemical calculation if necessary on a model molecular system.

1.3.2. Bond Length Distortion

The energy to distort a bond from its equilibrium length r_0 to a final length r can be specified by a harmonic function

$$\mathrm{V}(r) = \tfrac{1}{2}k(r - r_0)^2 \tag{2}$$

where k is the stretching force constant, or by a Morse function

$$\mathrm{V}(r) = A[1 - \mathrm{e}^{-B(r - r_0)}]^2 - A \tag{3}$$

Subtracting the bond energy A from the first term in Eq. (3) defines an energy scale that has as its origin the energy of the dissociated bond ($r = \infty$). At $r = r_0$ the bond energy contribution is then $-A$. The parameter B is defined in terms of the stretching force constant k and bond energy A as

$$B = \sqrt{k/2A}. \tag{4}$$

1.3.3. Bond Angle Deformation

The energy required to bend the angle between two bonds from its equilibrium value θ_0 to a final value θ may be described most simply by a harmonic function with bending force constant k_θ

$$V(r) = \tfrac{1}{2}\, k_\theta(\theta - \theta_0)^2 \tag{5}$$

1.3.4. Torsions about Bonds

The existence of several conformational states associated especially with organic molecules indicates the necessity of considering the energy variation incurred by twisting two portions of the system around a bond. A particularly simple torsional potential is

$$V(\phi) = \tfrac{1}{2}\, v_0[1 - \cos(n\phi)], \tag{6}$$

which exhibits n identical sinusoidal energy barriers of height v_0 in a complete rotation of ϕ. A more flexible function can be achieved by the addition of other angular terms to the right-hand side of Eq. (6).

1.3.5. Nonbonded Interactions

If the chemical system consists of an assembly of molecules or even a single flexible molecule possessing stable conformations, interactions must be considered between atoms that are not directly bonded. These model the van der Waals and dispersion phenomena that are responsible for the cohesion of molecular crystals.[36] Although usually weaker than other force field components, their summed effect is often critical in deciding the configuration of such systems.

In Section 4.2 we describe how nonbonded potentials are derived for polymer systems, but for now we list some of the isotropic functions used for their representation. Of these, two common forms are the Buckingham potential

$$V(r) = Ae^{-r/\rho} - Br^{-6} \tag{7}$$

and the Lennard-Jones potential,

$$V(r) = Ar^{-12} - Br^{-6} \tag{8}$$

The first term in Eqs. (7) and (8) increases steeply with diminishing distance r between the atom pair, reflecting the strong repulsion between the atoms at a short distance. In the Buckingham potential [Eq. (7)], the curvature of the potential in this region depends on the 'hardness parameter' ρ. The second term in Eqs. (7) and (8) represents several types of long-range attraction between atoms, sometimes

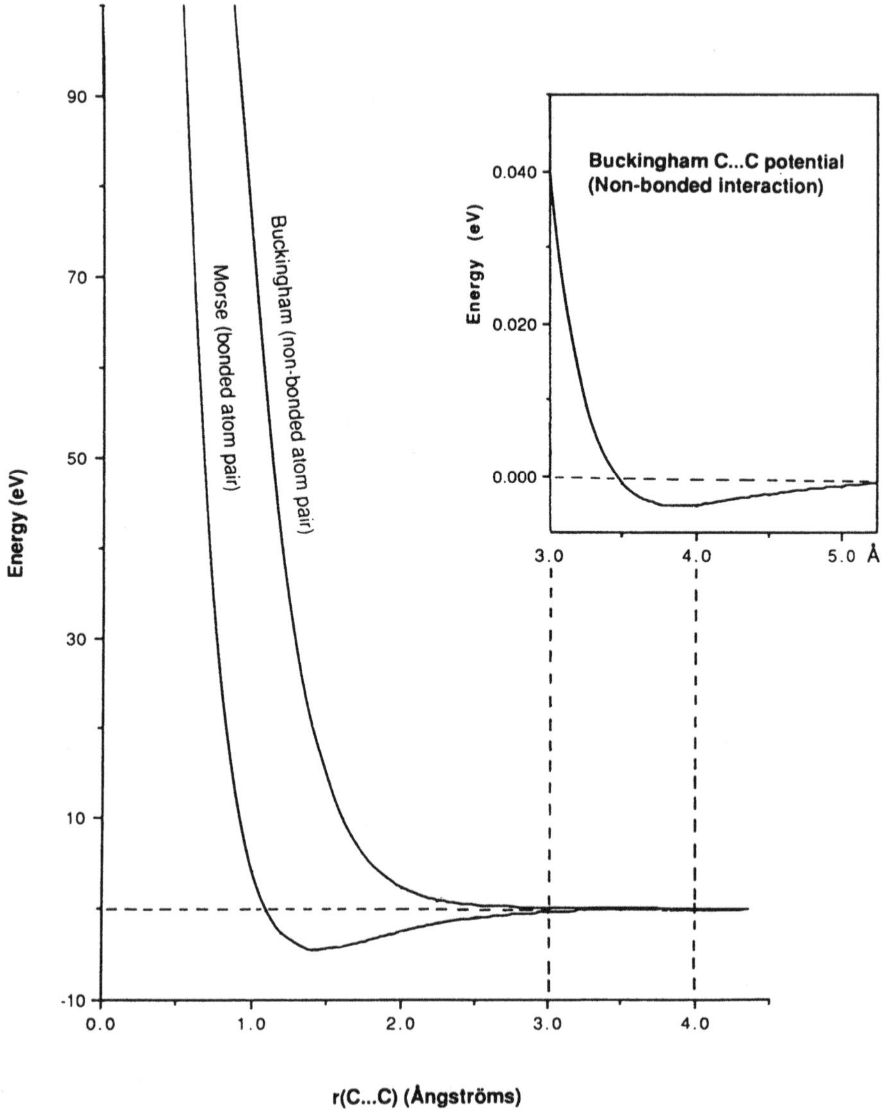

FIGURE 5.1. Comparison of two atom-pair potentials for C...C interactions. The Morse potential defined in Eq. (3) describes atoms joined by a covalent bond, while the Buckingham potential in Eq. (7) accounts for nonbonded interactions between a pair of carbon atoms. The much shallower energy well of the Buckingham potential is not visible when drawn to the same scale as the Morse, but it is revealed in the larger scale inset figure.

called the van der Waals interaction. The r^{-6} form arises from the leading terms in analytic expansions of dispersion interactions between nonbonded atoms and also of certain multipole moment interactions.[36]

Another way of representing atom-pair interactions, which lends itself particularly well to computational methods is the spline potential. A least-squares calculation fits various accurately known sets of numerical interaction energies at certain interatomic regions (perhaps from various theoretical and/or empirical investigations) into a functional form under conditions of matching first and second derivatives, so that a potential and its derivatives may be invoked at an arbitrary atom separation.

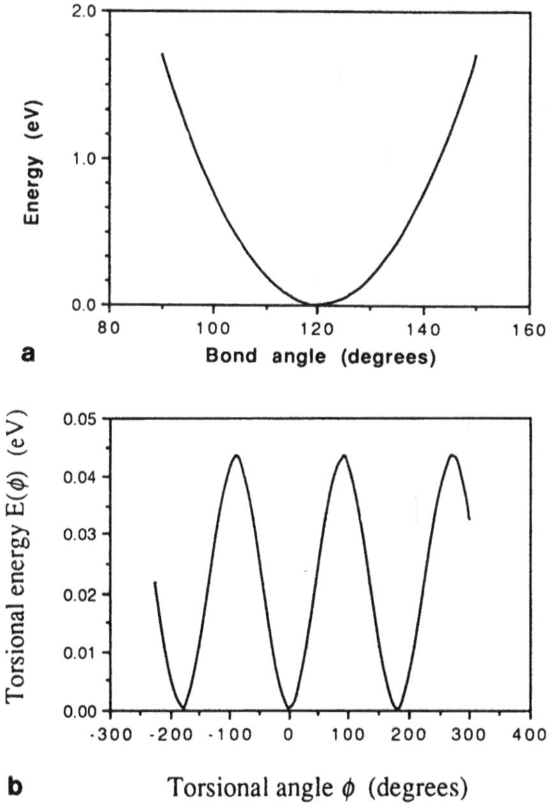

FIGURE 5.2. (a) Energy response predicted by the harmonic deformation equation [Eq. (5)] to a deformation of the bond angle. Data relate to three sp^2 carbon atoms making a natural bond angle of 120° at the central atom. (b) Torsional energy variation with rotation from the planar conformation ($\phi = 0°$) around two sp^2 carbon atoms C_1-C_2 in a conjugated molecule at the four-atom system A–C_1–C_2–B. Contribution is made by the π component of the molecule, and it does not include the steric interaction of A and B, which is normally provided by the nonbonding A...B potential.

1.3.6. Atom Polarization Potential

Various methods have been proposed for modeling the polarization of the charge distributions in an atom of the simulated system by the electric field produced by the charges on other atoms. In one such method, it is assumed that the modeled atom consists of a positively charged *core* coupled to a negative *shell* by a harmonic spring constant.[37] The core and shell charges, which sum to the net charge on the atom, and the spring constant are assigned by comparison with atomic polarizabilities.

To compare the energy changes accompanying various kinds of inter- and intramolecular distortion in carbon-based molecular systems, consider the set of typical potential functions plotted in Figs. 5.1 and 5.2.

When considering the bond-stretching, bond-angle-deformation, torsional, etc., potentials discussed here, it is important not to have the simulation program waste valuable computing time unnecessarily identifying atom pairs, triads, or quadruplets outside valid ranges for these potentials. Declaring appropriate cutoff distances for each potential ensures that only groups of atoms within the ranges of specified geometries are included.

1.4. Relaxing an Atomistic Configuration

From the following quantities:

Position vector that comprises $3N$ components specifying the spacial positions of the N atoms, and therefore the structure, of the system

Force field discussed in Section 1.2

a simulation program calculates the numerical value of the energy appropriate to the position vector (i.e., the energy of the supplied structure). The structure is then relaxed, or geometry optimized by adjusting the molecular configuration to minimize the energy of the system. Various relaxation procedures are appropriate in different circumstances. The method of steepest descents is best when the initial configuration is in a well-defined energy well, but not near the minimum. In this method the first derivative of the energy $\mathrm{V}(\mathbf{r})$ with respect to each of the $3N$ atomic coordinates x_i at a point in configuration space $\mathbf{x}$ is compounded into a gradient vector $\mathbf{g}$, with components defined by $g_i = \partial \mathrm{V}/\partial x_i$. The $\mathbf{x}$ vector is then modified as $\mathbf{x} - \lambda \mathbf{g}$, and the procedure is repeated with successively smaller steps λ until the energy minimum is attained ($\mathbf{g} = 0$).

Since convergence by steepest descents can be rather slow in the final stages of relaxation, the search for the minimum at this point can be switched to the Newton-Raphson method. In this procedure the second derivative at $\mathbf{x}$ is also calculated and compounded into a Hessian matrix $\mathbf{H}$, whose elements are

$H_{ij} = \partial^2 V/\partial x_i \partial x_j$ (i and j run from 1 to $3N$). The $\mathbf{H}_k$ is the curvature of the potential energy surface at the point $\mathbf{x}_k$ in configuration space, which with the slope vector $\mathbf{g}_k$, allows the energy V at a slightly displaced point $\mathbf{x}_{k'}$ to be expressed as a Taylor expansion

$$V(\mathbf{x}_{k'}) = V(\mathbf{x}_k) + \mathbf{g}_k^T \cdot (\mathbf{x}_{k'} - \mathbf{x}_k) + \tfrac{1}{2} (\mathbf{x}_{k'} - \mathbf{x}_k)^T \mathbf{H}_k (\mathbf{x}_{k'} - \mathbf{x}_k) \tag{9}$$

where the transpose of a matrix $\mathbf{A}$ is written $\mathbf{A}^T$. If the energies of the two configurations $\mathbf{x}_{k'}$ and $\mathbf{x}_k$ are nearly equal, so that they are near a minimum on the energy surface, then when the condition

$$\frac{d}{d\mathbf{x}_{k'}} V(\mathbf{x}_{k'}) \approx 0 \quad \text{(zero vector)} \tag{10}$$

as imposed on Eq. (9), we obtain

$$\mathbf{g}_k + \mathbf{H}_k \cdot (\mathbf{x}_{k'} - \mathbf{x}_k) \approx 0 \tag{11}$$

Note that using Eq. (10) does not eliminate the gradient vector $\mathbf{g}_k$, which provides the direction of the required displacement along the energy surface toward the minimum. An improved estimate of the configuration of the system is then

$$\mathbf{x}_{k'} \approx \mathbf{x}_k - \mathbf{H}_k^{-1} \mathbf{g}_k \tag{12}$$

The procedure is repeated iteratively until $\mathbf{g}_k$ is smaller than a preset discriminant, i.e., until an energy minimum is located.

Strictly speaking Eq. (9) is valid only in the region of a harmonic potential well. In other cases, particularly those associated with asymmetric minima, anharmonic terms are required. In systems containing shallow wells, showing high flexibility or fluctionality, convergence of the procedure described may be difficult, and other approaches to optimizing the configuration must be invoked. If the potential surface contains more than one minimum, the desired global minimum is not necessarily located by a single calculation, which stops after encountering the first (relative) minimum-energy lattice. The global minimum must be selected from the relative minima obtained by repeating the relaxation procedure with several initial configurations.

Using methods similar to those described here, materials in the gaseous, liquid, solid, and nematic states have been successfully simulated. The application of the simulation method to isolated molecules (modeling those in the vapor state) should be obvious from our discussion, and it forms a basis for the molecular mechanics series of computer codes, such as MM2,[38] and its extension to π-conjugated systems (MMP2)[39] and for the energy minimization sections of the CHARMM package.[40]

1.5. Monte Carlo Methods

Only when polymer chains form part of a crystal lattice can a precise structure be specified for the sample. In such a case the system can be defined by copying the contents of the unit cell along directions parallel to the lattice vectors. But polymers by their nature depart drastically from this ideal. Motions in noncrystalline media are subject to so many random or indeterminable factors that they cannot be treated by methods that require specifying a precise structure. Nevertheless we include a brief account of how randomness can be introduced into the simulation of polymer systems, since results have been shown to produce useful descriptions of noncrystalline polymer states. For more details about the methods, we refer the reader to books and specialist review articles[41–44] and descriptions of polymer simulation codes.[45,46]

Chains in the amorphous regions of a solid sample lack regular conformations, and since relaxation studies show that they are often dynamically unstable, a precise knowledge of the structure would, in addition to the complete set of conformations, also require time dependence. Since such information is inaccessible, polymer chains or assemblies of chains are generated that possess a known *statistical* distribution of conformations. One way that Monte Carlo methods do this is by representing a polymer chain as a set of beads separated by rigid or elastic rods, which are allowed to explore configuration space. In a coarse-grained representation each bead represents a polyatomic chain segment rather than a single atom. Alternatively by defining regular lattice points in coordinate space, the beads are placed at various nearest neighbor pairs, in a way that is random but subject to a set of structure-defining conditions.

A random-number-based algorithm is then applied in a sequence of steps of a Markov chain, attempting at each step to change the conformation of one section of the chain. The algorithm contains conditions with which to test the feasibility of the new structure at each step. In the simplest test, the move is rejected if it has placed two atoms on the same site. Alternatively if the energy change associated with the move is δU, calculated by using a force field as described in Section 1.3, then the move is accepted if the factor $e^{-\delta U/kT}$ exceeds a random number drawn from the range 0–1, otherwise it is rejected. Similar algorithms can be formulated that result in the growth or decomposition of chains, producing a molecular weight distribution.

A static assembly of chains resulting from a selection of random walks through conformation space is taken to constitute a time-averaged structure. Since Monte Carlo methods are quite demanding on computing time, simulating a highly complex polymer system is beyond the power of current methods, and simulated chains have to be much shorter than those of a true polymer. Despite these caveats the method can still provide a useful description of a polymer sample and permit the evaluation of such structural parameters as the mean square end-to-end distance or radius of gyration of the polymer $[M]_n$ in solution or melt and the glass transition temperature.[47] The latter temperate T_g is that heralding the onset of the glassy state when the fused polymer is supercooled below its melting point. The variations of these quantities with the degree of polymerization n at different temperatures are in accord with predictions from the classical statistical theories of polymers and their solutions.[48]

2. THE CRYSTAL LATTICE

In our description of simulation methods for conductive polymers, we pay particular attention to the crystal lattice. This is not only because it constitutes the only regions of the sample in which the (static) structure can be specified precisely, but also because these regions probably play a decisive role in the electrical phenomena characterizing conductive polymers.

2.1. Lattice Simulation

We describe an atomistic method based on the classic Born model[49] to simulate the lattice structure and its relaxation to a minimum energy configuration. Although the method was originally developed for ionic lattices, in which the dominating contribution to the energy is the Madelung term, derived from Coulombic interactions, the formalism can also be used for lattices consisting of predominantly covalent components, which possess smaller atomic charges.[50] The total energy is divided into valence and nonbonding contributions, which constitute the *short-range* interactions, and the summed Coulombic terms, which are *long-range*.

The procedures just described offer various approaches for simulating the transport of a particle in a crystalline medium. We describe static lattice methods in this section; molecular dynamics, which requires some additional specialized theory, is discussed in Section 3.

2.2. Short-Range Energy

A perfect crystal is completely defined by the contents of its unit cell. This is a region of the material that when repeated indefinitely in three dimensions, generates an infinite lattice. The *lattice energy*, E_{lat} is calculated by summing the interactions described in Section 1.2 for all atoms in the unit cell interacting with each other (in distinct pairs, triads, etc.) and with atoms in other parts of the lattice. The summed interactions between the cell and the rest of the lattice are then halved so that the total energy of a portion of lattice consisting of n cells is $n \times E_{lat}$. Failure to do this would include multiple accumulations of the same interactions.

Contributions from nonbonded interactions between atoms in the unit cell and those in other cells diminish rapidly with distance, for example, by r^{-6} for the Buckingham and Lennard–Jones potentials. However because of the large number of contributing atom pairs involving distant cells, it is not always practicable to seek convergence to a preset energy tolerance, and it is customary to impose a potential cutoff at about 10–15 Å to reduce computing times. Although interactions at larger distances beyond the cutoff may still make small negative contributions to E_{lat}, they are normally found to amount to small energy shifts that do not invalidate the relaxation procedures or the comparisons of stabilities. With the exception of the Coulomb energy (Section 2.3), remaining potential functions discussed in Section 1.2 involve small groups of covalently bonded atoms; thus convergence with respect to distance does not cause difficulty for these interactions.

2.3. Madelung Energy

The other interaction is the Coulomb potential, which is obtained from Eq. (1) by summing such contributions over all charged atom sites in the lattice. Decaying only as r^{-1}, the Coulomb terms decrease considerably more slowly with increasing distance than those described by the other potentials. The accumulation of slowly diminishing terms of alternating signs in E_{lat} as successively more distant lattice regions are considered may result in convergence problems.

Quantum chemical calculations show that as a member of the class of organic molecules, most polymers, even those containing oxygen, sulfur, and nitrogen atoms, possess atomic charges of rather small magnitudes. Consequently the Coulomb contributions to the lattice energy are usually slight enough to warrant the imposition of a simple potential cutoff, as used for the nonbonded contribution. On the other hand we are also concerned with polymers rendered conducting by the presence of *dopants*, which undergo electron charge transfer with the host polymer, resulting in bare ions, such as Li^+, K^+, ClO_4^-, AsF_6^-, etc., and probably in increased charges on atoms of the polymer. Precautions must therefore be taken to ensure that the calculated lattice energy is not sensitive to the cutoff distance used in the calculation.

For a regular lattice this can be achieved for the total Coulombic energy

$$E_M = \tfrac{1}{2} \sum_i \sum_j{}' \frac{q_i q_j}{4\pi\varepsilon_0 r_{ij}} \tag{13}$$

by performing an Ewald transformation of the right side of Eq. (13). The resulting expression consists of one summation series of exponentially decaying periodic terms in reciprocal space and a second strongly damped series in real space.[51] Both summations are convergent, and the result yields the Madelung energy E_M. This quantity is added to contributions from the short-range energy terms to give the total lattice energy.

2.4. Simulation Requirements

Calculating lattice energy as discussed in the forgoing sections is embodied in, for example, the perfect lattice section of the lattice-defect program CASCADE[52] and in the codes of the CERIUS[45] and BIOSYM[46] packages.* It should be clear that prerequisites for a successful application of these methods, in decreasing order of critical importance, are (1) reliable nonbonding and bonding atomistic potentials, (2) an accurate unit cell structure, and (3) a set of atomic charges, calculated by a good quantum chemical method. Of these requirements, a high-quality set of nonbonding potentials in (1) is essential and unfortunately the most difficult to obtain. Although using energy minimization methods to relax the lattice to a stable structure might be thought to reduce the critical status of (2), polymer chains often exhibit several structures of similar energies (polymorphism). As a result experimental data on the most stable structure are valuable. Interactions in doped polymer systems are usually dominated by nonbonded rather than coulombic interactions, and experience shows that structures and energy terms are often rather insensitive to the set of atomic charges required in (3).

We now describe two methods of static lattice simulation used to investigate the migration of a dopant.

2.5. Calculating Migration Energies

2.5.1. Supercell Method

Consider the motion of a large number of dopant ions in a crystalline region of a doped polymer. If this set of ions consists of those defined by each unit cell of the lattice, energy changes can be followed by subjecting the specified ion (and by implication its image in each unit cell) to a series of displacements and at each step applying the static lattice simulation method. Such a migration would imply a highly concerted motion in the dopants.

A less concerted dopant motion could be considered if the same displacement were not imposed on each translationally equivalent ion in the original lattice. By defining a supercell consisting of a sufficiently large assembly of elementary unit cells, the response of the lattice to the displacement of a single ion could in principle be simulated. However the usefulness of this procedure is limited by the demand that the supercell would make on computer storage requirements: This cell might need to be excessively large to achieve the desired independence of the selected ion.

**Note added in proof:* A powerful atomistic lattice simulation program GULP has recently been made available. Details may be obtained on application to Julian Gale, Dept. of Chemistry, Imperial College, South Kensington, London SW7 2AY, UK.

Since the supercell method relies on theory already discussed in forgoing sections, it need not be discussed further. It is referred to in an application to a lattice migration investigation in Section 5.3.

2.5.2. Defective Lattice Method

A local lattice defect is a configuration of atoms that interrupts the periodicity of the lattice, although within a few tens of Ångströms the periodicity may sufficiently recover for the effect of the defect on the rest of the lattice to be imperceptible. Such a defect might result from removing an atom from the lattice, leaving a vacancy, or from introducing an atom from outside the lattice to produce an interstitial defect. These fundamental defects can be used to define a general local defect as a configuration of vacancies and interstitials. For example a lattice defect consisting of an atom that has been displaced from its normal site can be regarded as the superposition of one lattice vacancy and one interstitial. Many ions are known to diffuse through a lattice by defect mechanisms in which the defect contains a net vacancy. It is therefore possible to simulate a defective lattice by methods similar to those described for a perfect lattice and thus to investigate ion migration through the lattice.

2.5.3. Relaxing a Lattice around a Defect

We outline the procedure for calculating the energy associated with a local defect, which is based on a method originally developed by investigating defects in ionic lattices and usually referred to as the two-region strategy.[9,53] In this model the lattice is divided into two regions (Fig. 5.3). Region I is usually a sphere with the defect at its center, and it is large enough to contain a few hundred atoms. Within this sphere all interactions are evaluated explicitly, and atoms are allowed to relax into their minimum-energy configurations, just as described for the perfect lattice. In Region II, which is the rest of the lattice, the relaxation is calculated by the Mott–Littleton approximation,[54] which is outlined in the sections that follow.

The defect energy can be written symbolically as

$$E^d = E^d{}_I(\mathbf{x}, \mathbf{x}') + E^d{}_{I,II}(\mathbf{x}', \zeta) + E^d{}_{II}(\xi, \xi + \zeta) \tag{14}$$

where E^d_I and E^d_{II} are, respectively, the defect energies in Regions I and II in isolation, and $E^d{}_{I,II}$ represents the interaction between Regions I and II. Vectors $\mathbf{x}$ and $\mathbf{x}'$, respectively, contain the coordinates of the perfect-lattice and defect-lattice atoms in Region I, while ξ and $\xi + \zeta$ contain those in Region II. Since interactions in Region I are considered explicitly the calculation of the defect energy $E^d{}_I(\mathbf{x}, \mathbf{x}')$ is straightforward. Assuming the defect energy in Region II depends quadratically on the *displacements* ζ of the atoms in the region, the ξ coordinates are not required, and we can write

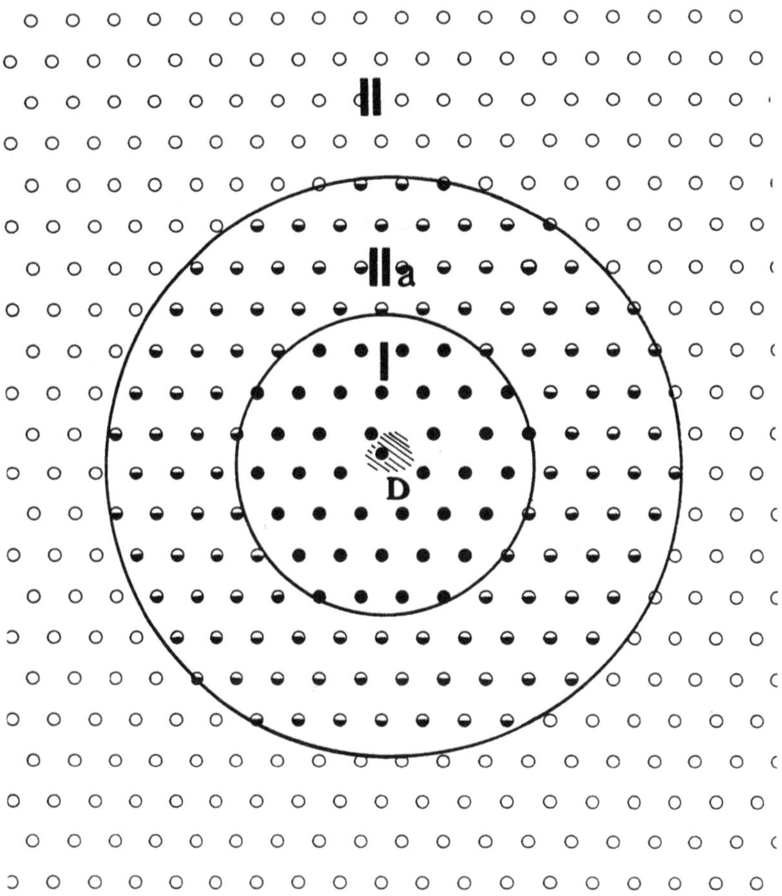

FIGURE 5.3. The model used in the two-region calculation of defect energies. In Region I, which contain a few hundred atoms forming the immediate environment of the defect D, interactions are evaluated explicitly. Between this and the outer Region II, whose contributions are treated by bulk dielectric methods, lies the interface Region IIa.

$$E^d_{II}(\zeta) = \tfrac{1}{2}\,\zeta \cdot \mathbf{A} \cdot \zeta \tag{15}$$

Equilibrium considerations allow Eq. (15) to be expressed as

$$E^d_{II}(\zeta) = \left(\frac{\partial E^d{}_{I,II}}{\partial \zeta}\right) \cdot \zeta \tag{16}$$

where the displacement vector ζ is calculated from the charge distribution in Region I resulting from the defect, which is now represented by an effective point charge at the center of the Regions. Coupling the charge with the Region II atoms is approximated using dielectric theory. The interaction term $E^d_{II}(\zeta)$ and its derivative are calculated in an inner part of Region II (the interface Region IIa; see Fig. 5.3), for which interactions are considered between atoms in this subregion and those in Region I.

The radii of the regions are normally increased until they have no further effect on E^d.

2.5.4. Static Simulation of Migration in a Lattice

Unlike the case in liquids and gases, where atoms and molecules are continuously moving, in a solid the motion is usually much more restricted. However weakly the migrant species is bound to neighboring atoms, it cannot readily move to a new position in the solid unless there is a space there that it can occupy. If this should be the case, the position vacated by the migrant species becomes available for a similar jump by a second species. When diffusion through a crystal lattice occurs by such a series of migrational transitions, it is said to occur by a vacancy mechanism.

In the simplest example of a jump mechanism, the species simply migrates to an adjacent vacancy. Whatever the relative energies of initial and final states of the transition, if the states are thermodynamically stable, the process is *activated*, i.e., energy must be supplied to accomplish the transition. The precise route between sites depends on the potential energy surface.

Applying this to the transport of dopant ions in a polymer lattice, a defect configuration can be defined, consisting of two or more dopant vacancies and one or more interstitial dopant ions. The interstitial species are fixed at various intermediate positions between the vacancies, though these positions need not be confined to a straight line. The procedure described in the first part of this section is then followed to calculate the defect energy for the series of defect configurations. This in effect monitors the potential energy surface and furnishes, among other things, the energy barrier for the minimum-energy pathway for the migrational transition.

3. MOLECULAR DYNAMICS

3.1. Equations of Motion

This method is widely used to simulate motions in liquids[55] and solids[56] and to study rapid diffusions in ionic lattices.[13] Although the application of molecular dynamics (MD) to studying ion transport in polymer systems is still in a seminal state, it is included in this chapter because of its correlation with the expansion of computing power, which will probably result in its prominence as an investigative method in this field.

The general procedure for applying MD to a solid or molecular system is to define a configuration vector **x** that comprises the $3N$ coordinates of the N constituent atoms of the system. Unlike the configuration vector described in Section 1.4 for static simulation, the present vector $\mathbf{x}(t)$ possesses a time dependence, which allows the atoms to explore the configuration space under the forces imposed on them. These forces are defined by the kinetic energies of the particles $\frac{1}{2} m_i v_i^2$ and by their mutual interactions, which are derived from potentials like those discussed in Section 1.3 and which are now expressed as a single function ϕ

$$\phi(\mathbf{x}_1,\mathbf{x}_2,\ldots,\mathbf{x}_N) = \sum_i \phi_i + (2!)^{-1}\sum_i \sum_j{}' \phi_{ij} + (3!)^{-1}\sum_i \sum_j{}' \sum_k{}'' \phi_{ijk} + \ldots, \tag{17}$$

where $\mathbf{x}_i = \mathbf{x}_i(t)$ consists of the three Cartesian components of atom i at time t and summations could represent (1) kinetic energies and core shell polarization terms, (2) bonding and nonbonding atom pair potentials and Coulombic terms, and (3) the bond angle deformation potentials, etc. The force $\mathbf{F}_i$ exerted on atom i by all the atoms of the system in configuration **x** is derived from ϕ by

$$\mathbf{F}_i(\mathbf{x}_1,\mathbf{x}_2,\ldots,\mathbf{x}_N) = -\nabla_i \phi(\mathbf{x}_1,\mathbf{x}_2,\ldots,\mathbf{x}_N) \tag{18}$$

In this way motions of the atoms can be made to emulate those of a real system resulting from thermal energy and mutual interactions.

The time evolution of the total conformation vector $\mathbf{x}(t)$ is implemented by allowing all N atoms to move in accordance with Newtonian dynamics, which couples their positions, masses, and momenta, and the forces acting on them. The general law describing the motion of particle i of mass m_i under the influence of the potential energy $V(\mathbf{x})$ derived from the remaining particles is

$$m_i \frac{d^2}{dt^2} \mathbf{x}_i(t) = -\nabla_i V(\mathbf{x}) \tag{19}$$

which leads to a set of $2N$ coupled equations:

$$m_i \frac{d\mathbf{x}_i}{dt} = \mathbf{p}_i \tag{20}$$

$$\frac{d\mathbf{p}_i}{dt} = \mathbf{F}_i \quad (i = 1 \text{ to } N) \tag{21}$$

With $\mathbf{x}(0)$ describing a configuration close to equilibrium, the dynamics are initiated by assigning to atom i a momentum vector $\mathbf{p}_i$, whose magnitude is consistent with a common kinetic energy assigned to all atoms of the system and initially corresponds to a low temperature. The direction of $\mathbf{p}_i$ is either chosen randomly or, provided the system is not in a precise equilibrium configuration, given that of the force $\mathbf{F}_i$ calculated from Eq. (18). After allowing the atoms to move for a very short time interval Δt of 10^{-15}–10^{-14}s, the equations of motion [Eqs. (20)–(21)] couple the new, slightly modified fields $\{\mathbf{F}_i\}$ to the set of momenta $\{\mathbf{p}_i\}$

that decide the new atomic trajectories. The particles now move for a second time step Δt under the influence of the modified field, and the sequence is repeated for $\sim 10^{-12}$s to equilibrate the system at low temperature.

Various algorithms have been proposed to update the elements of the configuration and velocity vectors; when Δt is sufficiently small, a commonly used pair is that of Verlet:

$$\mathbf{x}_i(t + \Delta t) \approx -\mathbf{x}_i(t - \Delta t) + 2\mathbf{x}_i(t) + \frac{\mathbf{F}_i(t)(\Delta t)^2}{\mathrm{m}} \tag{22}$$

$$\mathbf{v}_i(t) \approx \frac{[\mathbf{x}_i(t + \Delta t) - \mathbf{x}_i(t - \Delta t)]}{(2\Delta t)} \tag{23}$$

Particle velocities are then gradually increased until they correspond to the temperature of interest for the system, and the dynamics are continued for $\sim 10^3$ Δt intervals or until the distribution of velocities is Maxwellian. On achieving this condition, the dynamics describe a thermal equilibrium, and $\{\mathbf{v}_i, \mathbf{x}_i, t\}$ data are collected over an additional period of 10^3–10^5 time steps Δt. This maximum number of steps that is practicable thereby limits the total time interval of the dynamics to about 10^{-7}s. Accordingly no relaxations can be simulated with characteristic times longer than this. A limitation must sometimes be imposed at the shorter end of the time scale: to investigate dynamics with long characteristic times, it may be necessary to use a time step Δt that is longer than the period of a bond vibration ($\sim 10^{-13}$s). In such cases force constants are used that are diminished, typically by a factor of 7.

3.2. Simulation Box

If the solid being simulated possesses little or no long-range order, so that we are dealing with one that is amorphous, a large number of atoms must be included in the dynamics for an adequate representation of the system. This could be done firstly by applying Monte Carlo methods to randomize conformations of an assembly of polymer chains, followed by switching on the dynamics, as just described. However the caveat described in Section 1.2 on the danger of under-representing the disorder present in an actual sample indicates the necessity of considering an adequate number of atoms. Even if the sample being simulated is a crystal lattice whose periodicity is completely described by the crystallographic unit cell in the static lattice, the $\mathbf{x}(t)$ vector describing the dynamic lattice is not subject to such restrictions at a temperature above 0 K. This is similar to the theory of lattice vibrations, which requires a progressively greater number of cells for their description as the energy of the phonon mode decreases.

Ideally then the total number of particles described by Eqs. (17)–(23) should be as large as possible. Compromises with the limitations of computer core, disk, and computing time lead typically to the selection of $N \sim 1000$ atoms.

The region accommodating the N atoms is known as a simulation box; it is usually a cube of length L, containing a large number of unit cells. Since L is finite, conducting an MD procedure as we described would run the risk of evaporation, i.e., of having the particles (atoms, molecular fragments, or polymer chains) exit the box, thereby altering the composition. This effect is used in simulating the boundary between a condensed phase and the vapor, but it does not contribute to the processes of interest in this chapter. Most applications are performed under constant volume conditions in which the box is rigid, but if the simulation is to provide a full set of thermodynamic functions describing the dynamics, simulations would also have to be carried out at constant pressure and/or constant temperature. The resulting non-Newtonian environment would then require modifications of the equations of motion.[57]

To conserve constituent particles in the system, it is necessary to impose conditions on the simulation box. If the walls were rigid, so that they reflected the motion of the particles elastically, energy, momentum, and mass would be conserved. But unless the box were very large, reflection of the particles at the walls would result in serious boundary effects on the dynamics being simulated. The difficulty is overcome by the imposition of periodic boundary conditions. In a common model the cubic box described is surrounded at its faces, edges, and vertices by similar boxes, thus providing 26 images of the contents of the central domain. (A symbolic two-dimensional analog of the system is shown in Fig. 5.4.) Thus when dynamics require a particle to exit the central box by crossing a boundary, an identical particle with the same momentum vector enters the box via the opposite boundary.

Except for the Coulombic contributions, which are computed by an Ewald-type summation, as described in Section 2.3, a potential cutoff distance r_c is imposed to avoid unnecessary computing time calculating negligible contributions by short-range interactions from most of the $\frac{1}{2}N(N-1)$ atom pairs in the system. In the commonly used nearest-image convention, if the cutoff obeys the condition $r_c \leq \frac{1}{2}L$. Then atom pair interactions included are between atom i in the central box and either atom j in the same box or one of its images j' in an adjacent one, depending on whichever distance $|\mathbf{x}_i - \mathbf{x}_j|$ or $|\mathbf{x}_i - \mathbf{x}_{j'}|$ is least (see Fig. 5.4).

This periodic boundary model of a pseudoinfinite system does not provide the freedom that would be enjoyed by particles in a genuinely infinite space, but it renders the problem computationally tractable. Using the new generation of vector processor computers, which permit several independent interactions to be computed simultaneously, can lead to a more efficient solution of the equations of motion for a given N if the shape of the simulation box and its images are not a cube but a higher order polyhedron.[58]

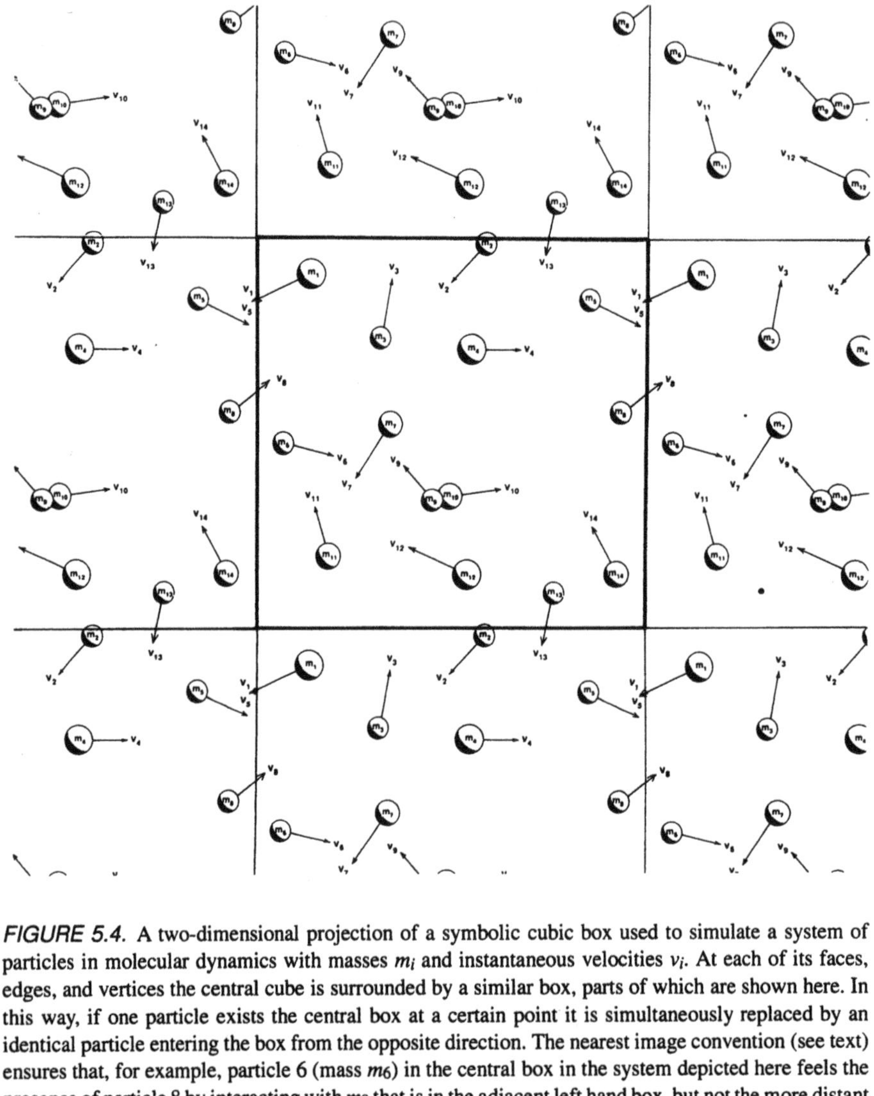

FIGURE 5.4. A two-dimensional projection of a symbolic cubic box used to simulate a system of particles in molecular dynamics with masses m_i and instantaneous velocities v_i. At each of its faces, edges, and vertices the central cube is surrounded by a similar box, parts of which are shown here. In this way, if one particle exists the central box at a certain point it is simultaneously replaced by an identical particle entering the box from the opposite direction. The nearest image convention (see text) ensures that, for example, particle 6 (mass m_6) in the central box in the system depicted here feels the presence of particle 8 by interacting with m_8 that is in the adjacent left hand box, but not the more distant m_8 in its own box.

3.3. Probability Functions

Although precise motions of individual atoms are computed in the MD simulation, the complex nature of their trajectories makes their interpretation difficult without statistical methods, which employ functions that express the probability of a given event or condition. Some of the commonest of these are now discussed, but for a more extensive collection, specialized works should be consulted.[59]

3.3.1. Mean Square Displacement $x_k^2(t)$

The mean square displacement of a particular species k between times t_1 and t_2 depends only on the interval $t_2 - t_1$. We can therefore write

$$x_k^2(t) = \langle[\mathbf{x}_k(t) - \mathbf{x}_k(0)]^2\rangle \tag{24}$$

where the angular brackets imply an average over all the k-species. The time variation of $\mathbf{x}_k(t)$ allows the calculation[13] of the diffusion coefficient D_k:

$$D_k = \tfrac{1}{6} d/dt[\mathbf{x}_k^2(t)] \tag{25}$$

Moments calculated from particle displacements can be used to elucidate the character of the diffusion process. For example the more closely the ratio

$$P_k(t) = \tfrac{3}{5}\left\{\frac{\langle[\Delta x_k(t)]^4\rangle}{\langle[\Delta x_k(t)]^2\rangle}\right\}^2 \tag{26}$$

approaches 1.0, the smoother or more continuous the diffusion. Departures from unity indicate the presence of abrupt motions or jumps.

3.3.2. Distribution Functions

In a dynamic system of volume V containing N particles, the probability of finding one particle at $\mathbf{x}_1$ while a second one is simultaneously at $\mathbf{x}_2$ is called the two-particle distribution function $\rho^{(2)}(\mathbf{x}_1,\mathbf{x}_2)$. If the particles moved independently of each other, $\rho^{(2)}$ would be a product of their independent probabilities N/V and $(N-1)/V$:

$$\rho^{(2)}(\mathbf{x}_1,\mathbf{x}_2)_{\text{uncorrelated}} = \frac{N}{V}\cdot\frac{N-1}{V} \approx \left(\frac{N}{V}\right)^2 \tag{27}$$

Now suppose the particles were allowed to feel each other's presence through the introduction of an interaction $V(\mathbf{x}_1,\mathbf{x}_2)$. For simplicity we suppose that the interaction is isotropic, as it would be if it were a $V(r)$-type potential like Eq. (7) or (8), which depends only on the separation $r = x_{12} = |\mathbf{x}_1 - \mathbf{x}_2|$. The distribution function $\rho^{(2)}$ must tend to Eq. (27) in the weak interaction limit (e.g., large x_{12}), but how would it behave at other separations? For the moment we evade the question and just modify the $\rho^{(2)}$ expression by introducing a dimensionless factor $g(x)$:

$$\rho^{(2)}(x) = g(x)\left(\frac{N}{V}\right)^2 \tag{28}$$

Since an MD calculation permits the calculation of $\rho^{(2)}(x)$, Eq. (28) is used to derive $g(x)$ which is termed the pair radial distribution function. This is a function that tends to 1.0 in the limit of large x; its departure from unity at other distances reflects the interaction between particles that correlates their motions.

The radial distribution function can be defined either for a pair of specified atomic species or alternatively for a pair consisting of a specific atom and any other atom in the MD box. At low x values, $g(x)$ is very small, reflecting the repulsive region of $V(x)$ at short interatomic distances, but it typically rises to a sharp peak when x corresponds to a particle diameter. In a highly mobile and homogeneous system, such as a liquid, Figs. 5.5(a) and (b) show that $g(x)$ exhibits a series of strongly damped peaks, corresponding to concentric shells of diminishing stability, which are in dynamic equilibrium with the bulk system. After several particle diameters, when the correlation is too weak to reveal local structure, $g(x)$ achieves its bulk value of unity. A more rigid system, such as a crystal lattice or an assembly of covalently bonded atoms, on the other hand, may offer a mobile particle several stable sites for its sojourn. The correlation in this case would have a longer range, resulting in sharper and more persistent $g(x)$ peaks [Figs. 5(c) and (d)]. The area under the $g(x)$ curve at a particular x is proportional to the cumulative coordination number characterizing the atom pair at that separation.

3.3.3. Correlation Functions

Since the dynamics are completely known throughout the simulation, the behavior of any (or all) of the particles can be selected for monitoring. But rather than try to describe the motions of all N particles in the simulation box by conventional dynamic variables, it is better to consider functions that are more appropriate for a bulk system.

For example it might be useful to know whether two particular dynamic variables f and h associated with the system are independent and if they are not, to have some measure of their degree of correlation. Suppose both f and h are functions of the configuration and momentum vectors $\mathbf{x}$ and $\mathbf{p}$ of the particles, which are themselves functions of time t measured from the instant that initial conditions were specified. The space correlation function $C_{fh}(\mathbf{x})$ and the time correlation function $C_{fh}(t)$ are defined as

$$C_{fh}(\mathbf{x}) = L^{-3}\int_0^\infty f^*(\mathbf{x}')\, h(\mathbf{x}+\mathbf{x}')d\mathbf{x}' = \langle f^*(\mathbf{x}')\, h(\mathbf{x}'+\mathbf{x})\rangle \tag{29}$$

and

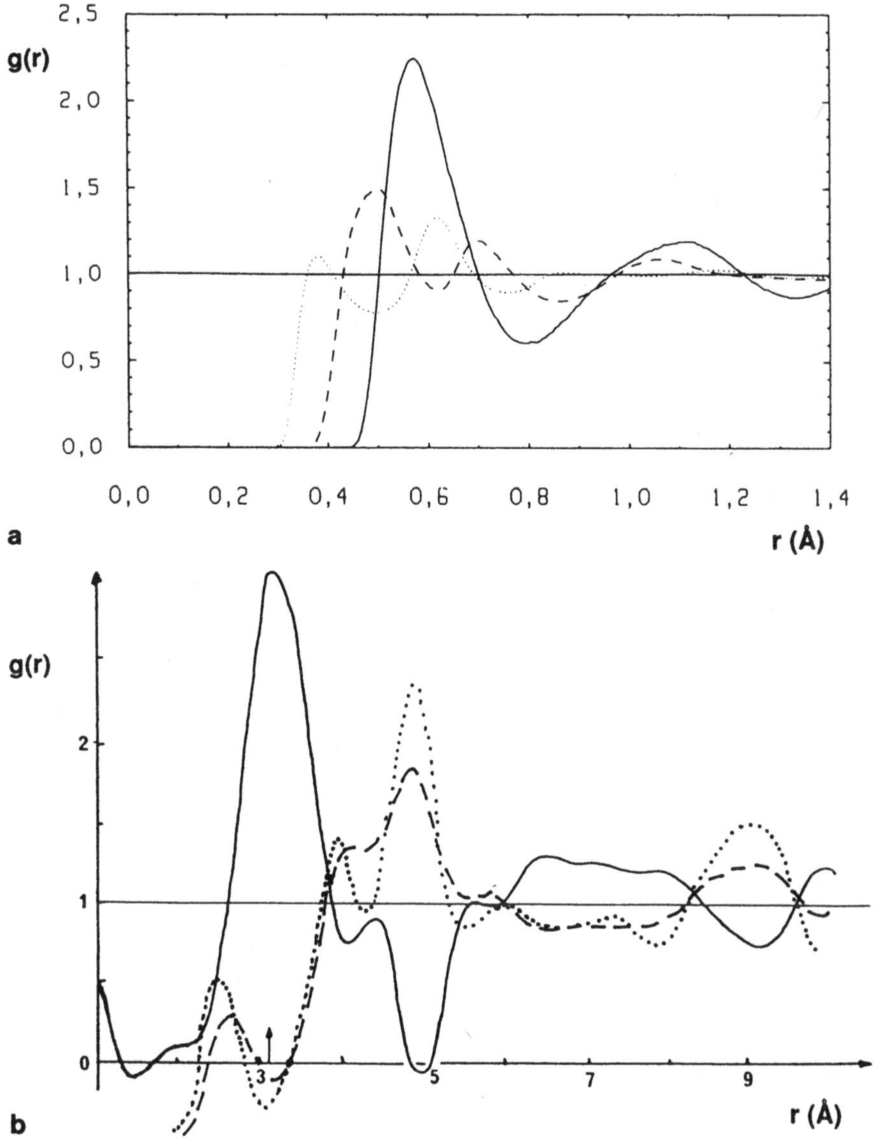

FIGURE 5.5. Radial distribution functions $g_{XY}(r)$ for (a) a molecular liquid (CCl_4), where the (X,Y) pairs are: —— (C,C), - - - (C,Cl), and ···· (Cl,Cl); (b) an ionic melt (KCl at 1073 K): —— (K^+,Cl^-), - - - (Cl^-,Cl^-), and ···· (K^+,K^+); (c) a solid, consisting of crystalline (curve c) and amorphous (curve a) silicon; and (d) a solvated polymer, where the pair consists of a polymer segment and a solvent molecule. The plotted function in (c) would be roughly compatible with the others if it were divided by $4\pi r^2$. Curves (b) and (c) are based on neutron-scattering data, while (a) and (d) are the result of MD calculations. [Curve (a) is from Ref. 60 (*Molecular Simulation*), (b) is from Ref. 61 (*Journal de Physique*), (c) is from Ref. 62 (US Atomic Energy Commission), (d) is from Ref. 63 (American Institute of Physics), and they appear by permission of the authors and publishers.]

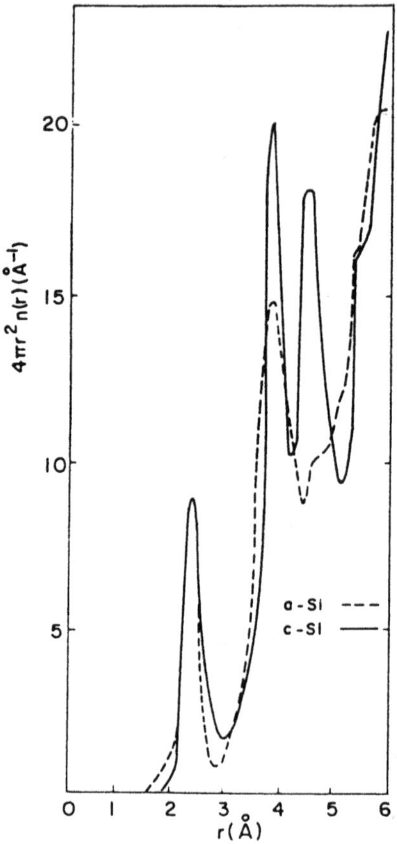

c

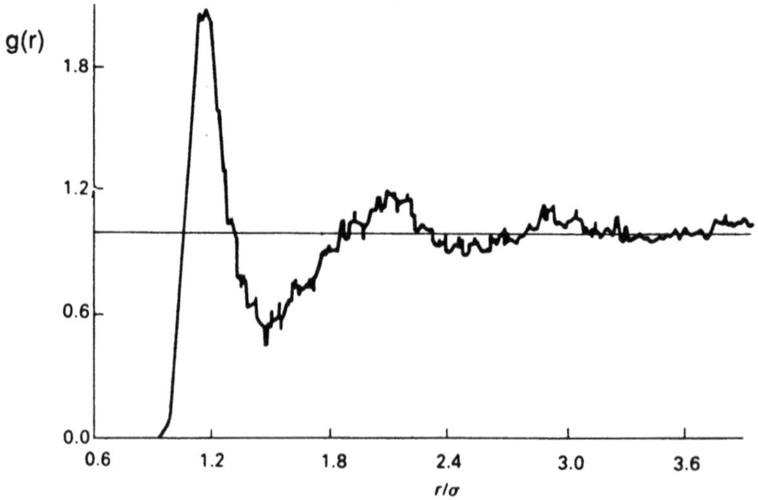

d

FIGURE 5.5. *(Continued)*

$$C_{fh}(t) = \tau^{-1} \int_0^{\tau} f^*(t')\, h(t'+t)dt' = \langle f^*(t')\, h(t'+t)\rangle \tag{30}$$

where angular brackets denote an average over all time events used in the simulation. If the system has reached dynamic equilibrium, the integrand should be independent of the latest time event t' to be included in the calculation of the right-hand side of Eq. (30), but since for practical reasons, this value is necessarily finite, the result is usually an average over several simulations with different time origins. Invariance to the time origin offers the freedom to rewrite Eq. (30) as

$$C_{fh}(t) = \langle f^*(0)\, h(t)\rangle \tag{31}$$

A special case of Eq. (31) occurs when h is the same function as f. The resulting autocorrelation function $C_{ff}(t)$ monitors the degree of correlation of f between the particles of the system. As an example consider the case when f is the velocity of the kth species $\mathbf{v}_k$. Then the velocity autocorrelation function (VAC) for species k is

$$z_k(t) = \langle \mathbf{v}_k^*(0)\, \mathbf{v}_k(t)\rangle \tag{32}$$

The VAC is closely related to the diffusion coefficient D_k of species k:

$$D_k = \int_0^{\infty} z_k(t)dt \tag{33}$$

Although this expression for D_k can be shown to be equivalent to Eq. (25), the method of summations over finite time steps results in small differences between values calculated from Eqs. (25) and (33).

A Fourier transformation into frequency space produces the power spectrum $f(\omega)$:

$$f(\omega) = \int_0^{\infty} z_k(t)\, e^{i\omega t}\, dt \tag{34}$$

When $f(\omega)$ is calculated from the dynamics of a given atomic species k, it shows sharp peaks at frequencies ω close to those characteristic of vibrational modes associated with atom k.

The relationship between $f(\omega)$ and $z_k(t)$ in Eq. (34) exemplifies an interesting duality between pairs of functions commonly encountered in structure-determining theory. Since some of the functions derived from MD investigations are related to quantities appearing in X-ray or neutron scatterings, the latter techniques may be used to calculate some of the quantities characterizing a dynamic system and thus to verify features of the MD simulation. The radial distribution function $g(\mathbf{x})$ and

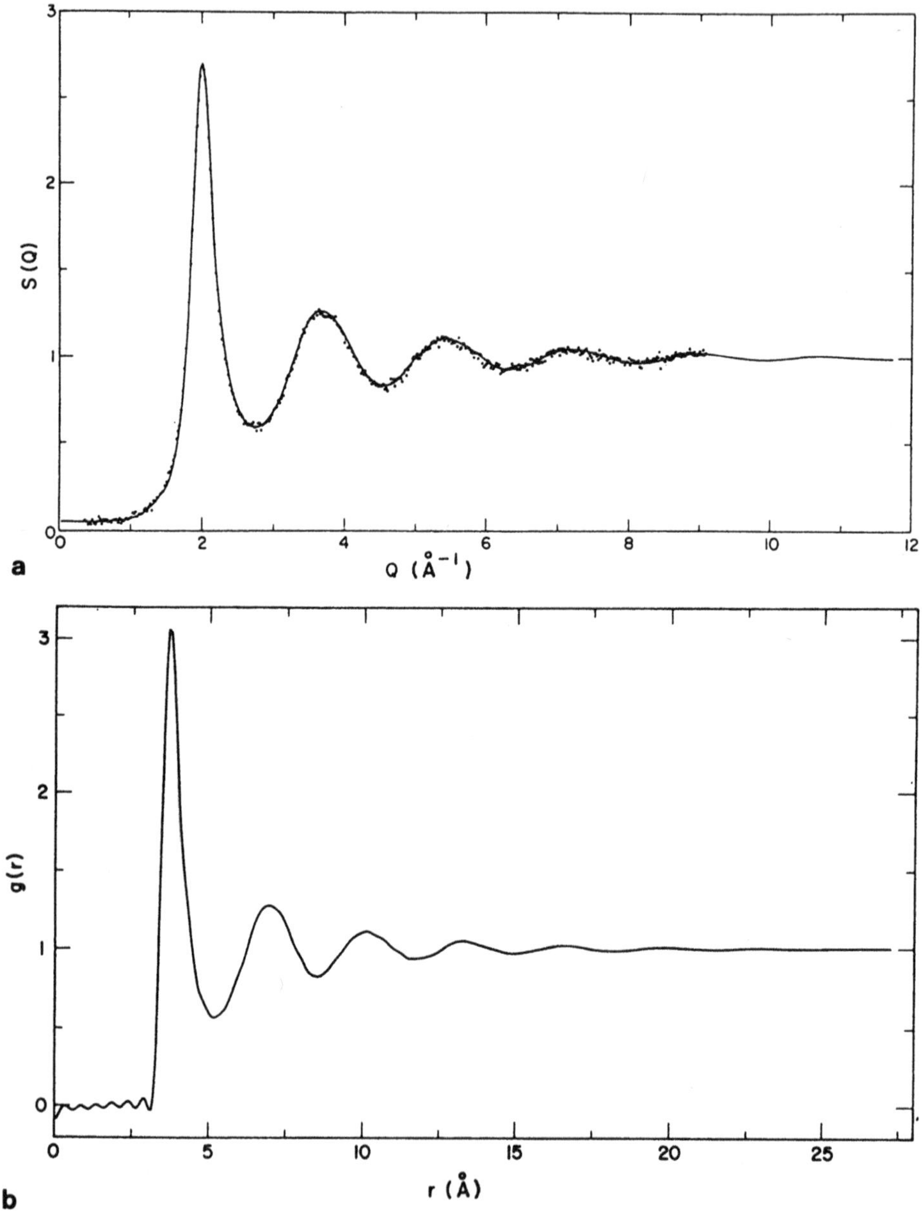

FIGURE 5.6. (a) The experimental structure factor $S(Q)$ for ^{36}Ar provided by neutron-scattering data at 85K. (b) The radial distribution function g(r) for ^{36}Ar at 85 K obtained from the Fourier transformation of the smoothed and extended solid line in (a). (From Ref. 64 by permission of the authors and publishers.)

the structure factor $S(\mathbf{q})$ constitute another conjugate pair related by a Fourier transformation:

$$S(\mathbf{q}) = 1 + \rho \int_0^{\infty} [g(\mathbf{x}) - 1] e^{-i\mathbf{q}\cdot\mathbf{x}} \, d\mathbf{x} \tag{35}$$

where ρ is the overall density of the system. Figure 5.6 shows an example of the transformation of structure factor data measured for liquid argon into the corresponding radial distribution function.[64]

3.4. Matter Transport

The dynamics of a particle k may be monitored by following its mean square displacement $x_k(t)$ defined by Eq. (24) or diffusion coefficient D_k from Eqs. (25) or (33). For our purpose these quantities are particularly useful when k is an ion or molecule that migrates through the polymer matrix; for example by following D_k as a function of imposed temperature T, an apparent activation energy E_D can be evaluated from the Arrhenius equation

$$D_k = D_k^0 \, e^{-E_D/kT} \tag{36}$$

If the polymer matrix is noncrystalline, the mechanism for the migration can be tested against various diffusion models. For example according to free volume theory, the migrant species in a stable site is trapped in a cage formed by the polymer strands. A suitable redistribution of vibrational energy enables the species to jump to a nearby lacuna or cavity in the matrix formed by random fluctuations in the local density. The probability of the occurrence of these cavities is related to the density of the sample through its free volume fraction v_f defined by

$$v_f = \frac{v_0 - v_c}{v_0} \tag{37}$$

where v_0 is the volume of the simulation box and v_c is the volume of the chains. The diffusion coefficient then depends on v_f according to

$$D_k = A_k \, e^{-B_k/v_f} \tag{38}$$

where A_k and B_k are constants. This equation may be used to test for a correlation between the diffusion coefficient and the free volume fraction—both furnished by the dynamics simulation—and hence to assess the importance of voids in the mechanism of the transport process being investigated.

4. SIMULATION DATA

4.1. Polymer Lattice Structures

A reliable knowledge of lattice structures of conductive polymers is unfortunately at present restricted to very few polymer species. This stems from the fact that the translational periodicity that characterizes the solid states of ionic and molecular matter, and which diffraction methods of structure determination normally demand, is present to a much smaller degree in polymer samples, particularly those discussed in this book. Even when polymer single crystals exist, notably for some insulating polymers, such as polyethylene, they may be contaminated with amorphous regions.[65] Low solubilities certainly contribute to the difficulty of preparing conductive polymers with sufficient crystallinity for diffraction studies. This precludes the usual methods of growing crystals from solution, but the problem is augmented by the existence of soft structures in which polymer chains can assume more than one orientation.

The unavailability of single crystals for diffraction studies appears to jeopardize one of the prerequisites listed in Section 2.4 for successful simulation—a precise knowledge of the structure of the unit cell. But rather than determining a complete structure from single crystal data, diffraction methods are often applied to stretched polymer films,[66,67] where the application of a strain (producing up to a fourfold length increase) results in a degree of alignment of the chains along the stretch direction. X-ray, neutron, or electron diffraction then provide data that permit the derivation of space groups and lattice vectors (i.e., unit-cell dimensions). The contents of the cells are determined by packing criteria for fitting polymer chains with assumed molecular geometries and van der Waals radii.

Other methods of determining chain structure in polymers include tunneling electron microscopy,[28–35] which has been used to reveal the presence of helical chains, and extended X-ray absorption fine structure (EXAFS),[68] which when applied to metal-ion-doped polymers can reveal the immediate atomic environment of the dopant ion.

4.2. Potentials

4.2.1. Valence Potential Functions

Although the largest contribution to the lattice energy of polymer systems comes from atom-pair bonding, often represented by the Morse potential in Eq. (3), it is not necessarily the most critical of the potentials. The transferable character of chemical bonds, and their rather small deformations during lattice relaxation, permit the use of bond-length and bond-angle deformation potentials [Eqs.(3) and (5)], which are based on averaged interactions derived from model molecules.

TABLE 5.1.
Morse Parameters for the Bond Stretching Potentials Defined by Eq. (3)

Bonded Pair	A (eV)	B (Å^{-1})	C (Å)	Ref.[a]
C–C	4.5873	1.9169	1.453	12
C–C	4.5873	1.9169	1.428	12
C=C	5.3067	2.1353	1.405	12
C–H	4.3051	2.1029	1.080	12
N–H	4.0331	2.1903	0.994	69, 70
C–N	3.1622	2.1537	1.400	69, 70
C–S	2.9963	2.0665	1.687	71, 72
B–F	6.6873	1.4604	1.390	73, 74
Cl–O	2.1254	3.3076	1.450	71, 75

[a]References in the last five rows are to thermochemical and force constant data for the calculation of the A, B, C parameters. (Adapted from Ref. 122 by permission of Elsevier Science Publishers B.V.)

Tables 5.1 and 5.2 list a selection of valence potentials that can be used for the simulation of polymer chains.

4.2.2. Nonbonded Functions

The interaction between nonbonded molecular systems can be expressed as a power expansion involving separation vectors and products of their electric multipole moments.[36] Much effort has gone into calculating intermolecular potentials,

TABLE 5.2.
Parameters for the Bond Angle Deformation Harmonic Potential Defined by Eq. (5)

Bond Triad	k_θ (eV rad^{-2})	Ref.
C–C=C	12.4200	12
H–C–C	5.2431	12
H–C=C	5.2431	12
C–N–H	2.3251	76
C–N–C	7.0953	76
N–C=C	4.9533	76
N–C–C	4.9533	76
C–S–C	8.4883	70
F–B–F	8.8032	74
O–Cl–O	2.8870	75

(Adapted from Ref. 123 by permission of Elsevier Science Publishers B.V.)

particularly for ionic species.[80] Experimentally they may in favorable cases be derived from pressure–volume investigations and from sublimation enthalpies and thermal conductivies of simple gases.[81] However despite the critical nature of these interactions, they are usually the least accurately known components of the total force field, and much effort is expended to reveal details of at least portions of the potential curves. Accurate energies are available only for systems of inert gas atoms[82] and rather less reliable interactions for gaseous mixtures of atoms and small molecules.[83] Moreover the practice of transferring the results to other systems is of doubtful validity.

If the dopant species are monatomic ions, the derivation of their mutual interactions fall into two classes. Using quantum chemical electron gas theory, electronic charge distributions derived from Hartree–Fock calculations on the ions are allowed to interact at various internuclear distances.[84,85] In another method lattice simulation calculations are performed to fit the constants in Eqs. (7) or (8) to measured bulk data, such as elasticity tensor components, dielectric constants and/or lattice structures,[80,86] thereby providing empirically derived potentials.

In covalent and partially ionic systems, for which these methods are not generally valid, it is usual to fit the potential functions to averaged sets of data derived from structural measurements or quantum chemical calculations on molecular clusters. Williams[77] for example derived a set of potentials that is now widely used to describe C...C, C...H, and H...H interactions. Using atomic orbital functions to fix the hardness parameter ρ in Eq. (7) for an interacting atom pair, he used the known crystal structures of a large number of hydrocarbons to calculate least squares values of the remaining Buckingham A and C parameters. Scott and Scheraga[78] used in an all-empirical method in their derivation of nonbonded atom pair potentials that stabilize known conformations of polypeptides.

The quantum chemical approach is exemplified by the work of Portmann *et al.*,[87] who performed *ab initio* computations on complexes of alkali metal ions with a large number of organic ligands. From their results the authors fitted the interactions E_{int} between the cations and the atoms H, C, N, O, and S to functions $V(r)$, using the relation:

$$E_{int} = \sum_{i>j} \sum V_{ij}(r_{ij}) = \sum_{i>j} \sum (A_{ij} r_{ij}^{-12} - B_{ij} r_{ij}^{-6} + C_{ij} r_{ij}^{-2} + D_{ij} q_i q_j r_{ij}^{-1}). \tag{39}$$

The parameters A, B, C, and D defining $V(r)$ were statistically optimized, so that they best reproduced the calculated interaction energies. Different parameter sets were proposed for the atom pair i, j (where i was in the organic ligand) to take account of the nature of the atoms to which i was bonded.

It is sometimes possible to calculate an ion–ligand potential using spectroscopic data on model molecular complexes. In their simulation studies on the migration of the lithium ion in intercalated graphites, Corish *et al.*[50] derived a

TABLE 5.3.
Parameters of Nonbonded Potentials in Eqns. (7) and (8)

Atom Pair	A	ρ or B	C	Ref.
		Buckingham Potential[a]		
C···C	3626.626	0.2778	24.6367	77
C···H	380.139	0.2725	5.4206	77
H···H	115.116	0.2674	1.1844	77
		Lennard–Jones Potential[b]		
C···N	9366.68	15.8713	—	78
N···N	6981.64	15.7412	—	78
N···H	1170.83	5.4205	—	78
S···H	13131.00	12.5845	—	76
S···C	116560.00	46.6608	—	76
S···S	599739.81	73.0703	—	76
C···F	3842.58	12.3800	—	76
N···F	2062.00	9.6444	—	76
S···F	46140.50	20.5854	—	76
F···F	1583.70	5.8370	—	76
F···H	405.94	3.4600	—	76
C···O	4978.32	20.0303	—	76
Cl···O	15919.34	32.6626	—	76
N···O	6626.19	15.8283	—	76
O···H	1085.43	5.3729	—	76
O···S	59063.31	33.8378	—	76
O···O	6283.61	15.9237	—	76

[a]Units for A, ρ, and C for Buckingham [Eq. (7)] are eV, Å, and eV Å^6, respectively.
[b]Those for A and B for Lennard-Jones [Eq. (8)] are, respectively, eV Å^{12} and eV Å^6.
(Adapted from Ref. 123 by permission of Elsevier Science Publishers B.V.)

Li^+...C potential from vibrational frequencies in Li^+ complexes with aromatic rings that was subsequently used in their study of lithium-doped polymers.[(12)]

Some nonbonded potentials that are useful in the simulation of polymer lattices are listed in Table 5.3.

4.2.3. Torsional Potentials

At present there are discrepancies between the description of internal rotation around bonds in the torsional functions used by various groups. Particularly in the case of π-conjugated chains, important contributions to the variation in torsional energy arise from the differing interactions between the σ and π components of

TABLE 5.4.
Selection of Barriers v_0 That Can Be Used in Eq. (6) to Describe Energy Variation during Torsional Rotation around Different Types of C–C Bonds in Conjugated Chains

Torsional C–C Bond	v_0 (eV)
Olefinic	1.957
Aromatic ring	1.087
Interaromatic ring[a]	0.435
Interolefinic[a]	0.217

[a]Second and third entries might apply for example to the interphenyl ring in polyparaphenylene and to the single C–C bond in polyacetylene, respectively. The values are taken from generative DREIDING force field of Mayo *et al.*[88]

bonding—features which are poorly described by the present state-of-the-art molecular simulation methods, since the π contributions to the molecular (or crystal) orbitals are sensitive to the topology of the molecular system. Now the remaining factors that decide the torsional conformation around a bond in a solid are the interactions (1) between the nonbonded atoms defining the four-atom torsional chain, and (2) between different polymer chains. The latter constitute the lattice forces, and are usually of a nature that tend to impose planarity on the chains. Perhaps because of these two important contributions the results of simulation of polymer lattices are often not inordinately sensitive to the choice of torsional potential.

A set which we have found to be satisfactory for conjugated polymers are those contained in the DREIDING generic force field,[88] in which atomic pair interactions depend more on features describing the bonding than on the identities of the atoms themselves. The energy barriers v_0 specified for a particular type of torsion may then be used in the simple function of Eq. (6), and data which may be useful to simulate common molecular chains are given in Table 5.4.

5. *STATIC LATTICE SIMULATIONS*

5.1. *Pristine Polymer Lattices*

Before using the methods described on ion-doped conductive polymers, we test their application on pristine (i.e., undoped) polymer lattices.

The paucity of experimental information on lattices with which to compare simulation results restricts the test to structural data. Even here comparisons are not always decisive, since various diffraction investigations have led to apparently conflicting structures for polyacetylene, polyparaphenylene, and polythiophene lattices. Although polyacetylene lattices are sufficiently well-characterized to distinguish trans from cis (in so-called cis-polyacetylene the C–C conformation alternates along the chains as cis and trans), the precise chain conformation in other polymer samples (e.g., polythiophene) has not been established with the same reliability. For this polymer some X-ray diffraction investigators have produced a planar all-trans structure,[89–91] while others, supported by tunneling electron microscopy, have proposed a helical conformation.[28–35] (Naturally the polymorphic character of polymer lattices does not permit a proposed structure to be labeled exclusive.)

Figure 5.7 shows the general lattice structure exhibited by common undoped conductive polymers—*trans*-polyacetylene,[22–24] *1,4* polyparaphenylene,[25,26,92] and *2,5* polythiophene[89–91]—which were obtained from diffraction investigations. In all cases unit cells are orthorhombic with slight monoclinic distortions, and polymer chains are aligned along the crystallographic *c*-axis. Figure 5.7 defines the setting angle θ for planar chains as the angle between the plane of the polymer and

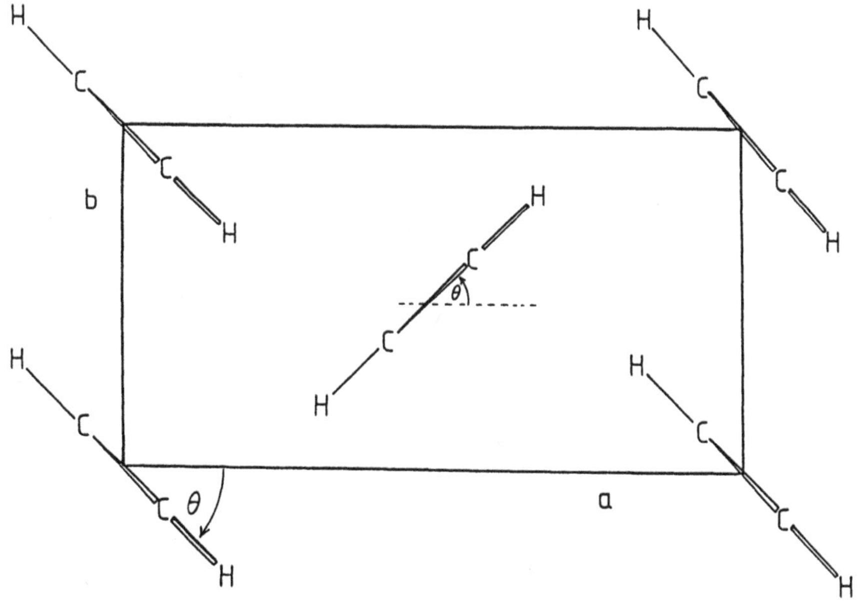

FIGURE 5.7. A common lattice structure found by diffraction methods for common *undoped* conjugated polymers. The lattice illustrated is that of polyacetylene, but a similar diagram describes polyparaphenylene and polythiophene. The unit cell is viewed along the direction of the chains, which in these lattices is parallel to the monoclinic *c*-axes. The angle θ, between the polymer planes and the *a*-axis. is the chain-setting angle.

TABLE 5.5.
Unit-Cell Data on Pristine Forms of Some Conjugated Polymers, All of Which Crystallize As Nearly Orthogonal Lattices

	a (Å)	b (Å)	c (Å)	ϕ (deg.)	Method	Ref.
Polyacetylene	7.4	4.1	5.0	51	X-ray diffraction	22
	7.3	4.2	5.0	55	X-ray diffraction	23
	7.3	4.2	5.0	24	Electron diffraction	24
	7.0	4.1	5.0	62	**LS**[a]	12
	7.1	4.6	5.0	22	**LS**[a]	12
Polyparaphenylene	7.8	5.5	4.3	57	Neutron diffraction	26
	7.8	5.5	4.3	45	Electron diffraction	92
	7.2	5.6	4.3	62	**LS**[a]	12
Polythiophene	7.8	5.5	—	69	X-ray diffraction (P)[b]	89
	10.8	4.7	—	—	X-ray diffraction (D)[b]	90
	7.4	6.1	7.9	68	**LS**[a]	125
	11.1	4.1	7.9	28	**LS**[a]	125

[a]**LS** denotes the results of lattice simulation.
[b]P and D are, respectively, obtained from the diffraction of powder and drawn film samples.

the *a*-axis, here taken to be the longer of the two axes that are normal to *c*. In each polymer, the simulation predicts more than one low-energy structure, often consistent with the results of apparently conflicting diffraction data. This can be seen in the two chain setting angles measured and calculated in polyacetylene in Table 5.5, and we shall see later that the structures derived from diffraction studies on both powder[(90)] and drawn-film[(91)] polythiophene samples appear as nearly energetically degenerate simulated lattices. This suggests that the plurality of the experimental structures might be due to the polymorphism and consequent inhomogeneity of the polymer samples. In fact there is evidence that the identity of the dominant structure depends on the method and conditions of preparation.[(11,92)]

5.2. Doped Polymer Lattices

When the dopant species enters the polymer matrix, it may form a sublattice of dopant ions, simultaneously undergoing charge transfer with the host lattice. Electron transfer supplies electronic charge carriers to the polymer sublattice, which now bears a net charge; n-dopants (alkali metals) place excess electrons in the conduction band, leaving the dopant species as Li^+, K^+, etc., while p-dopants (I_2, $FeCl_3$, AsF_5, etc.) create positive holes in the valence band, forming a dopant anion sublattice of I_3^-, $FeCl_4^-$, AsF_6^-, etc. In both cases the polymer sample becomes a conductor. A comparison of Figs. 5.7 and 5.8 shows that the chains, which now

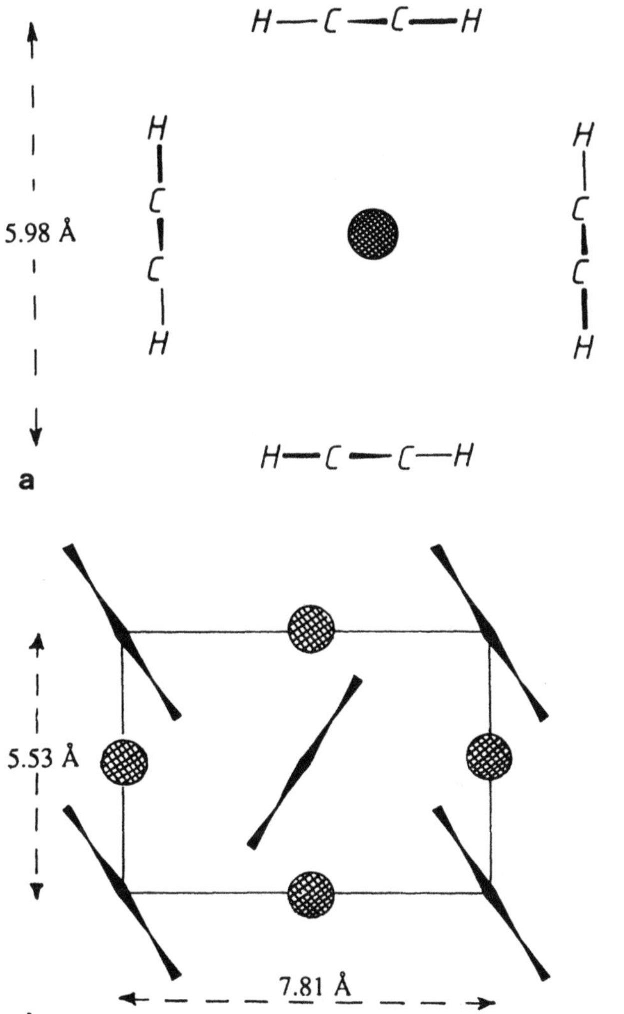

FIGURE 5.8. Structures of some common doped conductive polymers, each projected on a plane normal to the direction of the polymer chains. Cross-hatched circles indicate dopant ions. (a) The square channel structure of potassium-doped polyacetylene (other alkali-doped PA lattices show similar structures),[97,98] (b) lithium-doped polyparaphenylene,[26,92] (c) potassium-doped PPP,[113] and (d) AsF_5-doped PPP.[26] In (d) the dopant species, represented by ×, are probably AsF_6^- .

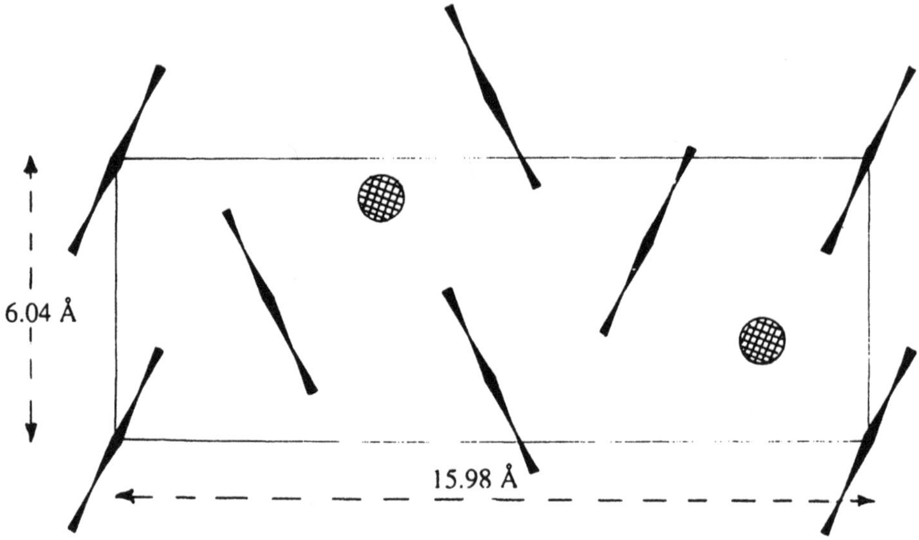

c

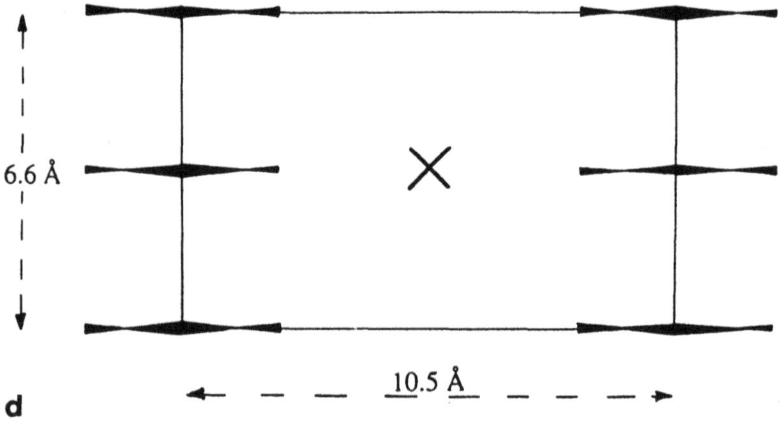

d

FIGURE 5.8. *(Continued)*

constitute the host sublattice, may exhibit a change of structure from that in the pristine state.

Quantum chemical calculations tend to confirm the intuitive chemical expectation that the excess charge conferred by dopants on the polymer is accommodated mainly by those atoms that constitute the polymer's π network. If the environment of the dopant ion has sufficient symmetry, atomic charges of the host sublattice can, in such an ideal case, be decided by the structure. This condition, which is met in first- and second-stage lithium-intercalated graphites C_6Li and $C_{12}Li$, respectively,[50,93] so that all carbon atoms are equivalent and may be assigned $q_C = -1/6\,e$ or $q_C = -1/12\,e$, is rare in doped polymer lattices. It is approached in fully doped polyacetylene (see Section 5.3), where each K^+ is surrounded by a serrated ring of eight equidistant, equivalent carbon atoms. This structure fixes the carbon charge at $q_C = -1/4\,e$. In other cases when atomic charges are determined by quantum chemical methods, it may be necessary to average calculated charges for some atomic species. This makes it possible to keep the number of distinct species (e.g., C atoms in slightly different environments, etc.), between which pair potentials must be specified to a practical minimum.

5.3. Doped Polyacetylene

5.3.1. Perfect Lattice

The *cis*-polyacetylene initially formed from the polymerization of acetylene monomer is often readily transformed into the thermodynamically more stable all-trans form on doping.[94] Although the pristine chain shows a Peierls distortion, in which C-C bond lengths alternate between partially single and double bonds,[95,96] one of the consequences of doping is a reduction in the degree of bond length alternation. This is because electrons or holes transferred between host and dopant sublattice occupy crystal orbitals with nodal distributions opposing, though not eliminating, the bond alternation pattern established by the occupied orbitals of the pristine polymer.

Doping may impose further structural effects that depend on the size and nature of the dopant species. Although there may still not be complete accord on the crystal structure of lithium-doped polyacetylene,[97,98] it appears that at low doping levels entropic factors are important in inserting a small nonaggregating ion, such as Li^+, into polyacetylene (PAc), with the dopant occupying sites with minimal strain or disruption of the host lattice. Iodine, on the other hand, in the form of I_3^- and I_5^- (and possibly higher polyiodides), produces structures in which anions cluster in columns or sheets to form intercalated layers between polymer chains.[99–101]

With the possible exception of lithium, which may interact chemically with organic chains, the small sizes of alkali metal n-dopants and high charge concentration compared with those of common p-dopants (I_3^-, $FeCl_4^-$, AsF_6^-, etc.) favor the creation of columnar arrays of sites surrounded by charge-compensating carbon chains of the polymer. According to independent X-ray and neutron diffraction studies, the crystal structure of alkali-doped PAc thus consists of near-tetragonal channels formed by parallel polymer chains, with metal cations occupying sites along the axes,[97,102–104] as shown in Fig. 5.8(a). The widths of the channels enclosing the Na^+, K^+, Rb^+, or Cs^+ columns are, respectively, 6.0, 5.98, 6.19, or 6.43 Å, showing that up to Rb^+, they are decided primarily by the van der Waals contact of the polymer hydrogen atoms.

The diffraction work does not specify the positions of the cations, nor does it provide an unambiguous measure of their separations along the channel axes. If all channels were uniformly occupied by dopant ions so that a M^+ ion occurred for every p CH units along any chain, the resulting overall composition would be described by $[CHM_{1/(2p)}]_x$. Elemental analysis of the most highly doped sodium samples gives the composition $[CHNa_{0.25}]_x$,[105] indicating that $p = 2$ or if the dopant is evenly distributed throughout the lattice, each –CH=CH– unit is flanked on both sides of its plane by a Na^+ ion, which is consistent with a commensurate host and dopant sublattice. Such a simple picture is not obvious for the potassium-doped PAc for which the composition $[CHK_{0.18}]_x$ of the most highly doped sample[105] implies $p = 2.78$; i.e., there is on average a K^+ site at every 1.39 –CH=CH– units. The composition $[CHK_{0.18}]_x$ must then be interpreted either as (1) an incommensurability of host and dopant sublattices (i.e., they show different periodicities), or (2) the result of heterogeneneous doping in which some regions of the polymer remain undoped while others sustain higher doping levels. Another possibility is of course that both (1) and (2) occur.

Since the lattice simulation method requires the presence of a structure that can be generated from a single unit cell, it cannot normally cope with an incommensurate potassium-doped PAc lattice in which the host and dopant sublattice are based on incompatible unit cells. For this system investigators simulated two commensurate structures—one considered to be a fully doped lattice $[CHK_{0.125}]_x$, which they termed PAK1, the other a semidoped species $[CHK_{0.125}]_x$ or PAK2.[106] These structures can be considered limiting cases of that described by experimental composition, which may perhaps be the result of a mixture of the two. A chain geometry was used in which it was supposed that doping made all the C–C bond lengths equal (1.40 Å). Then within the commensurability constraints defined by the two unit cells in Fig. 5.9, relaxed stable lattices were simulated for both structures in which the K^+ ions were sited at axial positions in the channel equidistant from the centers of four surrounding –CH=CH– units. Figure 5.9 shows that in PAK1, a dopant ion occupies a site over the center of each –CH=CH– unit, with a spacing of 2.48 Å, while in PAK2 only half of these positions is occupied, resulting in a separation of 4.96 Å.

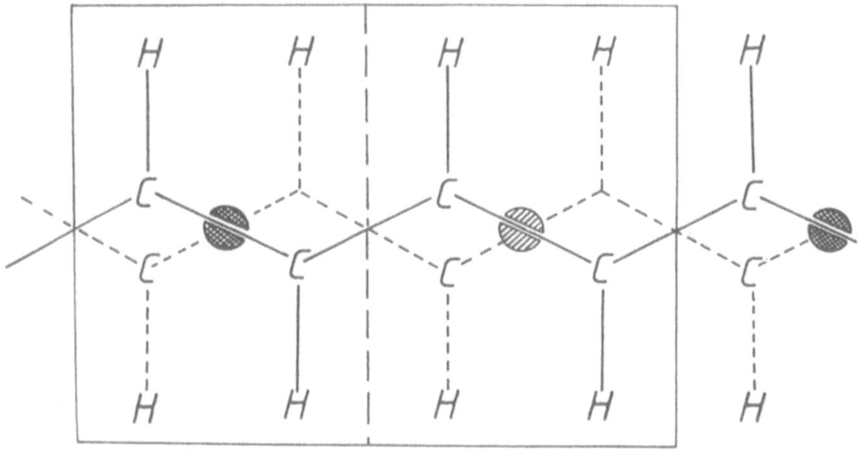

FIGURE 5.9. The potassium-doped polyacetylene channel in Fig. 5.8(a) viewed perpendicular to the channel direction for two assumed doping levels. If *both* K^+ ions shown are present, the lattice is described as fully doped, while the one with a cation at alternate segments is semidoped. (From Ref. 107 by permission of the publishers.)

5.3.2. Defective Lattice

When a local defect is created in the lattice, it is accompanied by a displacement of the surrounding atoms, thereby removing at least some of the translational and point group symmetry that is a property of the perfect lattice. Even if the effect of the defect is so localized that it is imperceptible on the structure a few tens of Å from the defect center (thereby permitting the application of the lattice-defect method described in Section 2.5), relaxing symmetry restrictions effectively introduces additional degrees of freedom into the permitted atomic displacements.

The results of applying the lattice-defect method to a single K^+ vacancy V_K in potassium-doped PAc show an interesting response by both the host (polymer) and dopant sublattices.[107,108] When the system investigated is PAK1, dopant ions to the left and right of V_K may either be attracted to, or repelled by, the vacancy, depending on the size of Region I and the cutoff declared for the potentials (see Section 2.5.3). The lack of convergence with respect to these parameters shows that calculation results are inconclusive. In all calculations conducted on the PAK2 lattice, on the other hand, dopant ions K_1 and K_2 on either side of V_K, which are initially separated by 9.92 Å, behave consistently by moving toward the vacancy until at equilibrium their separation is 5.73 Å, as shown in Fig. 5.10. Although the relaxation does not restore the K_1-K_2 distance to the PAK2 perfect-lattice repeat distance of 4.96 Å, the mobility of the K^+ ions enables them to close up to counter the effect of a gap in their ranks.

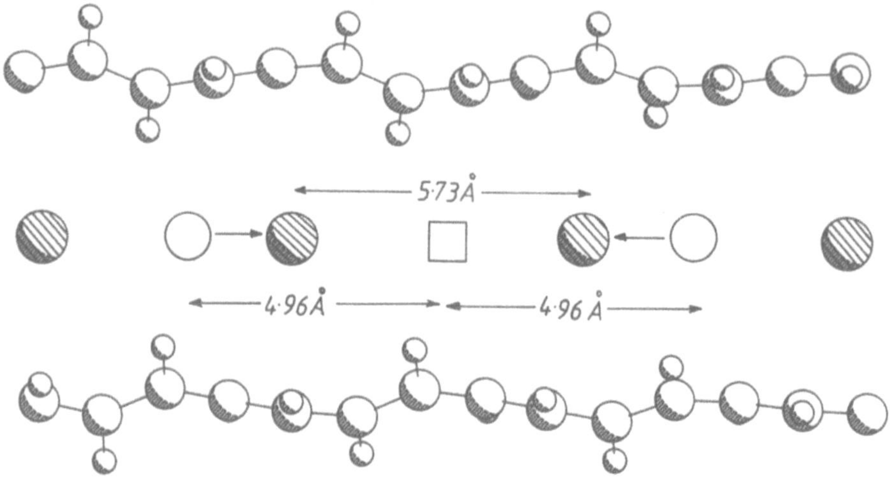

FIGURE 5.10. Response predicted by a lattice-defect calculation of a semidoped potassium-doped polyacetylene channel to a K^+ vacancy ○. The array of dopant ions ○ close ranks to eliminate the effect of vacancy. (From Ref. 107 by permission of the publishers.)

Mobility of the K^+ ions implies that their association with the PAc chains is weak (or at least not strongly directed). As the weak or undirected bonding limit of the two sublattices is approached, there is no reason to expect the periodicity of the K^+ ions along the channel to be compatible with that of the polymer. The resulting sublattices would be incommensurate, and as found in the simulations, they might exhibit a composition and dopant ion periodicities that need not be described by PAK1, PAK2, or any other commensurate structure.

This response of dopant ions to a vacancy in PAc differs from that in a conventional ionic lattice or even in graphite intercalated with alkali metal atom.[50,93] In these systems displacements exerted by a cation vacancy on its neighboring ions are usually so small (<0.1 Å) that cations can still be associated with their perfect-lattice sites; in such a case the vacancy retains its identity, and it can be described as stable. In the PAK2 lattice, on the other hand, behavior of the dopant ions shows that V_K is unstable under lattice relaxation. This is probably a consequence of the greater amount of space surrounding dopants in PAc than is usually experienced by K^+ ions in other lattices. The width of the channels (5.98 Å), which we explained owes more to the van der Waals contact of polymer chains than to the size of the cation, is significantly larger than the 5.4-Å layer separation in the graphite intercalate C_8K, for example,[109] in which a K^+ vacancy is stable.[110] A consequence of the instability of dopant vacancies in PAc is that it precludes estimating the energy barrier of a K^+ ion migrating via jump mechanism, as described in Section 2.5.2, which requires the two-vacancy, one-interstitial defect configurations to be stable.

Does the K^+ sublattice in potassium-doped PAc show a characteristic repeat distance? If so, does it depend on the doping level? Compositional and diffraction data are not unequivocal on this point,[97,102] but they do suggest that host and dopant lattices are incommensurate (as indeed X-ray diffraction has been established in the caesium sublattice of Cs-doped PAc).[100] This is also the conclusion of simulation calculations, which in addition predict a high mobility for the channel K^+ ions. Since the concept of lattice energy defined in Section 2.2 cannot be used to compare structures with different compositions, we must seek another method of predicting the optimum doping level.

5.3.3. Electrostatic Model

The lack of well-defined K^+ sites implicit in this description suggests that positions and the separation of dopant ions may be governed less by directed valence forces than by electrostatic interactions within and between the columns of K^+ ions and their surrounding sleeves of charge-compensating atoms. Although of course the actual channel structure is three-dimensional, it would be useful to examine to what extent the relationship between the two sublattices can be described by a one-dimensional model in which an array of positive ions is aligned along the axis of a hollow shell consisting of either (a) continuous or (b) discrete-compensating charges.

In case (a) the compensating negative charge consists of a continuous distribution on the walls of a coaxial cylinder of diameter 6 Å, while in case (b) the charge distribution, illustrated in Fig. 5.11, consists of a set of eight discrete negative charges ($q = \frac{1}{8}$), representing carbon atoms, around each positive charge, with geometry corresponding to the environment of each K^+ site in Figs. 5.8(a) and 5.9.

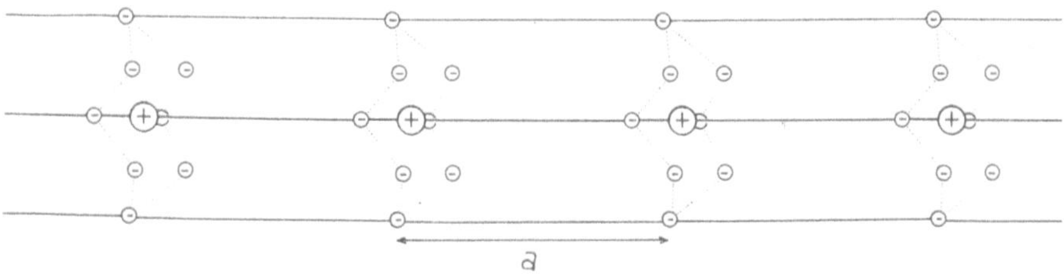

FIGURE 5.11. Model of a dopant channel used in the electrostatic treatment of potassium-doped polyacetylene, with a variable $K^+ \ldots K^+$ distance a to register the doping level. Each axial +1 charge is compensated by a serrated ring of eight negative charges. These are provided by two C atoms on each of the four surrounding polyacetylene chains, according to Figs. 5.8(a) and 5.9. (From Ref. 111 by permission of the publishers.)

The unit-cell energy E in case (a) can be expressed by applying electrostatics to the infinite chain.[111] The expression, which involves summations over the cells in the infinite linear lattice, was evaluated numerically. In the discrete negative charge model of case (b), the expression had to be transformed to reciprocal space for evaluation. Figure 5.12 shows the results of plotting the energy density E/a, which allows the stability of structures containing various ion separations a to be compared with different distributions of the compensating negative charge on the channel sides, simulating the polyacetylene chains (see Ref. 111). The energy minima obtained for the various models at $a \approx 4.2$ Å reveal rather a minor response from spacing in the positive ion array to quite a substantial change in the compensating charge distribution. This ion spacing is intermediate between the K^+ spacings

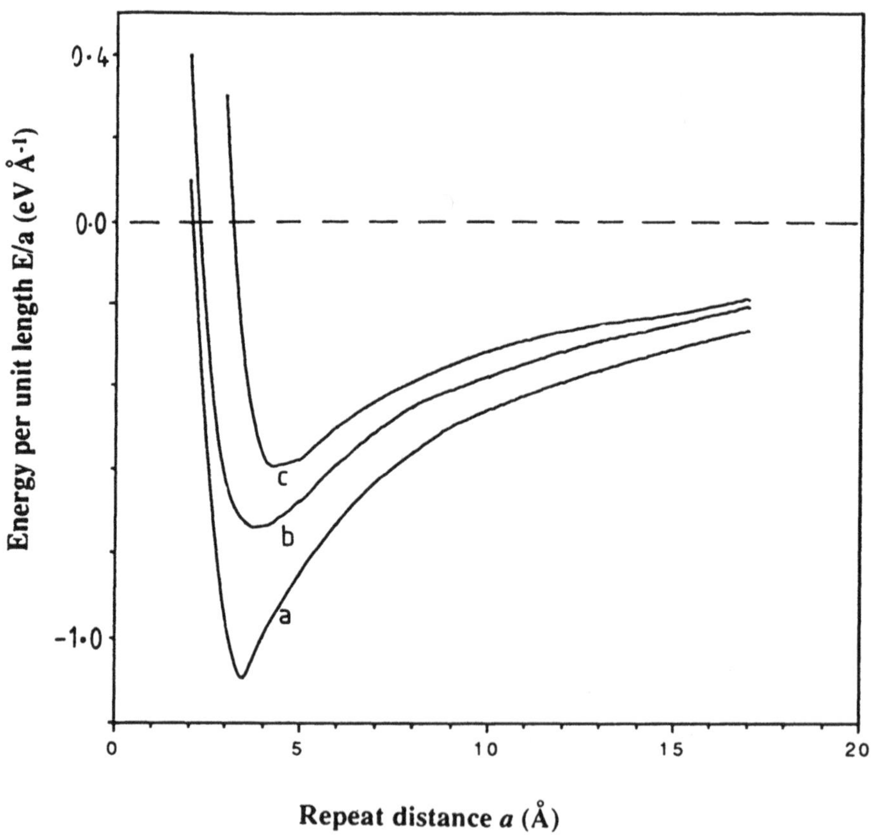

FIGURE 5.12. Energy density E/a of one-dimensional dopant channels in which the negative charge distribution on each polymer chain in the cells in Fig. 5.11 is spread over (a) one, (b) two, and (c) four C atoms. (From Ref. 111 by permission of the publishers.)

in the PAK1 (2.48 Å) and PAK2 (4.96 Å) structures, corresponding to incommensurate sublattices.

We have seen that the instability of a dopant vacancy in potassium-doped PAc does not allow the energy barrier to be calculated by standard static-lattice simulation methods. If however interactions between the host and dopant sublattices really are dominated by electrostatic effects, it is straightforward to calculate the energetics of the motion of a single K^+ in its channel through the charge distribution on the surrounding carbon atoms determined by static arrays of K^+ dopants in adjacent channels.[112] This calculation used the supercell method described in Section 2.5.1, where the unit cell was composed of 16 chains, each with 12 –CH=CH– segments, enclosing eight dopant channels. Of these channels seven contained the full PAK2 complement of three dopant ions in the unit cell, while the remaining channel was occupied by a single K^+ ion. The energy resulting from the motion of the K^+ ion in its channel showed a variation of ~0.1 eV, which can be regarded as an estimate of the barrier to the migration.

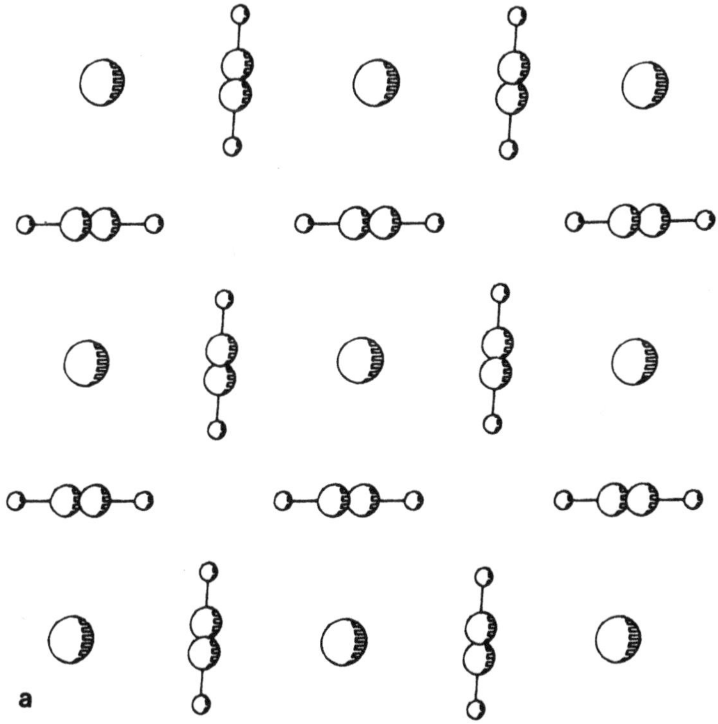

FIGURE 5.13. Lattice relaxation around a defect in potassium-doped polyacetylene, as predicted by the static simulation method. (a) The perfect lattice, (b) polymer chains in the environment of a K^+ vacancy. (From Ref. 92 by permission of the publishers of *Molecular Simulation*.)

5.3.4. Host-Dopant Interactions

Figure 5.13 shows another consequence of removing the symmetry constraints of the perfect lattice by creating a K^+ vacancy.[107,108] Whereas the unit cells in Figs. 5.8(a) and 5.9 generate PAK1 and PAK2 lattices that relax to a regular planar polymer backbone and an ordered arrangement of C–H bonds, additional degrees of relaxational freedom due to removing symmetry constraints enable distortions to occur that lower the energy still further. The result is a series of corrugations of the polymer backbone that when viewed along the chain direction (Fig. 5.13), are clearly revealed by the nonuniform directions of C–H bonds.

Distortions are also revealed in another defective-lattice investigation. Using the sum of mean square displacements in polymer atoms in the unit cell from their unrelaxed positions to monitor the distortion of the chains and declaring a single dopant ion to be simultaneously a vacancy and a mobile interstitial, the K^+ ion is displaced through the perfect-lattice site in the direction of the channel axis.

Results (Fig. 5.14) show the distortion rising sharply as the migrant K^+ ion shifts from its perfect-lattice site, thus lowering the lattice symmetry. The maximum reached by the distortion correlates with the minimum in the defect-energy curve,

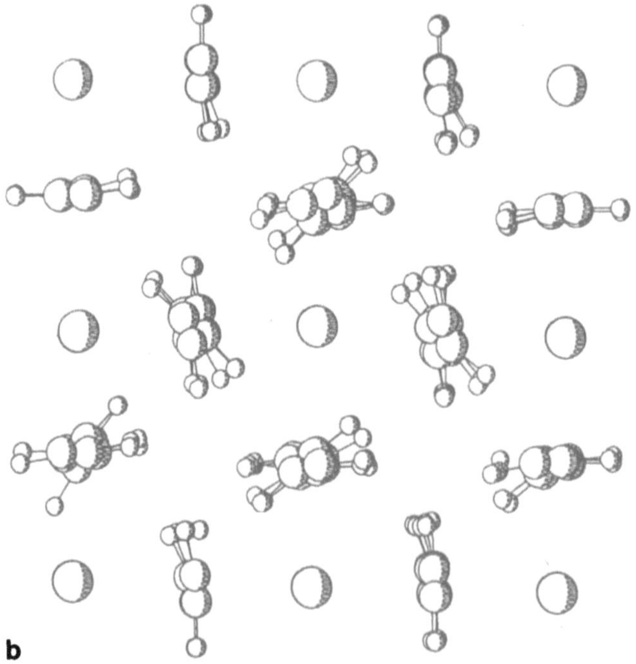

b

FIGURE 5.13. (Continued)

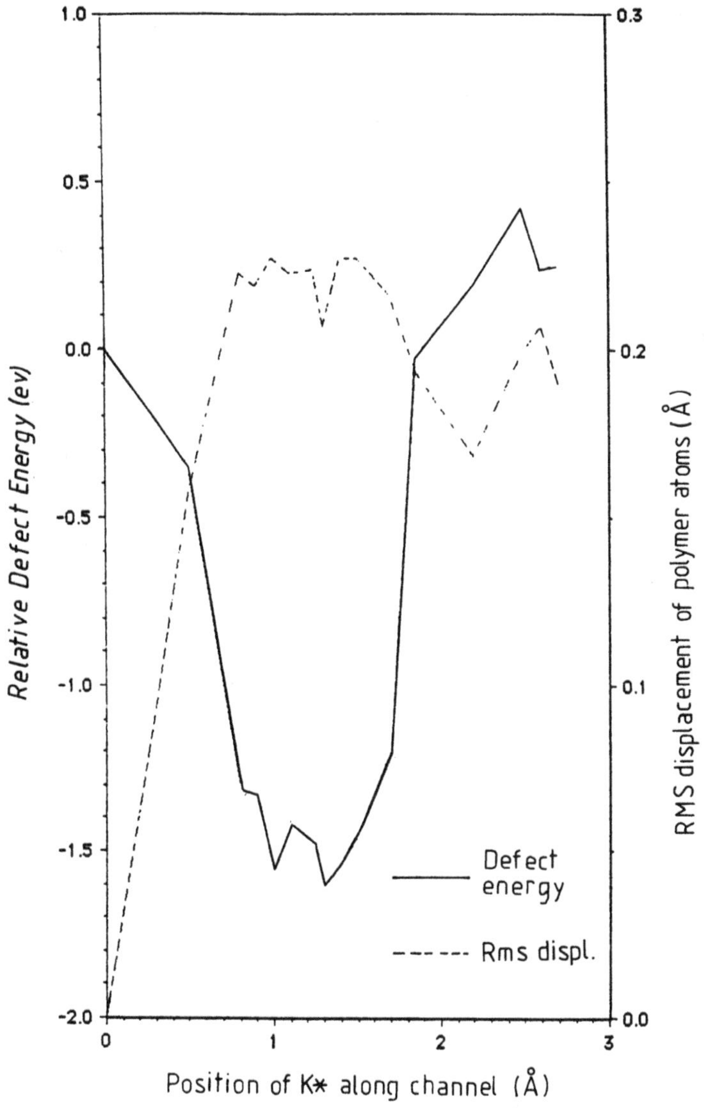

FIGURE 5.14. Energy variation and accompanying degree of lattice distortion to the displacement of a K^+ ion (labeled K^*) from its perfect-lattice site along the dopant axis in semidoped polyacetylene. (From Ref. 107 by permission of the publishers.)

revealing the tendency inherent in the simulated system to decouple the common translational periodicities of the host and dopant sublattices, i.e., to become incommensurate.

The subsequent behavior of the dopant ions is interesting. In Fig. 5.15 a section of the channel showing the positions of displaced K^+ ion (labeled K^*) and its nearest neighbors K_1 and K_2 is illustrated over the coordinate range in Fig. 5.14. As K^* is displaced to the right, it is followed by neighbor K_1, roughly maintaining the PAK2

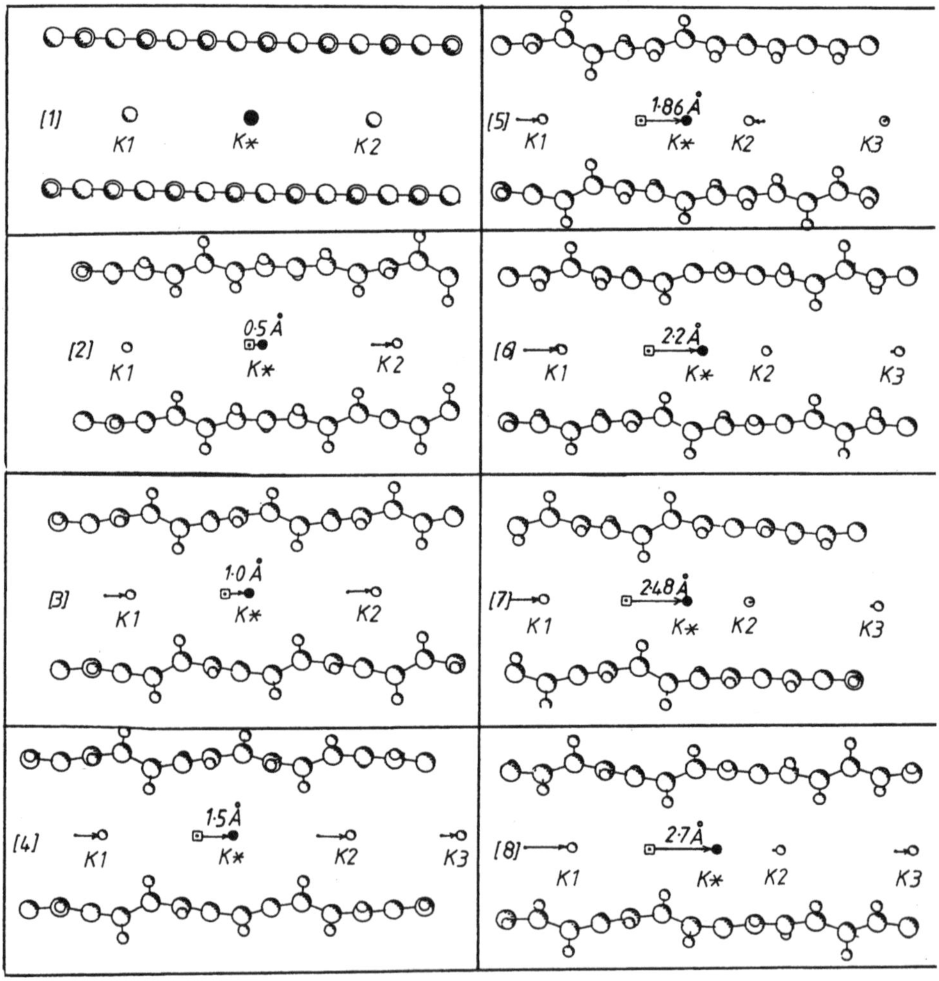

FIGURE 5.15. Response of host and dopant sublattices to the displacement of a K^+ ion in Fig. 5.14. (From Ref. 107 by permission of the publishers.)

repeat distance of 4.96 Å. At first, neighbor K_2 that is ahead of the displaced ion also keeps its distance from K^*, which is pushing it from behind. Then when K^* has been displaced by 1.86 Å, K_2 is abruptly attracted to its pursuer, the K^*-K_2 separation closing to the K^+ ion distance (2.48 Å) in PAK1, which is maintained as the train of ions advances.

The interpretation of all aspects of this calculation is not straightforward. The energy barrier of 1.5–2.0 eV in Fig. 5.14 does *not* apply to the natural migration of K^+ dopants along the channel, since such motion is expected to be limited without the presence of a vacancy in the dopant sublattice; it is merely the response of the column of K^+ ions when one of them is displaced a distance to which the application of the defective-lattice method is valid. The significant result of the calculation is that dopant migration seems to follow a cooperative mechanism.

5.4. Doped Polyparaphenylene

5.4.1. Perfect Lattices

The crystal structures of several doped species of Li^+ have been determined by diffraction methods. While uncertainty shrouds the structure of lithium-doped PAc, it has been shown that (probably because of the small size of the Li^+ ion) its accommodation in the polyparaphenylene (PPP) lattice occurs with very little modification to the position of the polymer chains.[26,92] The Li^+ ions occupy sites on the monoclinic (nearly orthorhombic) *a* and *b* axes, but although they thereby experience a channel-like environment, the walls of the channels are rhombic rather than tetragonal.

Corish *et al.*[106] simulated a Li^+-doped PPP lattice assuming planar polymer chains in, which all bond angles were 120° and ring and interring C–C bond lengths were, respectively, 1.4 and 1.5 Å. Researchers obtained a relaxed lattice only when the dopant ion was sited on either the *a* or *b* axis (but not on both, since such a trial structure failed to produce a stable lattice). This has the composition $[C_6H_4Li_{0.5}]_x$, which in fact corresponds to the highest reported doping level for the Li-PPP system.[92] As is also the case for PAc, larger dopant ions do modify the pristine PPP lattice (Fig. 5.8), with K^+ and AsF_6^- generating intercalationlike sites.[113,26]

5.4.2. Defective Lattices

Lattice simulations in Ref. 106 include structures of PPP doped with Li^+, K^+, and AsF_6^- ions, which led to fully relaxed structures with unit-cell parameters that were reasonably close to experiment.[26,92] Because of the simplicity of its unit cell, the Li-doped species was selected for a defective-lattice-type simulation.[107]

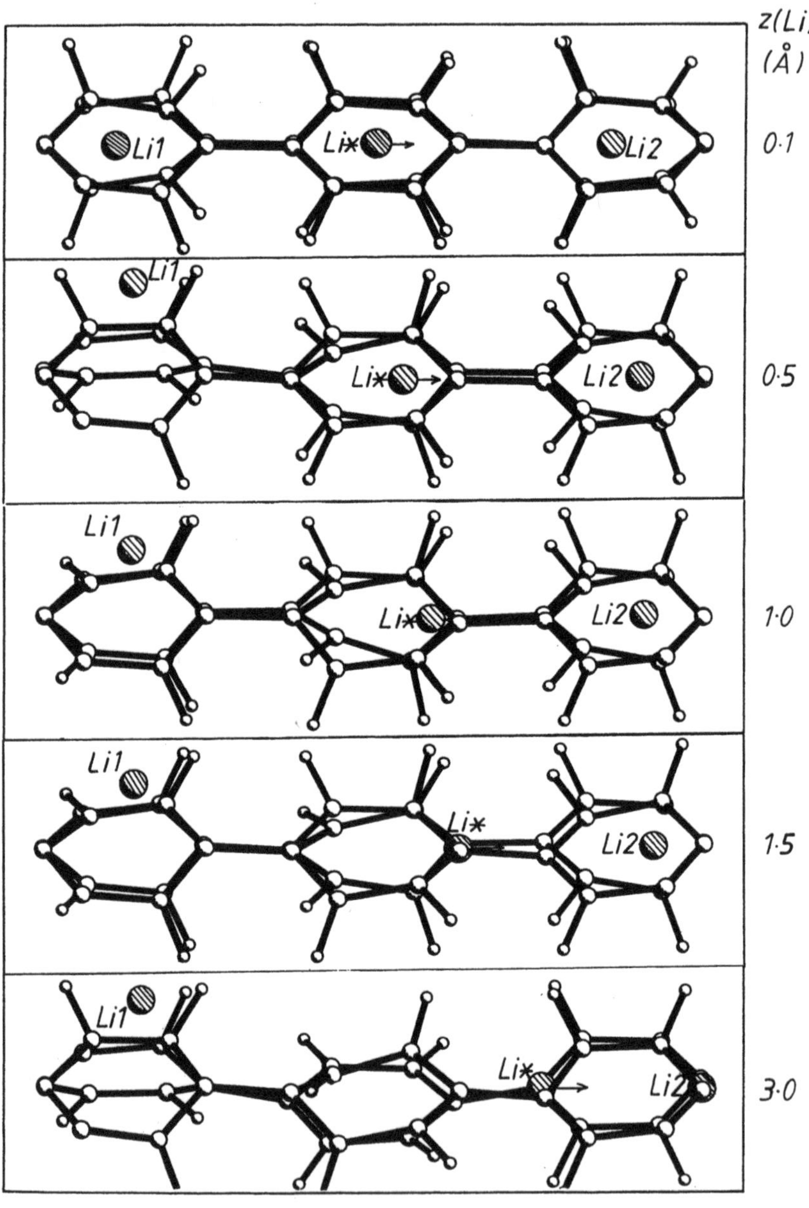

FIGURE 5.16. Response of host and dopant sublattices to various displacements $z(\text{Li})$ of a Li^{+} ion (labeled Li^{*}) from its perfect-lattice site in Li-doped polyparaphenylene. Dopant ions are aligned between two PPP chains. (From Ref. 107 by permission of the publishers.)

In PPP dopant ions Li^+ and K^+ appear to be more firmly held in their sites than in PAc. This is probably a consequence of energetically favorable distances and chain orientations of the phenylene rings in the host lattice, resulting in a greater number of short M^+...C interactions than occur between the M^+ ion and the –CH=CH– groups in PAc. As a result a vacancy in the dopant sublattice is not eliminated by adjoining ions moving in; instead such a vacancy destabilizes the simulated lattice. Forced migration of a single lithium ion Li^* using the method described for K^+ in PAc produces a very different effect on its nearest neighbors Li1 and Li2 than is encountered in potassium-doped PAc.[107] Figure 5.16 shows that displacing Li^* along the chain direction has a disruptive effect on its own sublattice. This should not be surprising because even if Li2 were not present, the passage of a Li^+ ion over the carbon atom of the interphenylene bond is energetically highly unfavorable, as it would bring the ion into a repulsive portion of two Li...C potentials. In the analogous migration of intercalated lithium in graphite, for example, the minimum-energy path of Li^+ takes the intercalant ion on a detour over an intermediate ring.[50,93] This has not been established with certainty in Li-doped PPP, but the simulation work strongly suggests that the dopant migration is less straightforward than it is in PAc.

5.5. Ion Migration in Polyacetylene and Polyparaphenylene

Simulation results show a high mobility of K^+ ions in PAc only for those ions in the channels. But since doped polymer samples are heterogeneous, obviously we cannot pretend that this description of dopant motion applies throughout the sample. The overall diffusion in the sample is, however, controlled by the slowest step, which it is therefore important to try to find. Now we must remember that:

(a) Electron microscopy reveals that the morphology of PAc consists of fibrils between which there is a large proportion of the volume of the sample (about two-thirds) containing empty space.[114] It therefore follows that a significant part of the migration of dopant ions involves transportation through a void.
(b) In amorphous regions where a precise structural knowledge is inaccessible, a detailed description of the migration mechanism is not possible.
(c) Even if crystalline regions of the polymer sample contain sufficient periodicity to produce a diffraction pattern, the long-term order is far short of that found in ionic and molecular crystals.
(d) Although lengths of perfect PAc chains are taken as infinite in the lattice simulation, in practice they are limited to ~1000 atoms.[115]

Taken together these facts imply that dopant ions encounter a significant degree of nonchannel environment during the migration through the sample, and they raise several questions. What is the structure of the channel ends, and what happens when dopant ions encounter them? Is migration hindered by the congesting effect of possible nonplanar chain end groups? Does this, and its passage through interchannel regions consistitute bottlenecks to the motion? Certainly the diffusion of some dopants at least appears to be much more rapid in environments in which they are not in close proximity to polymer chains: for iodine (probably as I_3^-), diffusion coefficients of D ~10^{-5}–10^{-6} $cm^2\ s^{-1}$ (comparable with values in fluids) have been reported for transport in the interfibrillar void of doped PAc samples, while those in the fibrils are ~10^{-16}–10^{18} $cm^2\ s^{-1}$.[116] With the possible exception of very small dopants, such as Li^+, we would expect migration to be significantly hindered in regions of the sample where the doping level is low, whether or not they are crystalline, since here the ions must provide the energy to produce structural changes in the host polymer lattice that are necessary to accommodate them in the doped regions.

Simulations reveal structural interactions between the migrating dopant ion and the host lattice in which polymer chains respond to the migrant by distorting from planarity. As would be expected, this interaction increases rapidly with the flexibility of the chains. In Section 6.2 we see that in nonconjugated polymers (i.e., those with no π electronic component to favor chain planarity), the host dopant interaction plays an essential role in the jump mechanism in migration.

5.6. Polypyrrole and Polythiophene Systems

5.6.1. Molecular Structures and Conformations

In the polypyrrole (PPy) and polythiophene (PTh) structures $(C_4H_2X)_n$ of principal interest in the study of conductive polymers, the rings (in which X is, respectively, NH and S) are linked at their 2 and 5 positions, as illustrated in Fig. 5.17. In this chain configuration, which involves the linear repetition of the antiparallel two-ring segment $-(C_4H_2X-C_4H_2X)-$, steric interactions between adjacent rings is minimized when the polymer is all-planar, which is the conformation expected to favor an electronically conducting state.

Some of the interest in these polymers follows from their chemical stabilities[117,118] and for PPy, from the solubilities and low-temperature fusibilities of the alkyl substitution products. Although crystallinities of 50% have been achieved in PTh,[118] as a rule they are significantly lower than in PAc and PPP; PPy, in fact, has very low crystallinity.[117] But while diffraction investigations have produced no structural description of PPy, those on PTh have led to the proposal of an orthorhombic lattice as shown in Fig. 5.18.[89–91] The unit cell contains two all-planar chains in an antiparallel conformation, making setting angles ϕ and $-\phi$, where $\theta = 68.8°$, thus sharing structural similarities with PAc and PPP.

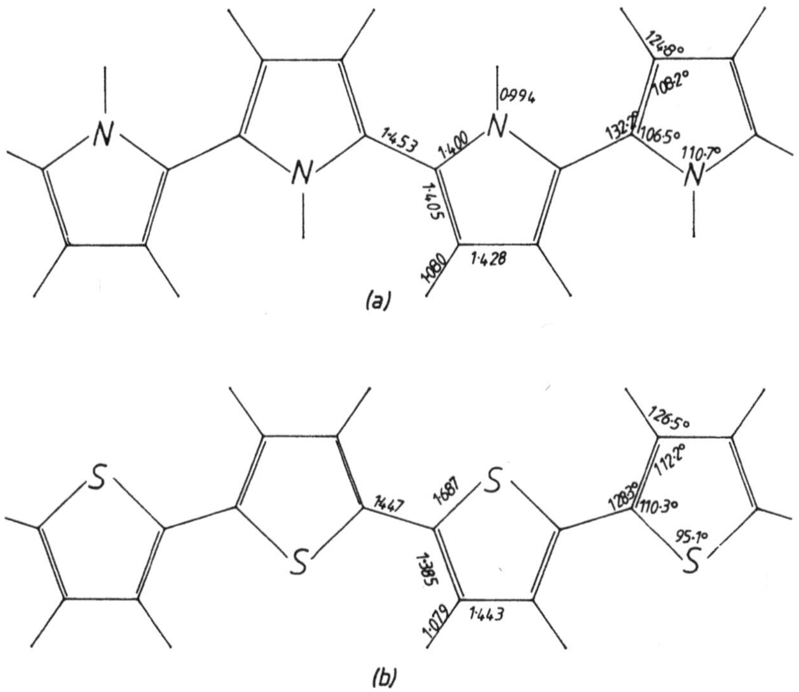

FIGURE 5.17. Chain structures and molecular geometries of (a) polypyrrole and (b) polythiophene. Bond lengths are in Å.

Since the synthesis of the polymers is usually accomplished by an anodic electrolytic polymerization the dopants are commonly anions of the electrolyte, which enter the chains *in situ*. They may subsequently be undoped electrolytically to yield the pristine polymer.[117,118] Many of the structure investigations have thus been on doped PPy and PTh samples, which appear to have morphologies that are more complex than those of PAc and PPP. For example, in their X-ray investigations on films of poly(3-alkylthiophenes) Mårdalen *et al.*[67] found that during the stretching process, some of the crystallites would first dissolve and then recrystallize. Diffraction studies[119] find the carbon skeleton of the chains to be all-planar, although the thermochromic properties of the material have been interpreted as a distortion of the regular conformation of the main chain above 190°C.[120]

A contributory factor at least in the morphological complexity of these polymer systems is the tendency of their chains to form helical conformations. The application of various electron microscopy methods[34] to PPy, PTh, and their substituted products has revealed the presence of helical and sometimes superhelical structures,[31] as well as (in one case at least) nearly parallel, straight-chain conforma-

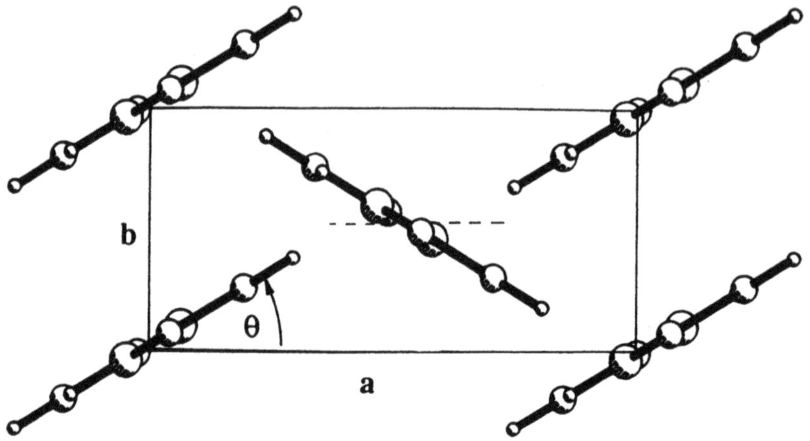

FIGURE 5.18. Unit cell of pristine polythiophene (also assumed for polypyrrole). (Adapted from Ref. 125 by permission of the publishers.)

tions.[33] The X-ray diffraction investigations have also reported the presence of helical lattices in $CF_3SO_3^-$ -doped PTh, and lattice vectors have been derived.[35] The number of thiophene rings per loop of the coil is about 11. The repeat distance $|\mathbf{c}|$ = 12.2 Å in the chain direction allows the pitch and the number of loops per translational repeat unit (~2) to be inferred.

In a theoretical approach Cui and Kertesz[121] conducted a series of semiempirical quantum chemical calculations on the conformational potential energy surfaces of single chains of PPy and PTh and their substituted products. Their results predicted distinct energy minima at a planar and helical chain conformation 0.08 eV higher, separated by a barrier of 0.2 eV.

Various experimental and theoretical studies seem to imply that samples of these two polymer systems probably contain various chain structures in both their crystalline and amorphous regions. A sensible approach would then be to conduct simulations separately on lattices consisting of straight-chain and helical configurations.

One problem associated with the simulations of helical chains is that most program codes generate the chain in the lattice from a precisely defined repeat segment along the *c*-axis of the helix. There is no certainty that such a purely translational repeat segment exists; even if it did, it might in general require an impossibly large unit cell. An alternative procedure is to specify a slightly modified chain structure defined by a repeat segment that can still model structural features of the actual helix. The unit cell defining the unrelaxed lattice would then be specified by (1) the repeat distance $|\mathbf{c}|$, (2) the number of basic structural segments (here the pyrrole or thiophene rings), and (3) the number of loops of the coil. Unlike the case of straight-chain lattices the maintaining of translational symmetry along the helical axis places severe restrictions on changes in (1)–(3) that are permitted to occur during lattice relaxation.

In a recent atomistic simulation of helical PTh, Corish *et al.* found agreement between the calculated and measured structural parameters of PTh helices as single chains and in lattices,(122) both of which have been investigated experimentally,(31,34,35) and which can apparently coexist in the same sample. Having defined the torsional angle ϕ around the inter-ring bond as zero for the *anti* conformation (Fig. 5.17(b)), they found a global energy minimum in the torsional energy profile for this conformation and a sharp relative minimum at $\phi = 177.3°$, i.e., near to the *syn* conformation (for which $\phi = 180°$, and which is precluded on steric grounds). The stabilizations of the planar $\phi = 0°$ and near-planar $\phi = 180°$ conformations are a consequence of the stabilization by the π-electron system (enhanced by doping,

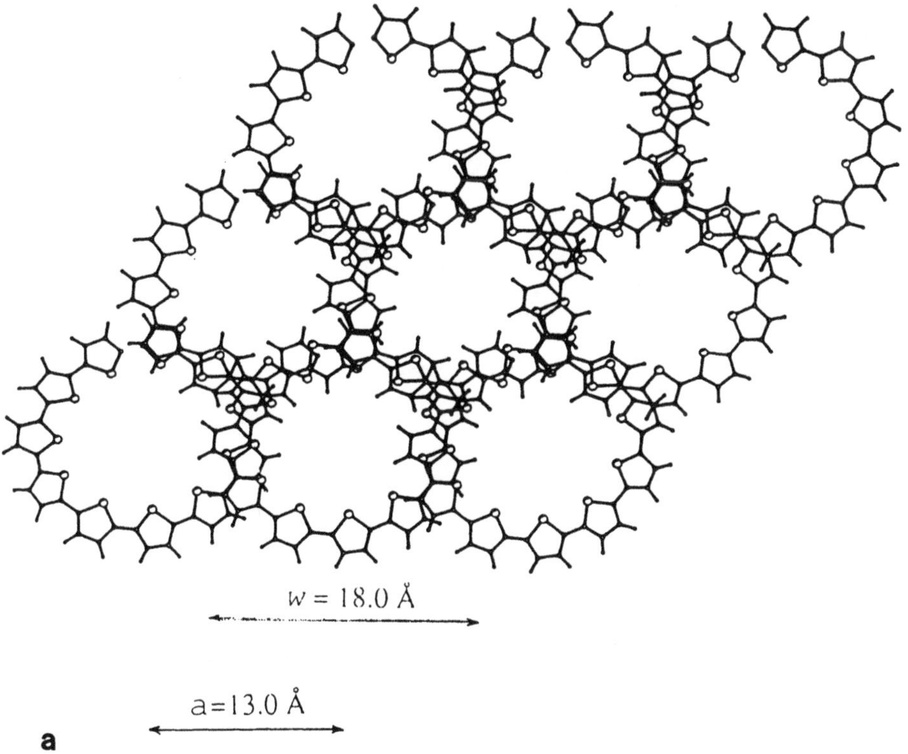

a

FIGURE 5.19. The lowest-energy relaxed lattice structure of helical polythiophene projected on planes (a) normal to, and (b) containing the helical axes, showing the cell length **a**, the external width **w** and the pitch **p** of the helices. In (a) three cell lengths along each axis have been taken to show the almost hexagonal environment of each chain. The pair of adjacent chains in (b) has been included to show the extent of their interleaving. (From Ref. 122, by permission of the publishers.)

as discussed above), and favorable stacking interactions between the stacked thiophene units, when the chains are planar. The calculations[122] predict that the helical chains in the lattice regions of the material interleave (interdigitate) in such a way that each turn of a given helix intercalates between two turns of the six surrounding helices (Fig. 5.19). As a result, the only lattice space available for use as sites for dopant ions is in the axial cavities of the polymer helices, which have a diameter of 11 Å.

Most of the current work on solid state PPy and PTh has been on materials exhibiting the conformational *anti* structure, whose planarities are facilitated by the transfer of electronic charge between the host and dopant sublattices.[123] Orchard

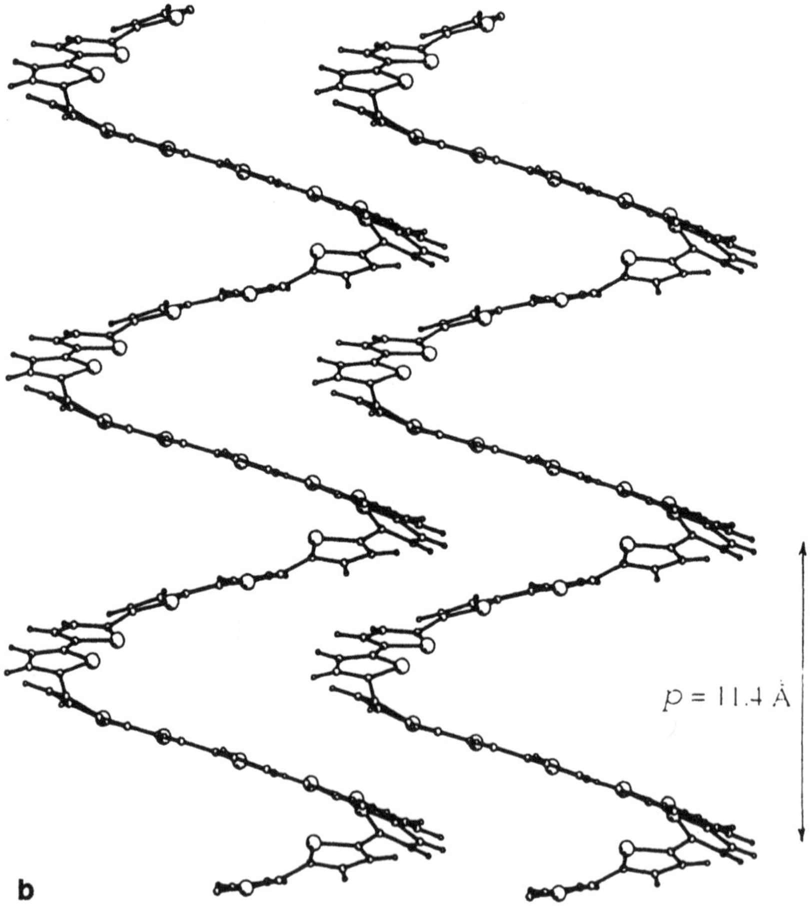

b

FIGURE 5.19. (*Continued*)

et al.[124] have conducted chain packing calculations on a large number of crystal systems, energy-optimizing the lattice vectors and chain setting angles. Recently, Corish *et al.*[125] simulated a set of fully-relaxed monoclinic lattices of PPy and PTh that were similar to those in pristine PAc and PPP (Table 5.5). Although no experimental data are as yet available to verify the result for PPy, their two lowest-energy PTh structures, with $|\mathbf{a}| = 7.4$, $|\mathbf{b}| = 6.1$ Å and $|\mathbf{a}| = 11.1$, $|\mathbf{b}| = 4.1$ Å, were quite close to the results of X-ray investigations on powder samples[90] ($|\mathbf{a}| = 7.8$, $|\mathbf{b}| = 5.5$ Å) and on drawn films[91] ($|\mathbf{a}| = 10.8$, $|\mathbf{b}| = 4.7$ Å). The drawn-film investigation did not provide a setting angle to compare with the predicted $\phi = 28°$ value,[125] but the angle $\phi = 68°$ calculated for the second simulation structure agrees excellently with the 69° derived from the powder studies.

If the three positions of the aromatic rings in PPy or PTh are substituted with alkyl (or alkoxy) chains, polymers are obtained that are structurally similar to the parent polypyrrole or polythiophene. An example of this substitution is the series of poly(3-alkylthiophenes) (P3AT) which exhibit the useful properties of solubility in common solvents and of enhanced environmental stability. Kawai *et al.*,[119] who investigated the changes in the lattice vectors of poly(3-alkylthiophenes) with doping levels of BF_4^-, AsF_6^- and toluene sulphonyl anions, conclude that the dopant ions assume sites between the chains and are coplanar with them. In relation to polymers in adjacent planes the anions constitute intercalants, although their sites are not between neighboring pairs of thiophene rings.

A simulation of the lattices of these materials was conducted recently[126] in order to investigate the interaction between the alkyl side chains on neighboring polymers and their response to pressure. The calculations required that in the most stable configuration of P3AT the alkyl chains were all-*trans*, with their carbon atoms coplanar with the main chain. The interleaving (interdigitation) predicted for the alkyl groups on adjacent, coplanar polymer chains is in accord with most of the X-ray investigations, although at least one diffraction group[127] favors an abutted structure, i.e., a non-interdigitated one in which the alkyl groups are tilted out of the plane of the main chain. Although the simulation[126] predicts a planar conformation in the undoped material, it appears that on doping, the alkyl groups readily assume an abutted conformation.[128]

The simulation studies on P3AT[126] concluded that when the alkyls were longer than C_4H_9 the polymers are expected to show several stable inter-chain separations as the alkyl chains pass through various degrees of interleaving (interdigitation). The authors related the effect to the lubricative properties of soft comb-like polymer chains, particularly at high pressures.

5.6.2. Simulating Doped Lattices

Among the species that dope PPy and PTh and render the polymers electronically conducting are BF_4^-, ClO_4^-, Cl^-, $CF_3SO_3^-$, $CH_3C_6H_4SO_3^-$, PF_6^-, and AsF_6^-.[117,118,129]

To date no systems containing n-dopants (cations) analogous to alkali-metal-doped PAc and PPP have been reported for PPy and PTh. The paucity of structural data makes simulation difficult, but guided by results of studies[67] on alkyl derivatives of PTh described in Section 5.6.1, it is reasonable to assume that polymer chains approximately retain their pristine lattice structures and anions take up interchain positions. By sweeping initial lattice vectors and chain-setting angles through the appropriate ranges, the process of lattice relaxation can then be used to yield the most energetically favored lattices.

Figure 5.18 shows the *c*-axis view of the monoclinic cell used to simulate the lattices of PPy and PTh doped with BF_4^-, ClO_4^-, and Cl^-.[125] Since structures of polymer chains were assumed to be the same as those in the pristine lattices, the unit cell contained two C_4H_2X rings (X = NH or S) in an antiparallel conformation in both chains. For these systems the highest doping level that appears to be sensitive to polymerization conditions occurs for $SO_3CF_3^-$ in PTh, reported as 50% (percentage dopant per C_4H_2S ring).[130] However most research groups refer to upper limits of 25% for the anions described here, implying a unit cell containing a single anion and the composition $[C_4H_2X.A_{0.25}]_x$.

In the same simulation investigation,[125] a dopant anion was placed near the midpoint of the *a* or *b* axes in Fig. 5.18 prior to energy minimization. As is usual for polymer systems, the relaxation produced several stable structures for each species and a fairly constant set of lattice vectors, which reveal only small expansions of the unit cell to accommodate the dopant anion. Structures obtained are mainly distinguished by the positions of the anion, which finds a number of stable sites. The principal change in the polymer sublattice is the setting angle, which in each case becomes zero. This indicates that doping transforms the chain settings from a herringbone arrangement to one that is parallel and anions intercalate into sites that Fig. 5.20 shows to be simultaneously between, and coplanar with, polymer chains.

Although to our knowledge no experimental confirmation of these predicted structures has yet been advanced, several lines of evidence suggest that the model provided by the simulation studies is a not unreasonable.

Diffraction work on alkyl-substituted PTh doped with the same anions[67] leads to an intercalation model similar to the simulated structures.

Although they do not present a structure of doped PTh, in their account of their X-ray investigations on the pristine polymer, Brückner and Porzio[90] describe the diffraction obtained from a ClO_4^- doped sample. From the similarity of certain peaks to those of the undoped polymer, they conclude that interplanar spacing is not greatly affected by dopants, but from the increased peak widths, they surmised that the ClO_4^- ions occupy random sites. While this does not necessarily imply intercalation by the anions, the predictions of several dopant sites with similar energies suggests the possibility of random occupations, and

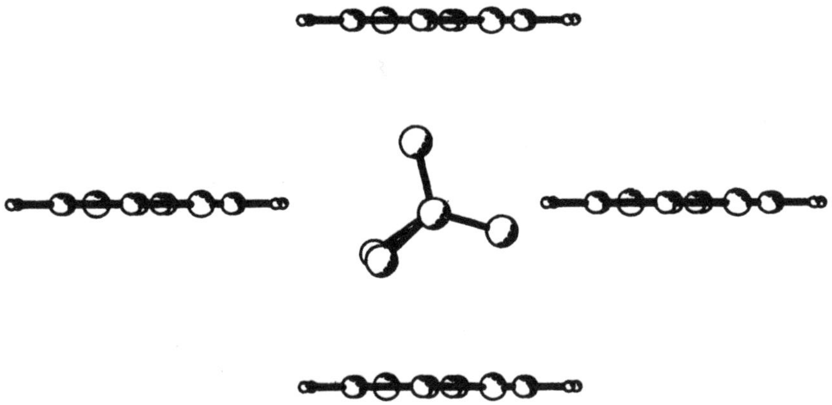

FIGURE 5.20. Unit cell adopted by the fully relaxed lattices simulated for polypyrrole and polythiophene doped with BF_4^- and ClO_4^-.

the rather small changes in lattice vectors on doping is consistent with sets of common repeat vectors.

Diffraction studies on the structures of doped polyparaphenylene show that although the polymer chains retain their nonzero setting angles in the presence of alkali metal cations, polyatomic ions such as AsF_6^-[113] may well impose an all-parallel arrangement on polymers [see Fig. 5.8(d) and Stamm and Hocker's comments[26] on the most favored structure of AsF_5-doped polyparaphenylene]. The latter structure is rather similar to that found in the simulated doped PPy and PTh lattices, with the ions occupying sites that are coplanar with the chains.

5.6.3. Ion Migration

As described for doped PAc and PPP, removing a dopant ion in lattices PP and PTh also results in chain distortions,[131] mainly due to the relaxation of translational symmetry restrictions. However now adjacent ions are pulled in only ~0.2 Å toward the vacancy site, and consequently an ion vacancy can be described as stable. As a result the motion of anions through the lattice via a jump mechanism involving a net cation vacancy can be investigated, as described in Section 2.5.4, by calculating lattice-defect energies associated with a series of defect configurations. In the simplest case—the two-vacancy/one-interstitial defect—each defect configuration consists of a vacancy at two adjacent anion sites and an interstitial anion, with the whole equivalent to a net vacancy. With the interstitial at various positions in the environment of the vacancy pair, a sequence of defect energies is calculated for a number of proposed pathways for the dopant ion. An exhaustive set of such

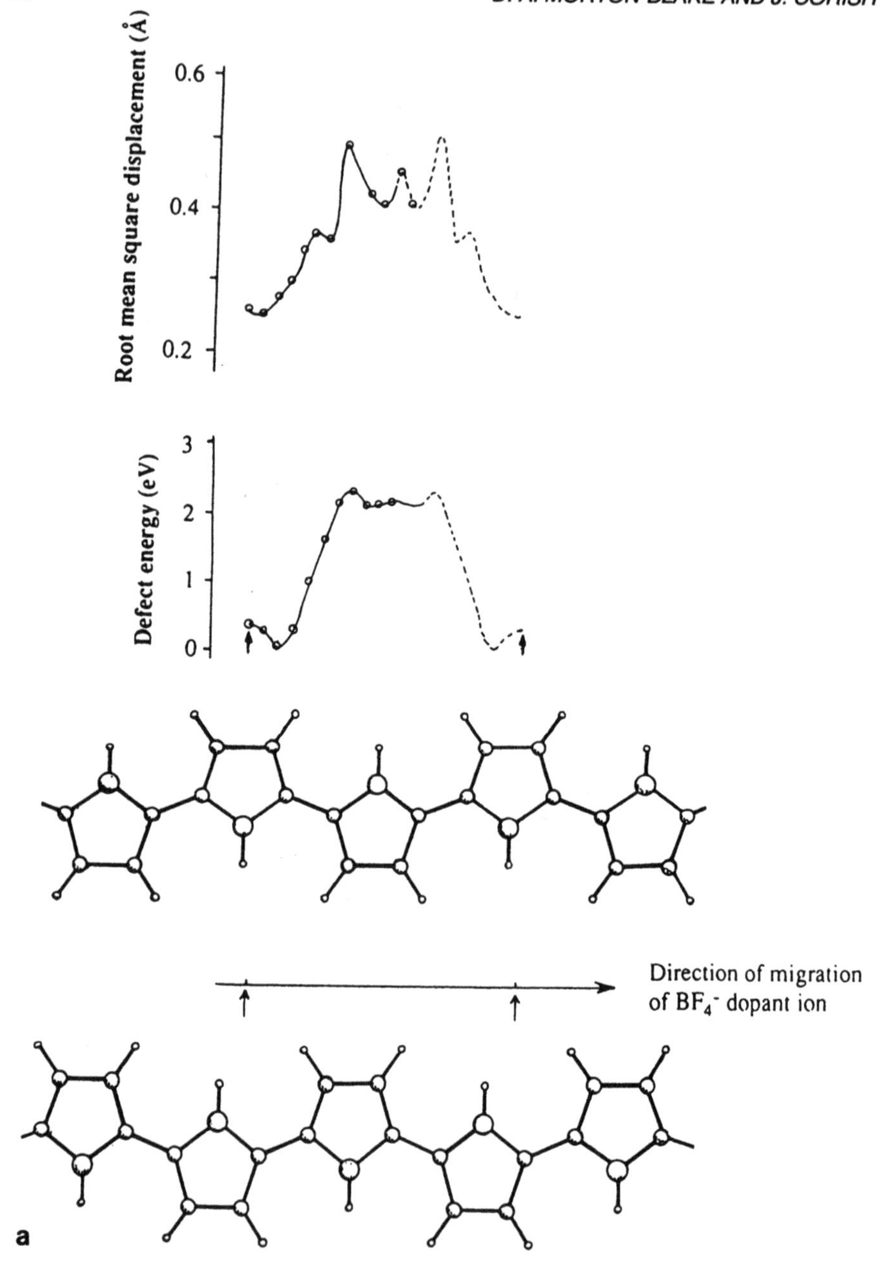

FIGURE 5.21. The minimum-energy migrational transition or jump of a BF_4^- dopant ion in (a) polypyrrole and (b) polythiophene. The short arrows show the positions of the structurally equivalent initial and final sites of the jump, and the defect energy curves and the r.m.s. distortions are drawn to the same scale for the two polymers in order to show their contrasting profiles. (From Ref. 132 by permission of the publishers.)

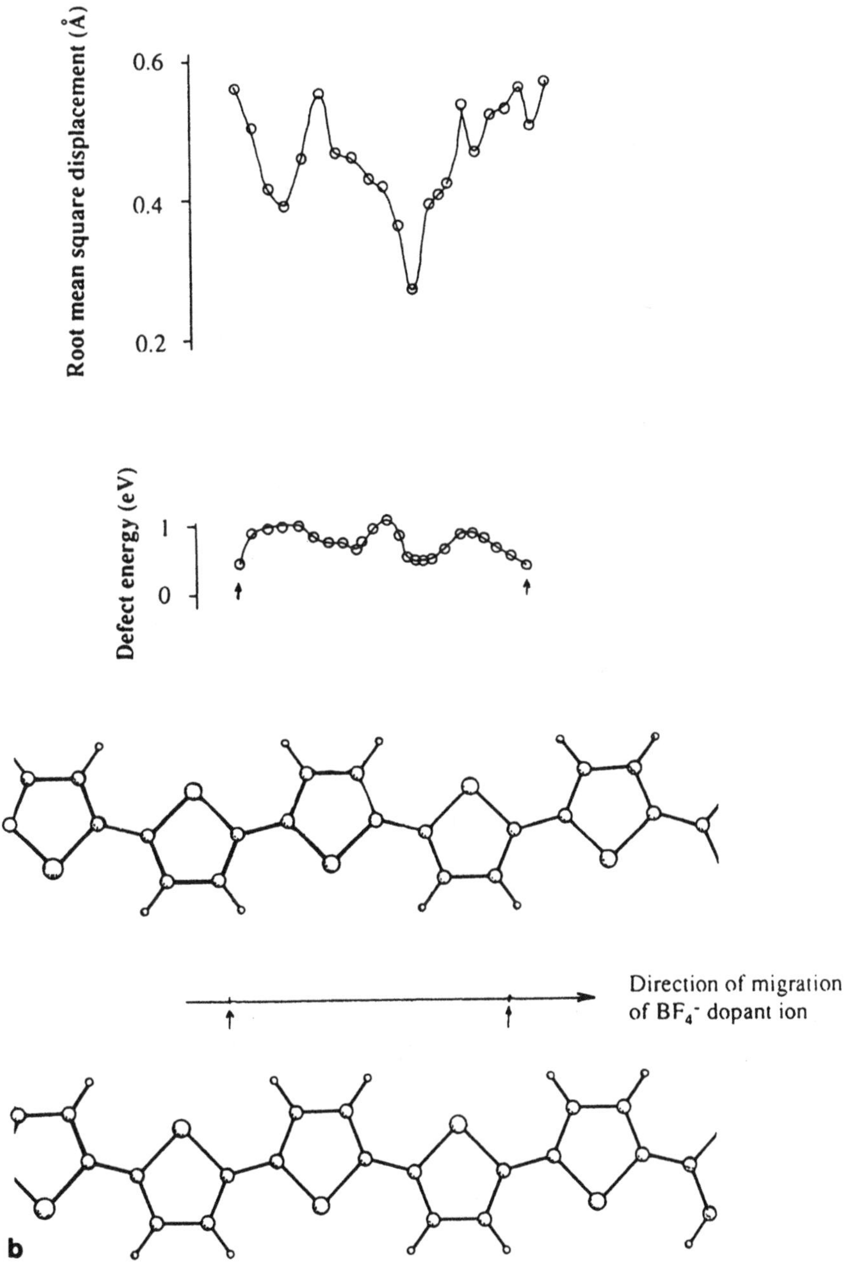

FIGURE 5.21. (*Continued*) Since the migrations result in slight structural modifications in the polymer chains, the positions of the atoms in the lower section of the diagrams are only approximate. For clarity, the changing orientation of the BF_4^- ion on its trajectory is not shown, neither is its precise trajectory which is almost, but not quite, a straight line.

quantities would provide a portion of the total defect energy surface and hence a dopant migration energy profile and trajectory.

Despite the structural similarities of PPy and PTh and dissimilarities (in size and shape) of the dopants BF_4^- and Cl^-, simulated migrations[131] show a greater sensitivity to the polymer lattice than to the particular dopant. During the course of its jump between a pair of equivalent vacancies in PPy, Fig. 5.21(a) shows that the Cl^- encounters two large barriers, thus defining an activation energy of 2.2 eV and a new (higher energy) dopant site. For the second leg of the jump, the Cl^- requires the subsequent expenditure of a rather smaller energy (1.5 eV). Simulation also predicts a second Cl^- site in PTh [shown in Fig. 5.21(b)], but in this case lattice relaxation around the defect makes this site more stable than the perfect-lattice site. Here barriers are in the range of 0.5–1.0 eV, i.e., significantly lower than in PPy.

Comparing the profile with the dopant position along the polymer direction (Fig. 5.22) shows that the double barrier arises from the ion passage between pairs of C–H bonds on adjacent polymer chains. Figure 5.21(a) shows that on either side of this constriction, there are two cavity regions coincident with the primary and secondary dopant sites shown in the energy profile. Molecular plots of Cl^- at positions corresponding to barrier summits in both polymers show an out-of-plane twisting in a ring in PPy that is absent in PTh.

A similar distinction between the two polymers is shown[132] when the dopant is BF_4^-. On leaving its primary site in PPy the dopant encounters a barrier of 2.1

a

FIGURE 5.22. The BF_4^--doped polypyrrole lattice and the distortion of the lattice during the migration of a dopant ion between two vacant sites. The sequence shows the environment of the dopant ion (a) in its 'perfect lattice' site, (b) when the migrant BF_4^- is at its initial (or final) site in the defective lattice and (c) when the BF_4^- is at the summit of the highest energy barrier in Fig. 5.19(a). (Adapted from Ref. 132 by permission of the publishers.)

eV, but the secondary site described for Cl^- is now replaced by a shallow crater, presumably because of the larger size of the BF_4^- ion. In PTh the BF_4^- ion (like Cl^-) encounters small energy barriers of ~0.5 eV.

The origin of the low barriers for PTh doped with either Cl^- or BF_4^- appears to be connected with the absence of the large cavities between adjacent coplanar chains that in PPy offer stable sites to the dopant ion.(131,132) As a result the dopant migration in PTh is, as shown in Fig. 5.21(b), associated with a more uniform potential energy profile than that in PPy. Although, to our knowledge, no experimental data are available to compare with the simulated ion migration in the

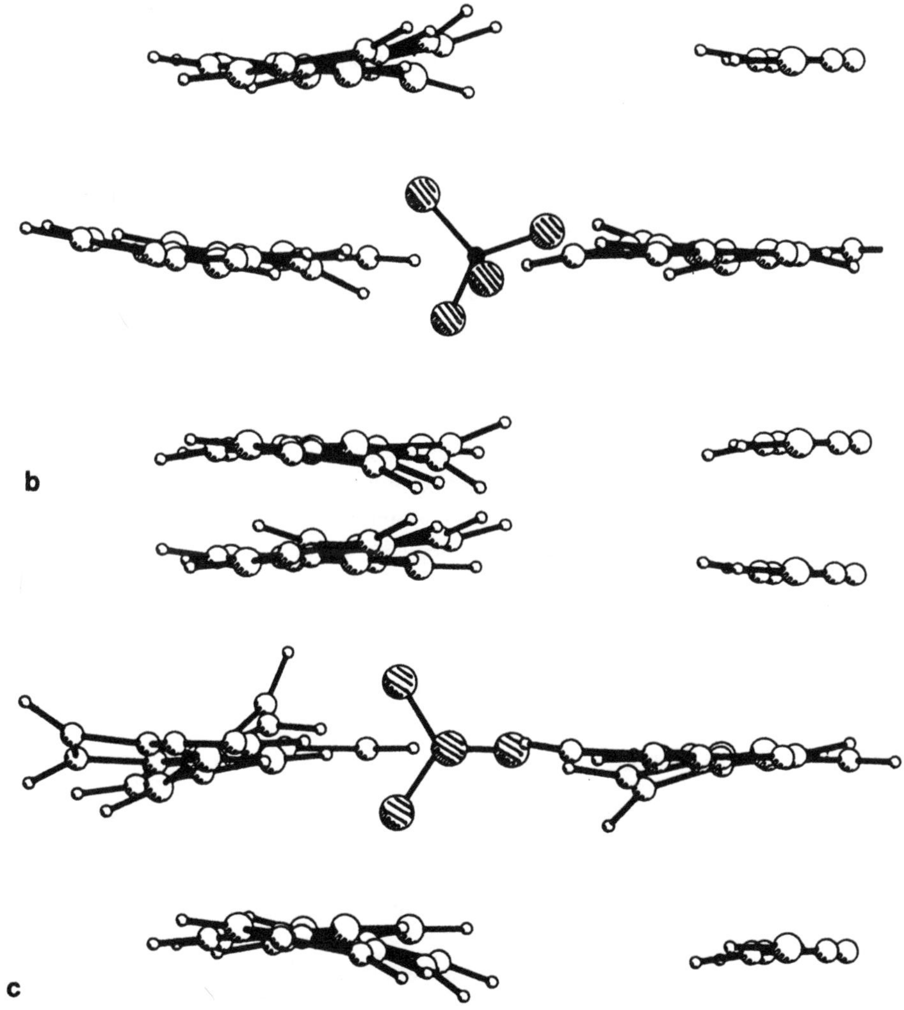

FIGURE 5.22. (*Continued*)

described 25% doped lattices of PPy and PTh, it is interesting that in a recent electrochemical investigation cf PPy doped to 50% with Cl^- ions,[133] the migration of the dopant was characterized by an energy barrier of 0.43 eV, which compares favorably with the 0.50 eV predicted in a simulation calculations on the 50% Cl^--doped polypyrrole.[131]

6. MOLECULAR DYNAMICS SIMULATIONS

Molecular dynamics has proved to be a powerful method for simulating and/or predicting several features of polymer systems. Properties on either side of the glass transition temperature T_g (see Section 1.5) have been successfully simulated, as has the solid-to-liquid transition, and provided descriptions of the dynamics (segmental motions, chain diffusion, conformational transitions, etc.) that are in accord with relaxation measurements and such bulk properties as shear viscosities and elastic moduli. The method may also provide a good description of the variation in heat capacity and other thermodynamic functions across a phase transition. Several collections of these investigations have recently been published.[125,126]

Although the literature contains several MD investigations of the association of ions with protein chains[136] and polyelectrolytes[137] and of the diffusion of small neutral molecules in polymer chains,[138–143] the transport of ions through conductive polymer lattices other than polyacetylene has not received the same attention. The current rapid expansion of powerful computing facilities, good quality calculations on nonbonded interactions and the interest in doped conductive polymers is likely to remedy this omission over the next few years, as also will the application of methods to amorphous phases of these systems. In this section we describe a seminal MD application to crystalline alkali-metal-doped polyacetylene and results of some MD investigations of the diffusion of small molecules (or penetrants) through amorphous flexible-chain polymer structures.

6.1. Crystalline Systems: Polyacetylene-Based Lattices

6.1.1. Pristine Polyacetylene

Sesé *et al.*[144–146] used a simulation box containing 16 polyacetylene (PA) chains, each consisting of six –CH=CH– monomers. Using a time step of 10^{-15}s and an initial equilibration period of 2 ps ($= 2 \times 10^{-12}$s), the authors applied the MD program FUNGUS[56,147] to calculate the dynamics of the system over a period of 150 ps at a simulated temperature of 300 K.

Their calculated mean square displacement $r^2(t)$ and the self-diffusion coefficient D for the carbon and hydrogen atoms defined in Section 3.3.1 [Eqs. (24) and (25)] show that PAc chains undergo diffusional motion mainly along their axial (c)

direction,[144,145] though of course boundary conditions ensure that this motion does not imply net matter transport. The mobility of the PAc chains can be verified simply from a static lattice calculation, which shows that the barrier to displacement of a chain is of the order of thermal energies. However detecting of such motions in actual PAc samples has not hitherto been reported, probably because of the presence of chain defects and cross-linking, which characterize PAc.

6.1.2. Lattice Lightly Doped with Lithium

The effect of doping was first considered by adding two Li^+ ions[144,145] to the simulation box containing the PAc chains. It was assumed that the electronic charge transferred from the dopant atoms was spread uniformly over the carbon atoms, thereby endowing each with a charge of $-(1/96)e$. To avoid uncertain electronic effects that moving dopant ions may exert on the C–C bond alternation on the chains, all C–C bonds were assumed to be identical, with the π bond order intermediate between that of a single and of a double bond.

With the two Li^+ ions in as widely separated positions as the simulation box allows, the dynamics were initiated and equilibrated for the same thermodynamic conditions as in pristine PAc. By artificially constraining the diffusive motion of polymer chains, the motion of Li^+ ions was monitored through the radial distribution function $g(r)$ [Eq. (28)]. The $g_{C,Li}(r)$ and $g_{H,Li}(r)$ were found to describe a liquid-like motion for Li^+ ions, while $g_{Li,Li}(r)$ was almost featureless except for a very small range of dopant separations. This indicates that in agreement with the results of the static lattice calculations (Section 5.3), the motion of the ions was rapid along the chain directions, but correlated by the dominant interionic Coulomb repulsions, so that they maintained (on average) the maximum distance apart permitted by the simulation box.

6.1.3. Potassium-Doped Polyacetylene

The highly doped structures referred to in the description of static lattice simulations (Section 5.3) as PAK1 and PAK2 have also been investigated by MD by Sesé and her colleagues.[144–146] As in the static studies the electronic charge transferred from the dopant atoms was distributed uniformly on the carbon atoms, which therefore bear net charges of $-0.25e$ and $-0.125e$, respectively in the two species.

The dynamics of PAK1 and PAK2 were investigated using a 432-atom and a 408-atom simulation box under the same 300K thermodynamic conditions that were applied to the pristine and lithium-doped lattices. Some of the liquid-like features of the Li^+ ions in lightly doped PAc also appear in the PAK systems, but as seen in Fig. 5.23 the $g_{K,K}(r)$ shape resembles the radial distribution functions in solids. In PAK1 there are peaks at r values corresponding to dopant site separations in the static lattice. This indicates that the dopant sublattice of the initial (static)

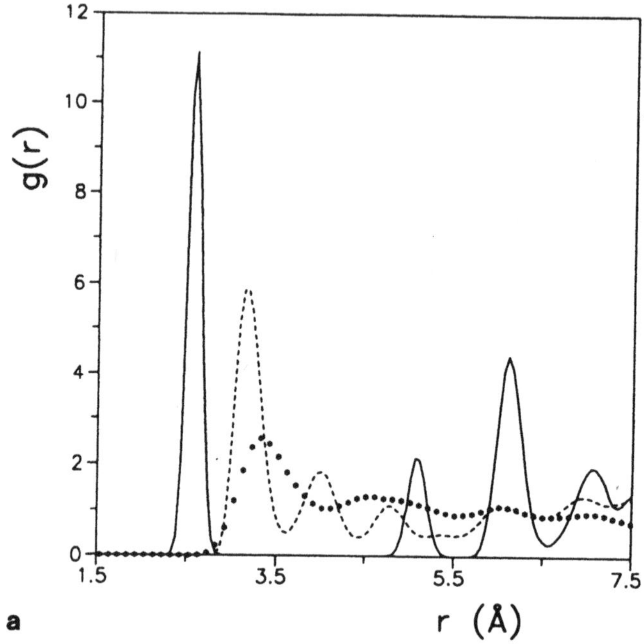

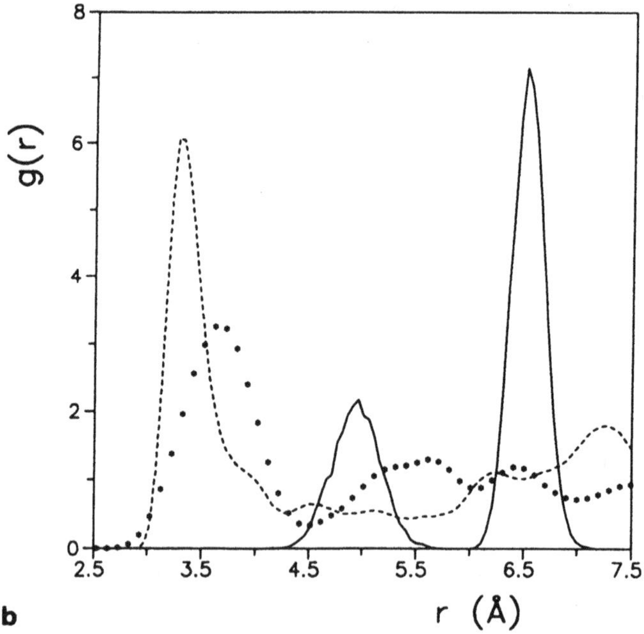

FIGURE 5.23. Radial distribution functions $g_{XY}(r)$ in potassium (a) 'fully-doped' and (b) 'semi-doped' polyacetylene lattices. The (X,Y) pairs are: —— (K^+,K^+), - - - (C,K^+), and * * * * (H,K^+). (Sesé *et al.*[144] by permission of the authors and of the publishers of *Molecular Simulation*.)

structure is preserved in the dynamics. Although in PAK2, the first $g_{K,K}(r)$ peak corresponds to the distance between nearest neighbor K^+ ions in the same channel, the position of the second peak (6.5 Å) is slightly greater than the one between the closest dopants in adjacent channels. This indicates an interchannel dopant correlation that tends to maximize dopant separations.

The K^+ displacement functions $\mathbf{r}_K(t)$ for both PAK species show greater amplitudes in the directions of the surrounding chains (*a* and *b* axes) than along the chain axis direction *c*, although as would be expected from the 'looseness' of the

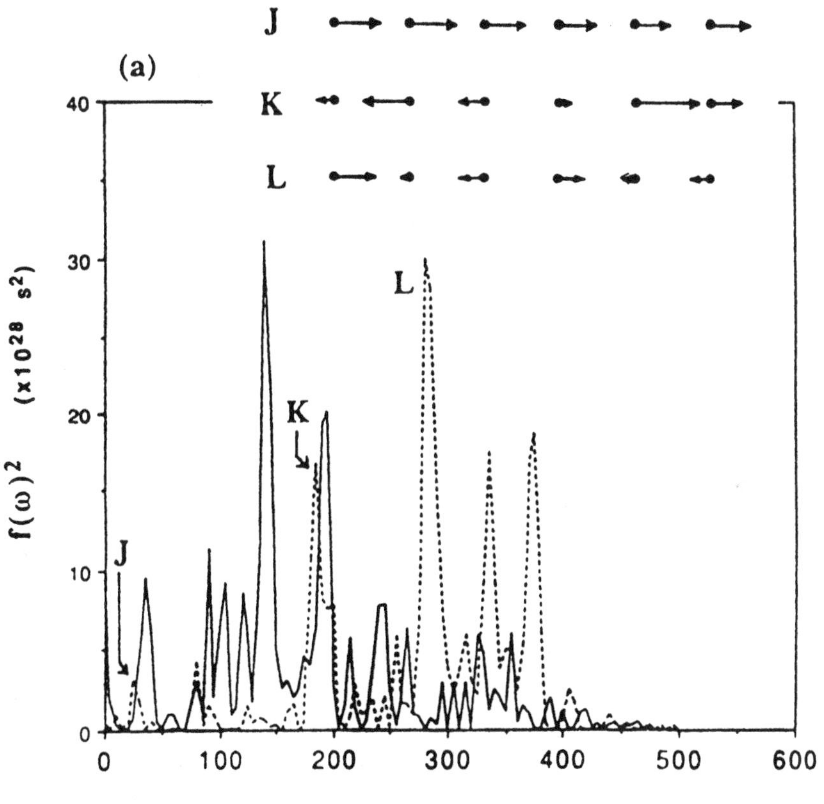

FIGURE 5.24. The power spectra of the dopant K^+ ions in (a) PAK1 and (b) PAK2 calculated the molecular dynamics investigation of Mori *et al.*[146] The inset in (a) shows the vibrational modes of the K^+ ions that give rise to the peaks indicated. (Reproduced by permission of the authors and of the publishers of *Synthetic Metals*.)

dopant sublattice in PAK2, the motions along the c direction are more pronounced than in PAK1.

In another MD investigation of doped PAc, the same group[146] calculated the effect of the K^+ ions on the power spectrum $f_k(\omega)$ furnished by the Fourier transform of the velocity autocorrelation function $z_k(t)$ via Eq. (32) in Section 3.3.3 (where k is one of the constituent atoms of the system, C, H, or K). They found that as well as increasing the frequencies of the polymer's out-of-plane acoustic modes, the dopant ions dampen the mobility calculated for the PAc chains in the pristine material. A comparison of the power spectrum function $f_K(\omega)$ of the dopant ions in PAK1 and PAK2 (Fig. 5.24) showed that the peaks associated with the dopant sublattice in PAK2 were confined to very low frequencies (primarily in the range close to zero, but extending to 100 cm^{-1}), while those in the more highly-doped

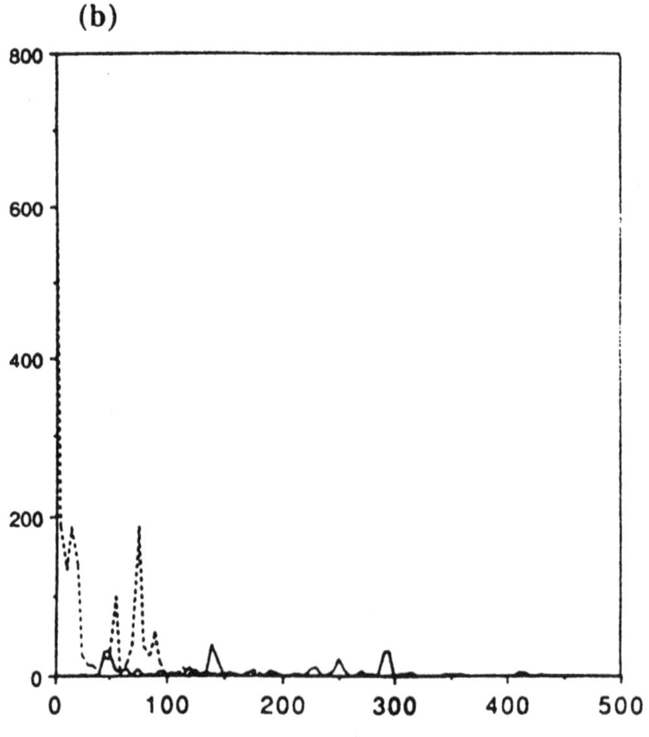

FIGURE 5.24. (*Continued*)

polymer PAK1 showed a more even distribution up to 400 cm^{-1}. One inference that may be drawn from the lattice frequencies (and their atom displacement vectors) concerns the greater mobility of the K^+ ions along the channels for the less highly-doped polymer. Also of interest is the sensitivity of the out-of-plane acoustic phonons of PAc to the doping level which supplements the host-dopant interaction predicted by the results of the static lattice calculations in Section 5.3.4 and depicted in Figs. 5.10 and 5.13–5.16.

6.2. Diffusion in Noncrystalline Systems

In the absence of molecular dynamics investigations of ion transport in noncrystalline phases of polymers, we describe investigations of the motion of small neutral molecules in amorphous polymers. The success of these studies in providing a description of the motion of (albeit) neutral diffusants encourages their extension to similar investigations of the transport of dopant ions through polymer materials.

A starting point to an MD study of amorphous polymer systems may be the application of Monte Carlo methods (Section 1.5) to randomize conformations of oligomer or endless chains in the MD box to which periodic boundary conditions (Section 3.2) are imposed. Among a series of papers reporting the computed dynamics of simple polymer systems, Takeuchi *et al.*[(138)] investigated diffusive motions of spherical particles through an amorphous matrix of an idealized polymer at temperatures above the glass transition temperature T_g. The polymer was modeled by a coarse-grained chain (Section 1.5), and dopant or penetrant particles were assigned masses equal to that of an oxygen molecule (32 amu).

In another investigation Takeuchi *et al.*[(139)] simulated an amorphous polyethylene system using a cell containing 600 CH_2 segments and 10 diffusant molecules. By varying the density of the simulated system up to the onset of crystallization, diffusion was shown to obey Eq. (38) over the complete density range and thus to be in accord with the predictions of free volume theory (Section 3.4). Fitting the temperature variation to Eq. (36) yielded an activation energy $E_D \approx 0.10$ eV, which is close to the barrier measured for the diffusion of O_2 in silicone rubber and moreover of the same magnitude as that of the torsional barrier associated with a chain link. Having investigated increasing diffusion rates with the flexibility of the polymer chain, the authors concluded that chain motions play a decisive role in penetrant diffusion. Sonnenburg *et al.*[(140)] came to a similar conclusion when they observed a severely reduced diffusant motion on damping thermal motions in the polymer chains.

More recently Takeuchi[(141,142)] reported details of a jump mechanism of a small diffusant molecule in a system of endless polymer chains that is quenched at the glassy state ($T < T_g$). For most diffusant species (simulating O_2 molecules) trapped in rigid cages formed by strands of the glassy polymer, the time evolution of the

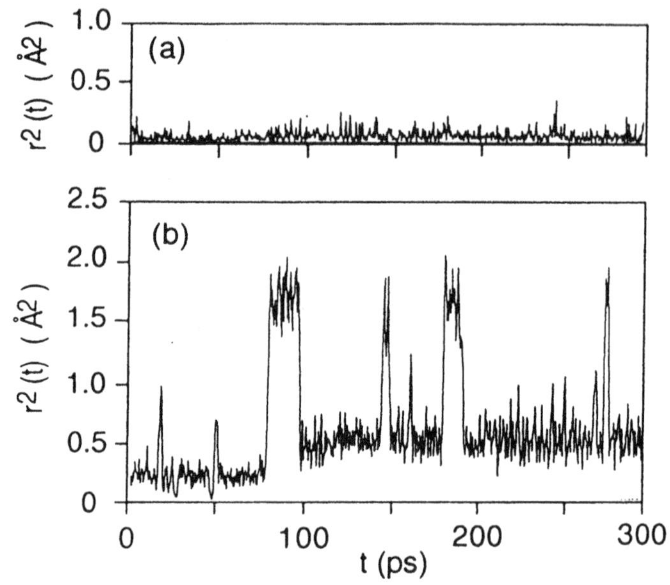

FIGURE 5.25. A comparison of the mean square displacement |$r(t)$| of a (a) trapped and (b) migrating oxygen diffusant in a glassy polymer, obtained by a molecular dynamics simulation. (Takeuchi,[141,142] by permission of the author and of the American Institute of Physics.)

mean square displacement $r^2(t)$, shown in Fig. 5.25(a), revealed only low-amplitude (0.1 Å) vibrational noise. However depending on cooling conditions (which could result in the persistence of slow relaxations even after reaching the quenched state), Fig. 5.25(b) shows that the $r^2(t)$ function of some O_2 molecules testify to a motional freedom by exhibiting 1.5–2.0 Å peaks superimposed on the vibrational noise.

Surprisingly, and contradicting expectations of free volume theory, the time behavior of the local free volume fraction curve LFVF(t) for jumping O_2 molecules does not monitor the jump by showing a change around the time of the jump. In fact this curve is practically indistinguishable from the LFVF(t) curve for trapped O_2 molecules. However plotting a sequence showing contours of the potential energy surface in the vicinity of the jumping molecule reveals a clear image of the transition of the molecule from one energy well to a second (Fig. 5.26). Opening a channel between the wells through which the molecule is transported implies a

→

FIGURE 5.26. A molecular dynamics time sequence of potential energy contours for a molecular oxygen diffusant in a glassy polymer. The time interval of 1 picosecond is that of a migrational transition or jump of the diffusant between two caged sites from the upper to lower part of the figure. The intervals of the energy contours are 50 kJ mol^{-1}. (From Ref. 142, by permission of the author and of the American Institute of Physics.)

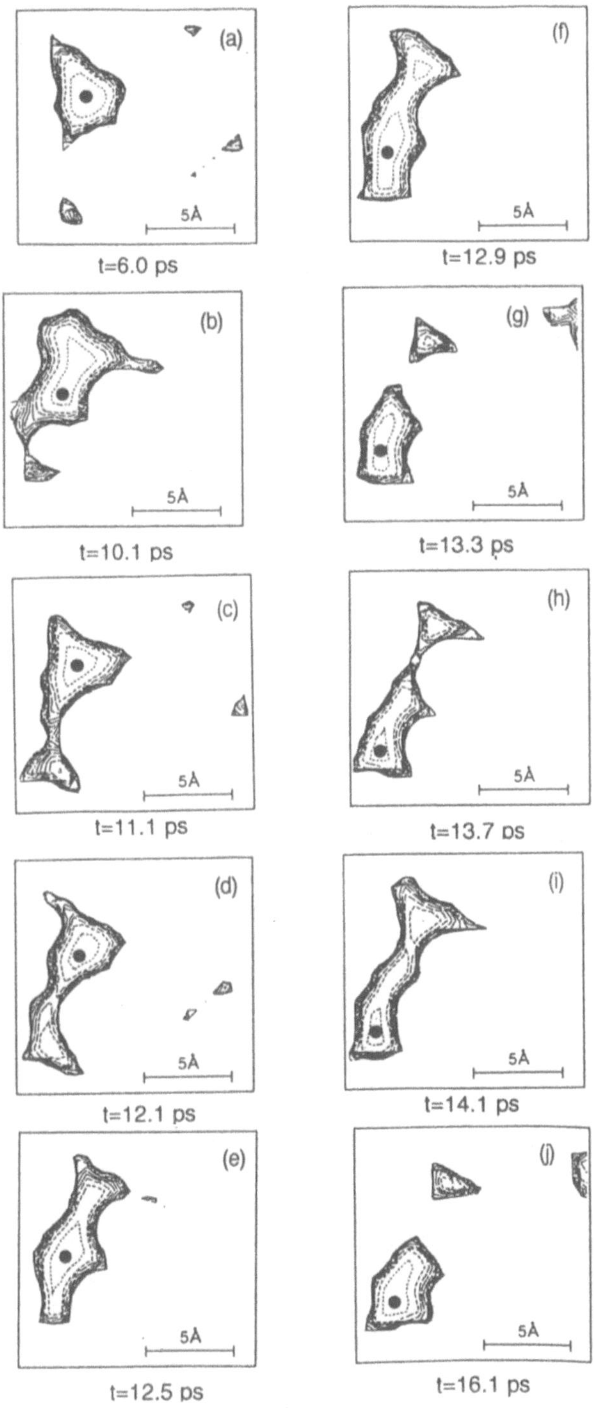
(a)
5Å
t=6.0 ps
(b)
5Å
t=10.1 ps
(c)
5Å
t=11.1 ps
(d)
5Å
t=12.1 ps
(e)
5Å
t=12.5 ps
(f)
5Å
t=12.9 ps
(g)
5Å
t=13.3 ps
(h)
5Å
t=13.7 ps
(i)
5Å
t=14.1 ps
(j)
5Å
t=16.1 ps

practically negligible energy barrier for the jump. Despite the glassy condition of the polymer system ($T < T_g$), torsional angle variations $\phi_a(t)$ of the chains in the immediate vicinity of the jumping O_2 molecule reveal significant changes near the time of the jump.

Recently Sok *et al.*[143] simulated the transport of He and of CH_4 through a polydimethylsiloxane (PDMS) membrane. The polymer was selected because of the almost invariably amorphous condition of its samples. Thus the interpretation of experimental diffusion data is uncomplicated by the presence of crystallites. The PDMS chains were represented by oligomers consisting of 30 monomer units, and the MD simulation was based on a fine-grained representation of polymer and diffusant, i.e., all atoms were included explicitly. Like the mechanism described by Takeuchi[141,142] here, too, transport occurred by a jump mechanism. Facilitated by fluctuations in the polymer chains, the hole surrounding the diffusant was observed to expand just prior to the jump, and a transient channel was opened, thus enabling the transition to occur. Diffusion coefficients derived for He and CH_4 (respectively, 18×10^{-5} and 2.1×10^{-5} cm^2 s^{-1}) are consistent with at least the magnitudes of experimental measurement (10×10^{-5} and 2.0×10^{-5} cm^2 s^{-1}).[148]

7. CONCLUSION

7.1. Statics or Dynamics?

Since a molecular dynamics simulation of a lattice provides a description not only of the structure but also of its behavior in time, it could be assumed that MD has more to offer than static simulations as an investigative method for the conductive polymers. But with a time step $\Delta t = 10^{-15}$ s (or even 10^{-14} s, which limits the vibrational description), computing costs restrict temporal events that can feasibly be simulated to those with characteristic times up to 10^{-9} seconds. This time limit coincides with those characterizing torsional and reorientational motions, and since many other types of relaxation mechanisms have time constants longer than this period, it is not possible to include in the dynamics all component motions that may play a role in the migration. Kremer and Grest[149] estimate that the 10^9 time steps required to simulate a motion with a time constant of 10^{-5} s would typically require 50,000 hours of computation time—a clearly impossible venture. Another limitation of MD is the application of periodic boundary conditions necessary to ensure chain continuity and particle conservation. As a result simulating an isolated local lattice defect as described for the static simulation method in Sections 2.5 and 5 has no parallel in MD. Finally it may sometimes be useful to investigate the response of the system being investigated to a perturbation (e.g., displacement) imposed by the investigator on a particular atom or group of atoms rather than restrict motions to the redistribution of thermal energy in the system. This facility, too, has no counterpart in MD.

On the other hand since the dynamics may be simulated at any chosen temperature or range of temperatures (rather than at 0 K, as is formally the case in static simulations), MD results may be processed to provide many experimentally interesting quantities—diffusion parameters, thermodynamic functions, radial distribution, and correlation functions—the last of which may be related to measured relaxation data. As a result the two techniques should be regarded as complementary. With the current rapid expansion in computing power, it is likely that at least some of the limiting caveats described will soon be eliminated.

7.2. Structure and Morphology

Solid structure formation is governed on one hand by a tendency toward a regular lattice, favored by enthalpic considerations, and on the other by entropic effects striving to introduce disorder. For simple ionic and molecular solids, one of these opposing effects is usually dominant, leading to either an amorphous or a crystalline state; the latter permits the use of diffraction methods to determine its structure. In polymers the conflict is resolved by the coexistence in the same sample of both amorphous and crystalline regions; however the crystallites are much too small for single-crystal diffraction, and often their paucity and low quality preclude deriving unambiguous structural information by diffraction techniques. It is here that the methods of atomistic simulation have in many cases proved to be the only ones capable of furnishing a description of structure and events occurring on an atomic level.

Static lattice simulation methods predict the tendency of polymers to avoid the strict regularity of a crystal by finding several lattice structures with similar energies but that vary in their chain-setting angle and cell vectors. The readiness with which the polymer sublattice in doped systems lowers its symmetry is echoed in both the defective static lattice simulations and the molecular dynamics treatment of lattice motions. In the former we observed the wide ranging response of the chains to even a small displacement of a dopant ion as described in Sections 5.3.2, 5.3.4, 5.4.2. and 5.6.3 while the dopant's power spectrum calculated in the course of the MD lattice simulations in Section 6.1 testifies to the important host-dopant interactions in potassium-doped polyacetylene.

Some polymers (e.g., polyacetylene[115] and polythiophene[34,35]) exhibit an additional morphological heterogeneity consisting of fibrils of dense material dispersed in a void, such that the polymer material and dopants occupy only about 50% of the volume. Although the sample is formally heterogeneous within the solid material, constituent phases are in such close proximity that not only may it be difficult to distinguish a boundary between them, but the constituents of the microdomains themselves are uncertain. Even single crystals which have been isolated in some polymers (polyethylene) possess dispersed amorphous regions.[65] Similarly, much of the noncrystalline component has been accorded by some authors[14] with a degree of order accruing from a general alignment of polymer chains. This is not to deny that regions devoid of order exist: Chain defects such as those resulting from *cis/trans* conversions, cross-linking, or C=C saturation lead to amorphous regions, which must be considered, by methods such as those described in Sections 1.5 and 6.2. But in comparison with flexible polymers, we expect the packing of the more rigid chains of conjugated polymers to result in at least some degree of chain alignment. The structural picture of conjugated polymers is thus one in which the chains are either completely disordered or occur in bundles in which the order varies from short-term regularity in one direction (more or less associated with that of the chains) to a full three-dimensional lattice.

7.3. Dopant Transport

From the preceding discussion, transport processes occur as migrant species move through many different environments, which may be broadly classified as (1) a crystalline lattice, (2) a reduced-order lattice (approximately parallel chains), (3) an amorphous polymer matrix, and (4) a void. In crystalline regions static-lattice methods show that migrational or diffusional motion depends on the lattice geometry surrounding the dopant sites. When dopants are small ions in a square channel environment, e.g., alkali metal ions in polyacetylene, simulation studies predict rapid and almost unactivated motion, but unusual features of the square-channel lattice perhaps make this a rather special case. Moreover unactivated motion occurs only over distances defined by channel lengths, which is probably no more than a few tens of Ångströms; the bulk diffusion of ions should be determined principally by their passage through the interchannel or amorphous regions. In fact although activation energies have not been reported for channel migration of alkali metal ions in PAc, the overall diffusion coefficient for Li^+ is quite low—about 10^{-17} cm^2 s^{-1},[150] suggesting that the motion is dominated by the pathways mentioned. Bénière *et al.*[116] measured different diffusional motion for I_3^- within and between the PAc fibrils. The overall diffusion, for which they find an activation energy of 0.2 eV, has a diffusion coefficient of 10^{-9} cm^2 s^{-1}; however more rapid diffusion between the fibrils (10^{-7} cm^2 s^{-1}) indicates that in certain circumstances the intrafibrilar regions could act as a bottleneck.

The polypyrrole and polythiophene matrices probably constitute more usual doped-lattice systems, with the dopant anions occupying sites between pairs of polymer chains which, in the lowest-energy structure, are coplanar with them. Here lattice migration should be much more restricted than in square channels, and barriers of 0.5–2 eV are expected as the rather rigid chains have to undergo distortion to allow passage to the anions.[131] To our knowledge no activation energies or diffusion constants have been reported for these systems.

Measurements on the transport of iodine in polyacetylene have been investigated by weight-uptake and electrochemical methods,[116] but because of the uncertainties relating to the proposed structures of the iodine-doped lattice,[99–101] it has proved difficult to simulate a stable lattice consistent with the various proposed structures.[151]

The wide-ranging response of polymer chains to a perturbation of the dopant that was recalled in Section 7.2 can be described as an interaction between host and dopant sublattices that assumes an extreme form for flexible polymers in the amorphous state. Here the dopant or penetrant species is trapped in a pocket formed by strands of a polymer chain, an environment that could exist in a chain-defect region of a conductive polymer. We have seen (Section 6.2) that in such a case, chains may undergo reorganization[140–143] in the form of torsional motions to release the migrant species for jumping to a second site. The model predicted a realistic activation energy of 0.1 eV in this process for flexible chains.[139] The lower flexibility of conductive polymers would be likely to increase this result in a rather higher barrier, probably to the 1-eV regime calculated for the lattice diffusion.

7.4. Future Methods

Frequent reference has been made to the rapidly expanding limits of computing power that continues to bring ventures from the realm of theory into the range of possibility. In static simulation for example, by including a larger number of species in defect regions and interactions between multiple defect centers it will be possible to investigate considerably more complex migration mechanisms than the simple, localized n-vacancy/m-interstitial type described for static methods.

In MD, too, demands for the future should be obvious from the discussion in Sections 3, 6, and 7.1. To obtain realistic simulations of the dynamics, not only is it essential to simulate considerably longer real-time relaxations but also to include an adequate number of particles in the simulation box. This is particularly important for modeling amorphous states of polymer systems for which the demands of computer storage and speed are critical and at present inadequately met.

A verdict on currently obtainable results from simulating of polymer systems is that it is at best semiquantitative, but with the ability to describe migration features. The most urgent problem in fundamental theory to be addressed using current methods is undoubtedly the lack of reliable potentials, particularly for nonbonded and torsional interactions. Good-quality investigations of intermolecular interactions on model systems may lead to some improvement, but this approach may be overtaken by the gaining view that using force fields based on transferable potentials should be abandoned in favor of calculating *all* interactions in an *ad hoc* quantum chemical manner during the course of simulation. Although such procedures that are applicable to polymer systems are not as yet developed, the results of *ab initio* molecular dynamics of simple molecular systems are very encouraging and pave the way to the next generation of simulation methods.[152,153]

REFERENCES

1. R. J. Bartlett, *Comparison of Ab Initio Quantum Chemistry with Experiment for Small Molecules: The State of the Art* (Reidel, Dordrecht, Holland, 1985).
2. E. Clementi, G. Corongiu, D. Bahattacharya, B. Feuston, D. Frye, A. Preiskorn, A. Rizzo, and W. Xue, *Chem. Rev.* **91**, 679 (1991).
3. J. Sadlej, *Semiempirical Methods of Quantum Chemistry* (Ellis Horwood, Chichester, England, 1985).
4. J. A. Pople and D. L. Beveridge, *Approximate Molecular Orbital Theory* (McGraw-Hill, Maidenhead, England, 1970).
5. J. A. Segal, ed., *Semiempirical Methods of Electronic Structure Calculations* (Plenum Press, New York, 1977).
6. J. J. Ladik, *Quantum Chemistry of Polymers as Solids* (Plenum Press, New York, 1987).
7. J. J. Ladik, J. M. André, and M. Seel, eds., *Quantum Chemistry of Polymers—Solid State Aspects* (Reidel, Dordrecht, Holland, 1984).
8. C. Pisani, R. Dovesi, and C. Roetti, *Hartree-Fock Ab Initio Treatment of Crystalline Systems* (Springer Verlag, Berlin, 1988).
9. C. R. A.Catlow, *Solid State Ionics* **8**, 89 (1983).
10. J. C. W. Chien, *Polyacetylene: Chemistry, Physics, and Material Science* (Academic Press, New York, 1984).
11. R. H. Baughman and E. Moss, *J. Chem. Phys.* **77**, 6321 (1982).
12. J. Corish, V. C. A. Hanratty, D. A. Morton-Blake, F. Bénière, and A. Morin, *J. Mol. Struct. (Theochem)* **207**, 41 (1990).
13. P. W. M. Jacobs, Z. A. Rycerz, and J. Moscinski, *Adv. Solid-State Chem.* **2**, 113 (1991).
14. R. H. Baughman, J-L. Brédas, R. R. Chance, R. L. Elsenbaumer, and L. W. Shacklette, *Chem. Rev.* **82**, 209 (1982).
15. P. Cukor, J. I. Krugler, and M. F. Rubner, *Macromol. Chem. Phys.* **182**, 165 (1981); W. Deits, P . Cukor, and M. Rubner, *Polymer Sci. Technol.* **15**, 171 (1981).
16. T. A. Skotheim, ed., *Handbook of Conducting Polymers*, Vols. 1 and 2 (Marcel Dekker, New York, 1986).
17. D. MacInnes, M. A. Druy, P. J. Nigrey, D. P. Nairns, A. G. MacDiarmid, and A. J. Heeger, *J. Chem. Soc. Chem. Commun. 1981*, 317.
18. F. Bénière, D. Boils, H. Canèpa, J. Franco, L. Lecorre, and J. P. Louboutin, *J. Electrochem. Soc.* **132**, 2100 (1985).
19. P. Burgmeier and R. W. Murray, *J. Amer. Chem. Soc.* **104**, 6139 (1982).
20. M. Gazard in T. A. Skotheim, ed., *Handbook of Conducting Polymers*, vol. 1 (Marcel Dekker, New York, 1986), p. 673.
21. F. Garnier, G. Tourillon, M. Gazard, and J. C. Dubois, *J. Electroanal. Chem. Interfacial Electrochemistry* **148**, 299 (1983).
22. R. H. Baughman, S. L. Hsu, L. R. Anderson, G. P. Pez, and A. J. Signorelli, *J. Chem. Phys.* **68**, 5405 (1978); R. H. Baughman, S. L. Hsu, L. R. Anderson, G. P. Pez, and A. J. Signorelli, in *Molecular Models, NATO Conference Series* (W. E. Hatfield, ed.), (Plenum Press, New York, 1979), pp. 187–201.
23. C. R. Fincher, C. E. Chen, A. J. Heeger, A. J. MacDiarmid, and J. B. Hastings, *Phys. Rev. Letts.* **48**, 100 (1982).

24. J. C. W Chien, F. E. Karasz, and K. Shimamura, *Makromol. Chem. Rapid Commun.* **3**, 655 (1982).
25. P. Kovacic, M. Feldmann, J. P. Kovacic, and J. P. Lando, *J. Appl. Polym. Sci.* **12**, 1735 (1968).
26. M. Stamm and J. Hocker, *J. Phys. Coll. C3* (Suppl. 6) **44**, 667 (1983).
27. G. R. Mitchell, F. J. Davis, R. Cywinski, and A. C. Hannon, *Polymer Comm.* **30**, 98 (1989).
28. R. Yang, X. R. Yang, D. F. Evans, W. A. Hendrickson, and J. Baker, *J. Phys. Chem.* **95**, 3765 (1991).
29. G. R. Mitchell, *Polymer Comm.* **12**, 346 (1986).
30. R. Yang, K. M. Dalsin, D. F. Evans, L. Christensen, and W. A. Hendrickson, *J. Phys. Chem.* **93**, 511 (1989).
31. R. Yang, D. F. Evans, L. Christensen, and W. A. Hendrickson, *J. Phys. Chem.* **94**, 6117 (1990).
32. R. Yang, X. R. Yang, D. F. Evans, W. A. Hendrickson, and J. Baker, *J. Phys. Chem.* **94**, 6123 (1990); R. Yang, X. R. Yang, and D. F. Evans, **95**, 3765 (1991).
33. G. Caple, B. L. Wheeler, R. Swift, T. L. Porter, and S. Jeffers, *J. Phys. Chem.* **94**, 5639 (1990).
34. F. Garnier, G. Tourillon, J. Y. Barraud, and H. Dexpert, *J. Mat. Science* **20**, 2687 (1985).
35. G. Tourillon and F. Garnier, *Mol. Cryst. Liq. Cryst.* **118**, 221 (1985).
36. A. D. Buckingham, in *Intermolecular Interactions: From Diatomics to Biopolymers* (B. Pullman, ed.) (Wiley, Chichester, England, 1978), pp. 1–67; P. Claverie, in *Intermolecular Interactions: From Diatomics to Biopolymers* (B. Pullman, ed.), (Wiley, Chichester, England, 1978) pp. 69–305.
37. B. G. Dick and A. W. Overhauser, *Phys. Rev. Letts.* **112**, 90 (1958).
38. N. L. Allinger, *J. Amer. Chem. Soc.* **99**, 8127 (1977).
39. J. T. Sprague, J. C. Tai, Y. Yuh, and N. L. Allinger, *J. Comp. Chem.* **8**, 581 (1987).
40. B. R. Brooks, R. E. Bruccoleri, B. D. Olafson, D. J. States, S. Swaminathan, and M. Karplus, *J. Comp. Chem.* **4**, 187 (1983).
41. A. Baumgärtner, in *Applications of the Monte Carlo Method in Statistical Physics* (K. Binder, ed.) (Springer Verlag, Berlin, 1984), p. 145.
42. G. G. Lowry, ed., *Markov Chains and Monte Carlo Calculations in Polymer Science* (Marcel Dekker, New York, 1970), p. 145.
43. K. Binder, in *Computational Methods of Polymers* (Marcel Dekker, New York, 1992), p. 221.
44. D. M. Hirst, *A Computational Approach to Chemistry* (Blackwell Scientific, Oxford, England, 1990).
45. CERIUS (Molecular Simulations, Cambridge, England).
46. BIOSYM Technologies (San Diego).
47. H-P. Wittmann, K. Kramer, and K. Binder, *J. Chem. Phys.* **96**, 6291 (1992).
48. A. Sariban and K. Binder, *Makromol. Chem.* **189**, 2357 (1988).
49. M. Born, *Atomtheorie des festen Zustandes* (Teubner, Leipzig, 1923); M. P. Tosi in *Solid-State Physics—Theory and Applications*, vol. 16 (F. Seitz and D. Turnbull, eds.), (1964), p. 1.
50. D. A. Morton-Blake, J. Corish, and F. Bénière, *Phys. Rev. B* **37**, 4180 (1988).
51. N. Anastasiou and D. Fincham, *Computer Physics Communications* **25**, 159 (1989).
52. M. Leslie, Daresbury Laboratory Report DL/SCI/TMB36T (Daresbury, Cheshire, U.K., 1984).
53. C. R. A. Catlow, R. James, W. C. Mackrodt, and R. F. Stewart, *Phys. Rev. B* **25**, 1006 (1982).
54. N. F. Mott and M. J. Littleton, *Trans. Faraday Soc.* **34**, 485 (1938).
55. H. J. C. Berendsen and W. F. van Gunsteren, in *Molecular Liquids—Dynamics and Interactions* (A. J. Barnes, W. J. Orville-Thomas, and J. Yarwood, eds.) (Reidel, Dordrecht, Holland, 1984), p. 475; D. J. Tildesley, *Molecular Liquids—Dynamic and Interactions* (Reidel, Dordrecht, Holland, 1984), p. 519.
56. J. R. Walker, in *Computer Simulation of Solids* (C. R. A. Catlow and W. C. Mackrodt, eds.) (Springer Verlag, Berlin, 1982), p. 58.
57. D. J. Evans and G. P. Morriss, *Comp. Phys. Rep.* **1**, 297 (1984).
58. D. J. Adams, *Information Quarterly for Computer Simulation of Condensed Phases (SERC)* **10**, 30 (1983); W. Smith, *Information Quarterly for Computer Simulation of Condensed Phases (SERC)* **10**, 37 (1983).

59. H. J. C. Berendsen, in *Molecular Dynamics Simulation of Statistical Mechanical Systems* (G. Cicotti and W. G. Hoover, eds.) (North-Holland, Amsterdam, 1986).
60. H. Stassen, J. Samios, and T. Dorfmüller, *Molecular Simulation* **8**, 215 (1992).
61. J. Y. Derrien and J. Dupuy, *Journal de Physique* **36**, 191 (1975).
62. S. C. Moss and J. F. Graczyk, in *Proc. 10th Int. Conf. on the Physics of Semiconductors, Cambridge, Mass.* (S. P. Keller, J. C. Hensel, and F. Stern, eds.) (Atomic Energy Commission, Washington, DC, 1970), p. 658.
63. W. Bruns and R. Bansal, *J. Chem. Phys.* **74**, 2064 (1981).
64. J. L. Yarnell, M. J. Katz, R. G. Wenzel, and S. H. Koenig, *Phys. Rev. A* **7**, 2130 (1973).
65. A. Keller and S. Sawada, *Makromol. Chem.* **74**, 190 (1964).
66. M. Satoh, H. Yamasaki, S. Aoki, and K. Yoshino, *Mol. Cryst. Liq. Cryst.* **159**, 289 (1988).
67. J. Mårdalen, E. J. Samuelsen, O. R. Gautun, and P. H. Carlsen, *Solid State Commun.* **80**, 687 (1991); J. Mårdalen, E. J. Samuelsen, O. R. Gautun, and P. H. Carlsen, *Synth. Met.* **48**, 363 (1992).
68. H. Shirakawa and T. Kobayashi *J. Phys. Colloque* **C3**, 3 (1983); F. Bénière, N. Bertru, C. R. A. Catlow, M. Cole, J. Simonet, and L. Angély, *J. Phys. Chem. Solids* **53**, 449 (1992).
69. R. H. Petrucci and R. K. Wismar, *General Chemistry with Qualitative Analysis,* 2nd ed. (Macmillan, London, 1987), p. 240.
70. N. Isaacs, *Physical Organic Chemistry* (Harrow Longman Scientific and Technical, New York, 1967), p. 130.
71. J. C. Bailar, T. Moeller, J. Kleinberg, C. O. Guss, M. E. Castellion, and C. Metz, *Chemistry* (Harcourt Brace Jovanovich, San Diego, 1989), p. 223.
72. G. Bauer and H. Mikosch, *J. Mol. Struct.* **142**, 121 (1986).
73. G. M. Bodner and H. L. Pardue, *Chemistry, an Experimental Science* (Wiley, New York, 1989), table A15.
74. R. L. Hunt and B. S. Aule, *Spectrochim. Acta* **37A**, 63 (1981).
75. H. Schulze, N. Weinstock, A. Müller, and G. Vandrish, *Spectrochim. Acta* **29A** 1705 (1973).
76. A. J. Hopfinger, *Conformational Properties of Macromolecules* (Academic Press, New York, 1973).
77. D. E. Williams, *J. Chem. Phys.* **47**, 4680 (1967).
78. R. A. Scott and H. A. Scheraga, *J. Chem. Phys.* **45**, 2091 (1966).
79. M. Clark, R. D. Cramer III, and N. van Opdenbosch, *J. Comp. Chem.* **10**, 982 (1989).
80. See for example *Molecular Simulation* **4**, (1990), special issue: "Practical Calculation of Interionic Potentials in Solids."
81. G. C. Maitland, M. Rigby, E. B. Smith, and W. A. Wakeham, *Intermolecular Forces* (Clarendon Press, Oxford, England, 1981), chap. 3.
82. J. A. Barker, R. O. Watts, J. K. Lee, T. P. Schafer, and Y. T. Lee, *J. Chem. Phys.* **61**, 3081 (1974).
83. R. Shafer and R. G. Gordon, *J. Chem. Phys.* **58**, 5422 (1973).
84. R. G. Gordon and Y. S. Kim, *J. Chem. Phys.* **56**, 3122 (1972).
85. P. T. Wedepohl, *Proc. Phys. Soc.* **92**, 79 (1967).
86. See for example S. M. Tomlinson, C. R. A. Catlow, H. Donnerberg, and M. Leslie, *Molecular Simulation* **4**, 335 (1990), special issue: "Practical Calculation of Interionic Potentials in Solids."
87. P. Portmann, T. Maruizumi, M. Welti, M. Badertscher, A. Neszmelyi, W. Simon, and E. Pretsch, *J. Chem. Phys.* **87**, 497 (1987).
88. S. L. Mayo, B. D. Olafson, and W. A. Goddard, *J. Phys. Chem.* **94**, 8897 (1990).
89. Z. Mo, K-B. Lee, Y. B. Moon, M. Kobayashi, A. J. Heeger, and F. Wudl, *Macromolecules* **18**, 1972 (1985).
90. S. Brückner and W. Porzio, *Makromol. Chem.* **189**, 961 (1988).
91. M. Satoh, H. Yamasaki, and S. Aoki, *Mol. Cryst. Liq. Cryst.* **159**, 289 (1988).
92. M. Stamm, J. Fink, and B. Tieke, *Mol. Cryst. Liq. Cryst.* **118**, 281 (1985).
93. J. Corish, F. Jugie, D. A. Morton-Blake, M. Leslie, and F. Bénière, *Phys. Rev. B* **41**, 10870 (1990).

94. J. C. W. Chien, *Polyacetylene: Chemistry, Physics, and Material Science* (Academic Press, New York, 1984), p. 160.
95. S. Suhai, *Chem. Phys. Lett.* **96**, 619 (1983).
96. E. J. Mele, in *Handbook of Conducting Polymers,* vol. 2 (T. A. Skotheim, ed.) (Marcel Dekker, New York, 1986), p. 795.
97. R. H. Baughman, N. S. Murthy, and G. G. Miller, *J. Chem. Phys.* **79**, 515 (1983).
98. S. Flandrois, C. Hauw, and B. François, *Mol. Cryst. Liq. Cryst.* **117**, 91 (1985).
99. J. C. W. Chien and J. A. Hirsch, *Makromol. Chem. Rapid Commun.* **3**, 269 (1982).
100. R. H. Baughman, N. S. Murthy, G. G. Miller, and L. W. Shacklette, *J. Chem. Phys.* **79**, 1065 (1983).
101. G. G. Miller. R. H. Baughman, and N. S. Murthy, *J. Chem. Phys.* **89**, 2523 (1988).
102. J. P. Aimé, M. Bertault, P. Delannoy, R. L. Elsenbaumer, G. G. Miller, and M. Schott, *J. Phys. Lett.* **46**, L379 (1985).
103. J. Ma, D. Djurado, J. E. Fischer, N. Cousel, and B. Bernier, *Phys. Rev. B* **41**, 2971 (1990).
104. F. Saldi, M. Lelaurin, and D. Billaud, *Solid State Commun.* **76**, 595 (1990).
105. Quoted in Ref. 95.
106. J. Corish, V. C. A. Hanratty, J-P. Margrita, D. A. Morton-Blake, F. Bénière, and A. Morin, *J. Mol. Struct. (Theochem)* **207**, 53 (1990).
107. J. Corish, V. C. A. Hanratty, D. A. Morton-Blake, and F. Bénière, *Radiation Effects and Defects in Solids* **119–21**, 499 (1991); J. Corish, V. C. A. Hanratty, D. A. Morton-Blake, and F. Bénière, *Molecular Simulation* **9**, 65 (1992).
108. J. Corish, *Phil. Mag. A* **64**, 1073 (1991).
109. T. Trewern, R. K. Thomas, G. Naylor, and J. W. White, *Faraday Trans. 1* **78**, 2369 (1982).
110. D. A. Morton-Blake and C. Shudell, unpublished.
111. D. A. Morton-Blake, *Synthetic Metals* **35**, 281 (1990).
112. J. Corish, V. C. A. Hanratty, D. A. Morton-Blake, and F. Bénière, *Synthetic Metals* **39**, 311 (1991).
113. R. H. Baughman, L. W. Shacklette, N. S. Murthy, G. G. Miller, and R. L. Elsenbaumer, *Mol. Cryst. Liq. Cryst.* **118**, 253 (1985).
114. J. C. W. Chien, *Polyacetylene: Chemistry, Physics and Material Science* (Academic Press, New York, 1984), chap. 4.
115. J. C. W. Chien, *Polyacetylene: Chemistry, Physics and Material Science* (Academic Press, New York, 1984), chap. 2.
116. F. Bénière, J. Corish, M. E. G. Lyons, T. McCabe, and A. Morin, *J. Electrochem. Soc.* **137**, 2181 (1990).
117. G. B. Street, in *Handbook of Conducting Polymers*, vol. 1 (T. A. Skotheim, ed.) (Marcel Dekker, New York, 1986), p. 265.
118. G. Tourillon, in *Handbook of Conducting Polymers*, vol. 1 (T. A. Skotheim, ed.) (Marcel Dekker, New York, 1986), p. 293.
119. T. Kawai, M. Nakazomo, and K. Yoshino, *J. Mater. Chem.* **2**, 903 (1992).
120. W. R. Salaneck, O. Inganäs, B. Thémans, J. O. Nilsson, B. Sjögren, J-E. Österholm, J-L. Brédas and S. Svensson, *J. Chem. Phys.* **89**, 4613 (1988).
121. C. X. Cui and M. Kertesz, *Phys. Rev. B* **40**, 9661 (1989).
122. J. Corish, D. A. Morton-Blake, K. Veluri, and F. Beniere, *Molecular Simulation* **14**, 381 (1995).
123. J. L. Brédas, B. Thémans, J. G. Fripat, and J. M. André, *Phys. Rev. B* **29**, 6761 (1984); J. L. Brédas, *J. Mol. Struct. (Theochem)* **107**, 169 (1984).
124. B. J. Orchard, B. Freidenreich, and S. K. Tripathy, *Polymer* **27**, 1533 (1986).
125. J. Corish, D. A. Morton-Blake, K. Veluri, and F. Bénière, *J. Mol. Struct. (Theochem)* **283**, 121 (1993).
126. J. Corish, D. A. Morton-Blake, F. Bénière and M. Lantoine, *Trans. Faraday Soc.* (1996, in press).
127. T. J. Prosa, M. J. Winokur, J. Moulton, P. Smith, and A. J. Heeger, *Synthetic Metals 55-57*, 370 (1993), and references therein.

128. J. Corish, D. A. Morton-Blake and F. Bénière, to be published.
129. G. Tourillon and F. Garnier, *J. Pol. Sci. Phys. Edn.* **22**, 33 (1984).
130. T. Yamamoto, K. Sanechika, and A. Yamamoto, *Bull. Chem. Soc. Jpn.* **56**, 1497 (1983).
131. J. Corish, D. A. Morton-Blake, K. Veluri, and F. Bénière, *J. Mol. Struct. (Theochem)* **334**, 109 (1995).
132. K. Veluri, J. Corish, D. A. Morton-Blake, and F. Bénière, *J. Mol. Struct. (Theochem)* (in press).
133. C. H. Lyons, Ph.D. diss., University of Dublin (1993); C. H. Lyons and M. E. G. Lyons, *The Analyst* (1996, in press).
134. R. J. Roe, ed., *Computer Simulation of Polymers*, Prentice Hall, Englewood Cliffs, NJ (1991).
135. J. Bicerano, ed., *Computational Modelling of Polymers*, Dekker, New York (1992).
136. H. Einspahr and C. E. Bugg, in *Metal Ions in Biological Systems*, vol. 17 (H. Sigel, ed.) (Dekker, New York, 1984), pp. 51–97.
137. L. Xie and G. C. Farrington, *Solid State Ionics 53-56*, 1054 (1992).
138. H. Takeuchi and K. Okazaki, *J. Chem. Phys.* **92**, 5643 (1990).
139. H. Takeuchi, R-J. Roe, and J. E. Mark, *J. Chem. Phys.* **93**, 9042 (1990).
140. J. Sonnenburg, J. Gao, and J. H. Weiner, *Macromol.* **23**, 4653 (1990).
141. H. Takeuchi, Ph.D. diss., Tokyo Institute of Technology (1992).
142. H. Takeuchi, *J. Chem. Phys.* **93**, 2062 (1990).
143. R. M. Sok, H. J. C. Berendsen and W. F. van Gunsteren, *J. Chem. Phys.* **96**, 4699 (1992).
144. G. Sesé, C. R. A. Catlow, and B. Vessal, *Molecular Simulation* **9**, 99 (1992).
145. G. Sesé and C. R. A. Catlow, *Phil. Mag. B* **68**, 397 (1993).
146. Y. Mori, G. Sesé, and C. R. A. Catlow, *Synthetic Metals 55-57*, 4493 (1993).
147. E. Hernandez, G. Mills, and B. Vessal (unpublished).
148. Y. Ichiraku, *J. Membr. Sci.* **34**, 5 (1987); R. M. Barrer and H. T. Chio, *J. Polym. Sci. C* **10**, 111 (1965).
149. K. Kremer and G. S. Grest, in *Computer Simulation of Polymers* (R. J. Roe, ed.) (Prentice Hall, Englewood Cliffs, NJ, 1991), p. 167.
150. F. Rachdi, P. Bernier, E. Faulques, S. Lefrant, and F. Schué, *J. Chem. Phys.* **80**, 6285 (1984).
151. V. C. A. Hanratty and D. A. Morton-Blake (unpublished).
152. R. Car and M. Parrinello, *Phys. Rev. Lett.* **55**, 2471 (1985).
153. J. C. Greer, R. Ahlrichs, and I. V. Hertel, *Z. Phys.* **18**, 413 (1991).

6

Numerical Simulation Methods Applied to Electroactive Polymer Films

John F. Cassidy

1. INTRODUCTION

This chapter introduces in a tutorial manner some numerical methods for solving mass transport equations for specific electrochemical systems. These are applied to polymer coatings on electrode surfaces containing redox species. Unlike some theoretical approaches, numerical methods are considerably more straightforward and require only some knowledge of programming. For a typical mass transport situation, the equation to be solved is

$$\frac{\partial c}{\partial t} = \mathrm{D}\frac{\partial^2 c}{\partial x^2} \quad + \quad \text{other} \tag{1}$$

where partial derivatives correspond to diffusion, while the other term corresponds to concentration changes with time due to migration, convection, or chemical reactions. In Eq. (1) c is the concentration, t is time, x is distance from the electrode, and D is the diffusion coefficient. In particular migration may be a problem within redox and especially conducting polymers. However in the first instance, the other terms are neglected. The goal is to solve Eq. (1) and determine an expression for the concentration, which is generally a function of distance from the electrode and time $c(x,t)$. If an expression for concentration is found, then it is possible to determine the current readily. Analytic techniques can be used in a variety of

John F. Cassidy • Chemistry Department, Dublin Institute of Technology, Dublin 8, Ireland.

Electroactive Polymer Electrochemistry, Part 2: Methods and Applications, edited by Michael E. G. Lyons. Plenum Press, New York, 1996.

instances to determine explicit expressions for concentration profiles. Information on analytic techniques can be found in a separate chapter. Generally analytic methods can be applied for simple mechanisms, but for more complicated problems, numerical methods are required. The key to these numerical methods is discretization of the system. For example in Fig. 6.1, which shows expected concentration profiles for a species in a polymer layer that reacts at an electrode surface, there is initially a concentration c^0 within the film. Then the cell is switched on, and the concentration at the electrode surface is forced to zero. Because of this species from a little further out diffuse toward the electrode to remove this difference in concentration. This diffusion is characterized by Eq. (1), and if Eq. (1) can be solved, the series of concentration profiles $c(x,t)$ (see Fig. 6.1) may be found. For these methods concentrations are evaluated at specific points in time and positions in distance. Finite difference methods therefore involve applying a grid with distance and time increments h and k, respectively. So rather than ending

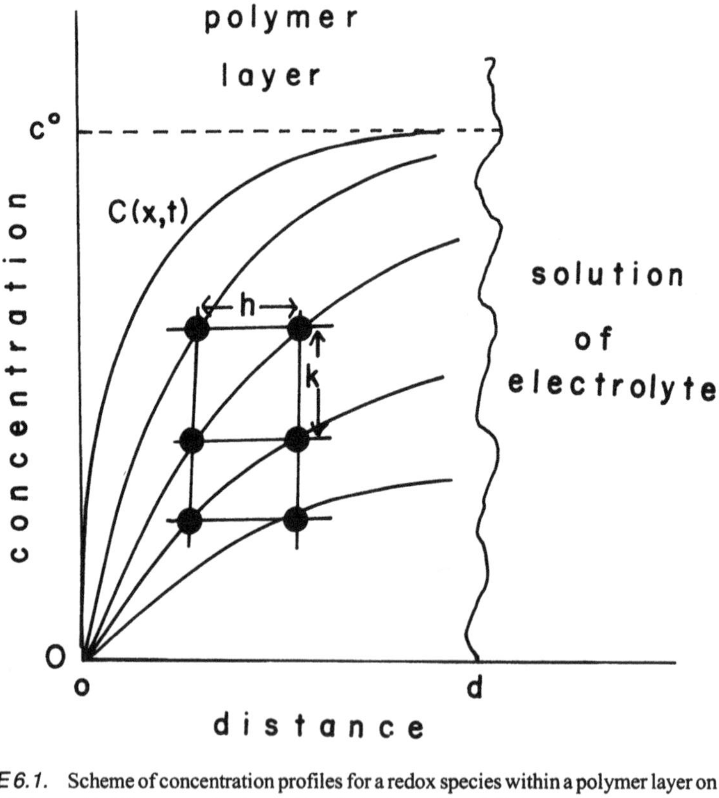

FIGURE 6.1. Scheme of concentration profiles for a redox species within a polymer layer on applying a potential step so that the concentration at the electrode surface falls to zero.

up with the complete profile, a digitized profile at particular distances and times is obtained. This incurs an error that can be quantitated.

The numerical technique depends on a Taylors expansion that states that concentration at position $i + h$ can be determined if the concentration at point i, and various derivatives of that concentration at i are known

$$c_{i+h} = c_i + hc_i' + \frac{h^2 c_i''}{2!} + \frac{h^3 c_i'''}{3!} + \ldots \tag{2}$$

where h is the distance away from i to the point $i+h$ and dashes represent derivatives. This expression can be applied either for distance (the x-axis of the grid in Fig. 6.1) or time (the y-axis of the grid in Fig. 6.1, since for a particular distance, concentration decreases as a function of time). An expression related to Eq. (2) is also possible for a distance in the other direction:

$$c_{i-h} = c_i - hc_i' + \frac{h^2 c_i''}{2!} - \frac{h^3 c_i'''}{3!} + \ldots \tag{3}$$

These expressions are derived from a theory that says it is possible to extrapolate a function to a position $i + h$ if it is known at i and if its tendency to change (derivative terms) is also known. An expression similar to Eq. (2) can be written for the time axis as:

$$c_{t+k} = c_t + kc_t' + \frac{k^2 c_t''}{2!} + \frac{k^3 c_t'''}{3!} + \ldots \tag{4}$$

where k is the time interval. By algebraically combining Eqs. (2–4) it is possible to obtain a digital form of Eq. (1).

From Eq. (4), it is possible to obtain an approximate expression for $(dC)/(dt)$ by neglecting the third and remaining terms:

$$c_t' = \frac{(dc)}{(dt)} \cong \frac{(c_{t+k} - c_t)}{k} \tag{5}$$

We see that this is not a true equality; however at least it is an approximate expression for the derivative, and the error can be quantified. Furthermore by adding Eqs. (2) and (3), it is possible to obtain a second derivative, since:

$$c_{i+h} + c_{i-h} \cong 2c_i + h^2 c_i'' \tag{6}$$

and therefore

$$c_i'' = \frac{(d^2 c_i)}{(dx^2)} \cong \frac{(c_{i+h} + c_{i-h} - 2c_i)}{h^2} \tag{7}$$

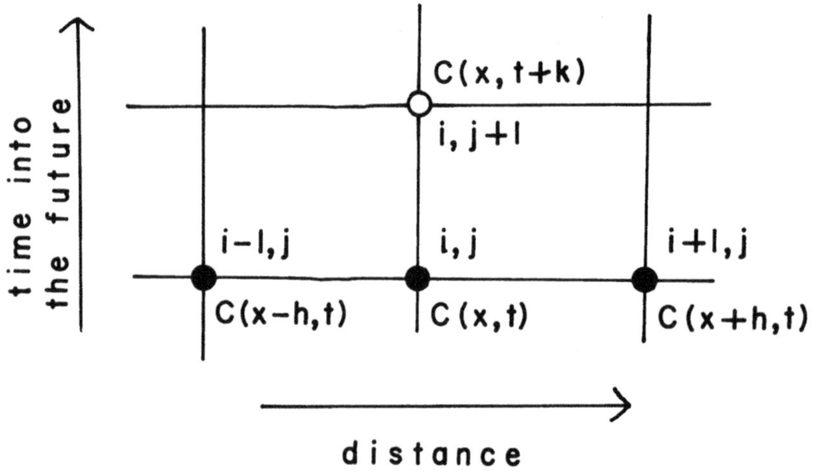

FIGURE 6.2. The distance-time grid for the finite difference method.

By a further approximation, it is possible to substitute Eqs. (5) and (7) into Eq. (1) by assuming that the full derivatives are equal to the partial derivatives. In this manner Eq. (1) becomes:

$$\frac{(c_{t+k} - c_t)}{k} = \frac{D(c_{i+h} + c_{i-h} - 2c_i)}{h^2} \tag{8}$$

Since concentration is both a function of time and distance, let us generalize the notation according to Fig. 6.2, where i,j represents concentration at position i and time t and c_{i+1},j represents concentration at position $i + h$ and time t. Thus Eq. (8) can be written as:

$$c_{i,i+j} = c_{ij} + \frac{Dk(c_{i+1,j} + c_{i-1,j} - 2c_{i,j})}{h^2} \tag{9}$$

Looking at Eq. (9) and Fig. 6.2, we see that information at a particular time j can be used to obtain information at a future time $j + 1$. In the diagram concentration at the full circles are known, and concentrations at the open circle can be evaluated. To use this method, we must have somewhere to start, and that is a boundary condition. For example at the electrode surface in Fig. 6.1, we can set a potential so that concentration of the redox species is always zero; thus it is possible to move from this. A nice aspect of Eq. (9) is that the units work out: Since k is measured in seconds and h in centimeters, the parameter $Dk/h^2 = D_m$ is unitless. Furthermore it is possible to ratio all concentrations to the initial bulk concentration c^o to obtain dimensionless concentrations $C = c/c^o$. Thus:

$$C_{i,j+1} = C_{i,j} + D_m(C_{i+1,j} + C_{i-1,j} - 2C_{i,j}) \tag{10}$$

Using this technique to probe into the future one point at a time is called the explicit method. It is inadequate because the value of D_m cannot be greater than 0.5, otherwise the method is unstable.[1]

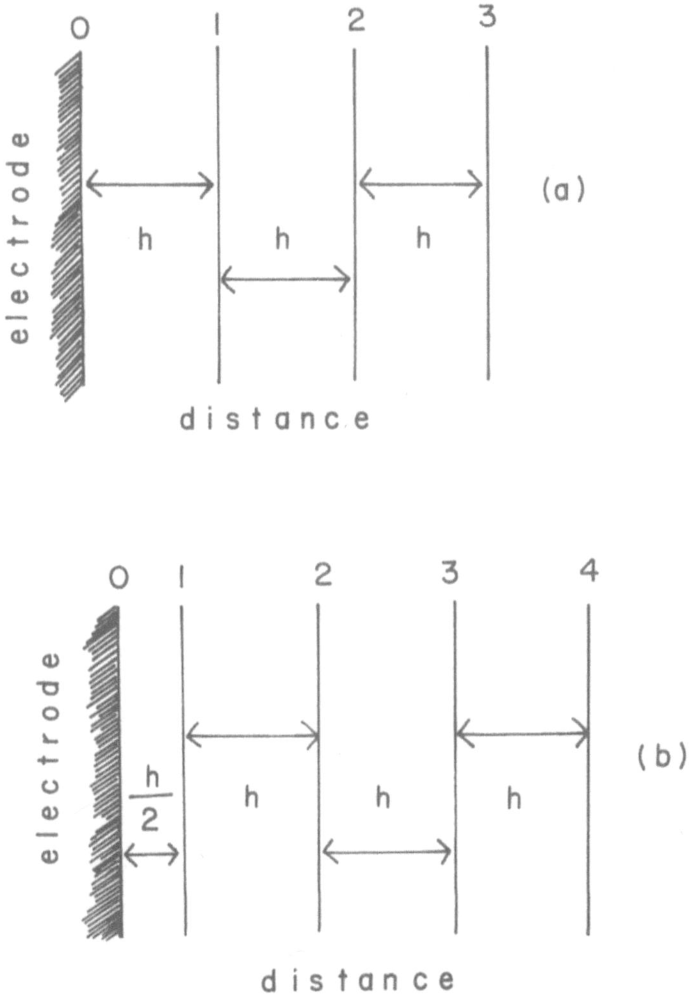

FIGURE 6.3. (a) Equal and (b) unequal distance spacings used in the finite difference method.

Since the preceding method is widely used in electrochemistry, and since the concentration profile at the electrode surface is more important than in the bulk solution, a trick is used to characterize the profile better close to the electrode. By decreasing the distance interval closer to the electrode, as shown in Fig. 6.3(b), we hope to obtain a better indication of the concentration profile and also its derivative, which is related to the current. An estimate of Eq. (9) applied to the situation closer to the electrode, namely, the derivative evaluated at point 1 in Fig. 6.3(b) is[2,3]

$$\frac{C_{1,j+1} - C_{1,j}}{k} = \frac{D}{h}\left(\frac{C_{2,j} - C_{1,j}}{h} - \frac{C_{1,j} - C_{0,j}}{h/2}\right) \tag{11}$$

or

$$C_{1,j+1} = C_{1,j} + D_m(C_{2,j} - 3C_{1,j} + 2C_{0,j}) \tag{12}$$

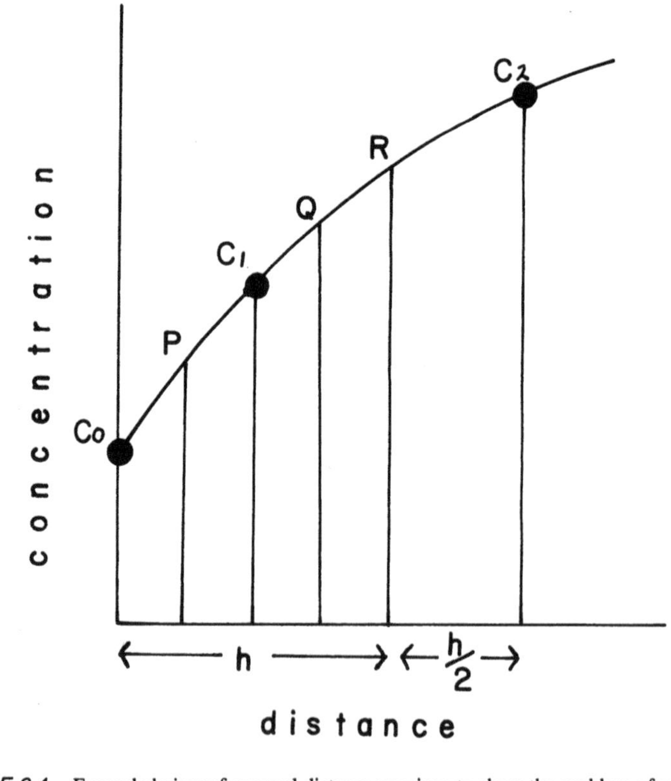

FIGURE 6.4. Expanded view of unequal distance spacings to show the problem of asymmetry.

where the right-hand side of Eq. (11) represents the average of two slopes. We run into a problem at this point. Although Eq. (12) is quoted widely and used frequently in the literature,[4] it suffers from the problem that it is unsymmetric.[5]

Expanding Fig. 6.3(b) yields Fig. 6.4. The slope at point P located at a distance $h/4$ is $2(C_1 - C_0)/h$. The slope at point R located at a distance h from the electrode is $(C_2 - C_1)/h$, and the slope at point Q located at a distance $3h/4$ from the electrode is $2(C_2 - C_0)/(3h)$. It is possible to obtain two different estimates of the second derivative. The difference in slopes between R and P (located at $5h/8$ from the electrode) is

$$C'' = \frac{4(C_2 - 3C_1 + 2C_0)}{(3h^2)} \tag{13}$$

while the difference between slopes between Q and P located at the point C_2 is also given by Eq. (13)

There is a four-thirds difference between Eqs. (12) and (13), a fact that has not received a great amount of attention. Britz[1] comments that although Eq. (12) is wrong, better results are obtained from it than the technically correct Eq. (13). By leaving out the factor 4/3, a fudge factor 0.75 is implicitly included for which there is no justification. One is inclined to recoil from these methods because of this ambiguity. Britz comments that it is better to use equal intervals close to the interface, i.e.,

$$C_{1,j+1} = C_{1,j} + D_m(C_{2,j} + C_{0,j} - 2C_{1,j}) \tag{14}$$

than to use a $h/2$ interval close to the electrode.

2. Potential Step in an Infinite Solution—Explicit Method

Consider a reaction at an electrode in a solution of oxidized species O of concentration c^o.

$$O + ne^- \rightleftharpoons R \tag{15}$$

At a certain time ($t > 0$), a potential step is applied so that O is reduced at the electrode and the concentration of O at the electrode is zero, i.e., in terms of dimensionless concentrations, $C^o_{0,j} = c^o_{0,j}/c^o = 0.0$, where the superscript o represents oxidized species, and subscripts represent a distance $o \times h$ away from the electrode and a time $j \times k$ after the potential step is applied. Once the reaction is proceeding, the rate at which O is consumed at the electrode is equal to the rate at which R is produced, i.e.,

$$\left.\frac{dC^o}{dx}\right|_{x=0} = -\left.\frac{dC^r}{dx}\right|_{x=0} \tag{16}$$

Let us use equal spacings to begin with; then from Fig. 6.3(a), Eq. (16) becomes:

$$\frac{(-C^o_{1,j} + C^o_{0,j})}{h} = \frac{(C^r_{1,j} - C^r_{0,j})}{h} \tag{17}$$

and since $C^o_{0,j} = 0.0$, then

$$C^r_{0,j} = C^r_{1,j} + C^o_{1,j} \tag{18}$$

When Eq. (18) is substituted in Eq. (14) for r, we obtain Eq. (19)

$$C^r_{1,j+1} = C^r_{1,j} + D_m\,(C^r_{2,j} - C^r_{1,j} + 2C^o_{1,j}) \tag{19}$$

Similarly for O we obtain

$$C^o_{1,j+1} = C^o_{1,j} + D_m\,(C^o_{2,j} + 0.0 - 2C^o_{1,j}) \tag{20}$$

Though superscripts and subscripts are cumbersome, they help identify parameters.

To solve the potential step problem, a differential equation of the form of Eq. (1) must be solved for both species. These equations can be approximated using either Eq. (9) for equally spaced panels [Fig. 6.3(a)] or Eq. (12) when the spacing of the panel close to the electrode is $h/2$ [Fig. 6.3(b)].

It is possible to set the number of iterations on a time scale to some constant value L. Based on this, if the time scale of the experiment is t_c, then $k = t_c/\text{L}$. For this particular method D_m may not take values greater than 0.5, so D_m is set to 0.45; then:

$$h = \left(\frac{Dt_c}{D_m\text{L}}\right)^{1/2} \tag{21}$$

Based on an approximation of the thickness of the expansion of the diffusion layer after a time t_c equal to $6(Dt_c)^{1/2}$. If there are J_{MAX} points in the x direction, then:

$$(J_{MAX})h = 6(Dt_c)^{1/2} \tag{22}$$

Substituting Eq. (21) into Eq. (22), $J_{MAX} = 6(D_m\text{L})^{1/2}$. Following this we can use the following expression for the current:[6]

$$i = -nFAD\,\frac{dC}{dx}\Big|_{x=0} \tag{23}$$

If the distance divisions are all the same, Eq. (23) becomes:

$$i = \frac{-nFADc^o(C^o_{1,t} - C^o_{0,t})}{h} \tag{24}$$

If we define a dimensionless parameter Z as:

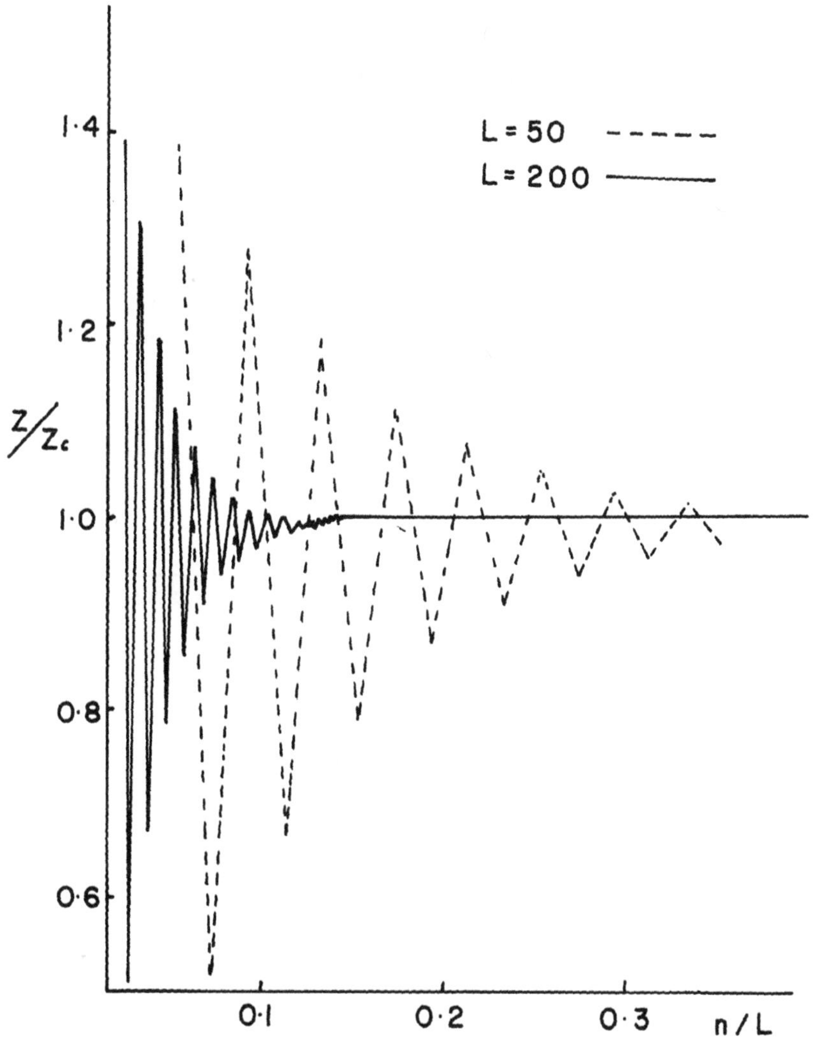

FIGURE 6.5. The output from Program 1 for the potential step method. The value of Z is ratioed with that of the analytic solution. The number of time divisions are taken as 50 and 200. n is the counter for L, and it represents dimensionless progress in time.

$$Z = \frac{i(t_c)^{1/2}}{[nFA(D)^{1/2}c^o]} \tag{25}$$

Ratioing Eq. (25) compared to the expected dimensionless Cottrell result $Zc = (L/\pi k)^{1/2}$, we obtain

$$\frac{Z}{Zc} = \frac{i(t_c\pi k)^{1/2}}{(LD)^{1/2}nFAc^o} \tag{26}$$

Program 1 (see page 115) gives the FORTRAN solution of Eq.(26). Initial concentrations are defined in the DO loop labeled 10. The diffusion equation is described in the DO loop labeled 30. It is possible to change the program to handle the situation when spacing close to the electrode is *h*/2 by changing the following lines:

```
FONEW(1) = FOOLD(1)+DM*(-2.D0*FOOLD(1)+FOOLD(2))
FRNEW(1) = FROLD(1)+DM*(FOOLD(1)-FROLD(1)+FROLD(2))
W = DSQRT(FLOAT(L)*DM)*FONEW(1)
```

to

```
FONEW(1) = FOOLD(1)+DM*(-3.D0*FOOLD(1)+FOOLD(2))
FRNEW(1) = FROLD(1)+DM*(2*FOOLD(1)-FROLD(1)+FROLD(2))
W = 2.D0*DSQRT(FLOAT(L)*DM)*FONEW(1)
```

Figure 6.5 shows the output of this program for two different values of L. We see that the output Z/Zc settles at 1 more quickly when L is large.

3. Cyclic Voltammetry in a Thin Layer of Redox-Containing Electroactive Species

The next problem concerns sweep voltammetry, which involves applying a slowly varying linear waveform starting at a potential where no electrochemistry occurs. The potential is swept past the formal potential of the couple. Once again we consider the reduction of the species O confined within a thin layer [Eq. (15)].

A potential profile as shown in Fig. 6.6 is applied to the electrode for the purpose of examining current potential characteristics. Behavior depends on the slope of the ramp: If the rate of change of potential is sufficiently slow then on the forward sweep, it is possible to reduce all O within the layer. Thus the current initially increases as the potential is swept toward $E^{o\prime}$ and afterward the current decays to zero, since all the O is reduced. A similar situation occurs during the reverse sweep for the oxidation of R, where *a* (mirror image) peak occurs for its oxidation. If the sweep rate is faster, then the first scan may not capture all the species O for reduction. In other words the diffusion layer may not extend to the edge of the polymer layer. In this case a slightly asymmetric cyclic voltammogram

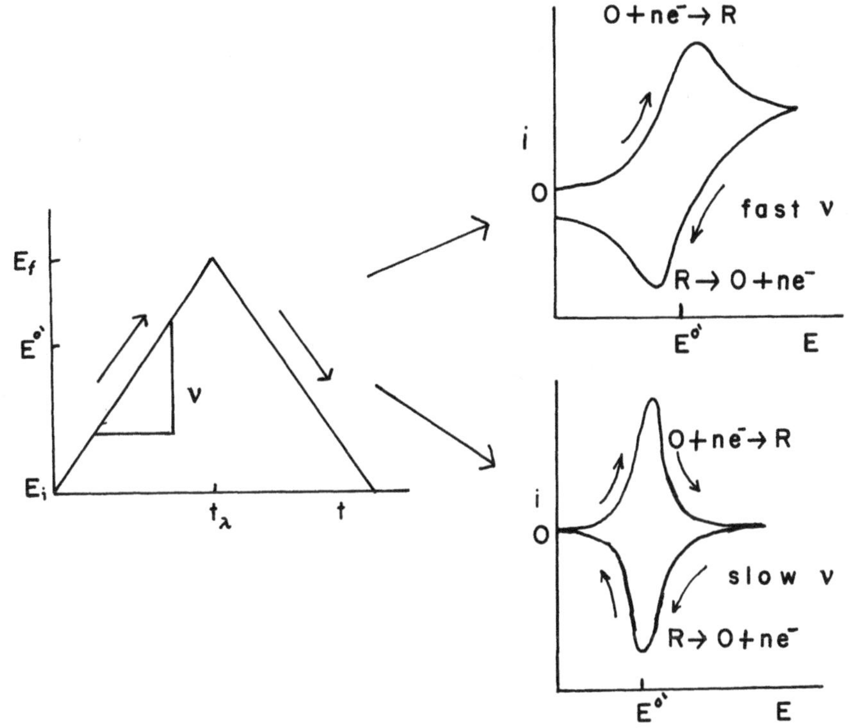

FIGURE 6.6. A potential waveform applied to a redox species confined to a thin layer at a rapid sweep rate and a slow sweep rate; v is the sweep rate; $E^{0\prime}$ is the formal potential of the couple.

is obtained, characteristic of bulk solution behavior, as can be seen in Fig. 6.6. The two equations to be solved are

$$\frac{\partial C^o}{\partial t} = D\frac{\partial^2 C^o}{\partial x^2} \tag{27}$$

$$\frac{\partial C^r}{\partial t} = D\frac{\partial^2 C^r}{\partial x^2} \tag{28}$$

The system is assumed to be at equilibrium, and therefore the Nernst equation follows, so the surface concentrations are given by:

$$\frac{C^o_{0,j}}{C^r_{0,j}} = \exp\left[\frac{nF(V - V^{o\prime})}{(RT)}\right] \tag{29}$$

where $V^{o\prime}$ is the formal potential of the couple and V is the applied potential. This potential can be made dimensionless by defining $E = V/0.0256 = FV/(RT)$ at $T =$

298 K. The scan rate v determines the time of the experiment t_c, since $t_c = (V_f - V_i)/v$, where V_f is the final potential and V_i is the initial potential of the sweep.

The other boundary conditions for the situation shown in Fig. 6.6 are

$$\left.\frac{dC^o}{dx}\right|_{x=0} = -\left.\frac{dC^r}{dx}\right|_{x=0} \tag{30}$$

or

$$\frac{(C^o_{1,j} - C^o_{0,j})}{h} = -\frac{(C^r_{1,j} - C^r_{0,j})}{h} \tag{31}$$

Furthermore since no material leaves the layer, a similar expression can be written at the outer edge, i.e., when x equals d.

$$\left.\frac{dC_r}{dx}\right|_{x=d} = -\left.\frac{dC_o}{dx}\right|_{x=d} = 0.0 \tag{32}$$

or

$$\frac{C^o_{N+1,j} - C^o_{N,j}}{h} = \frac{C^r_{N+1,j} - C^r_{N,j}}{h} = 0.0 \tag{33}$$

For O, $C^o_{N+1,j} = C^o_{N,j}$; the diffusion equation at the edge for O is

$$C^o_{N,j+1} = C^o_{N,j} + D_m(C^o_{N-1,j} - C^o_{N,j}) \tag{34}$$

A similar equation exists for R. At the electrode surface:

$$\frac{C^o_{0,j}}{C^r_{0,j}} = \exp(nE_m) \tag{35}$$

where E_m is the potential increment, i.e., $(E_f - E_i)/L$, and n is the number of steps of time up to a particular time; L is the number of time divisions and $E^{o\prime}$, the formal potential, is zero. These relations can be substituted into Eq. (31) to give

$$C^r_{0,j} = \frac{C^r_{i,j} + C^o_{1,j}}{[1 + \exp(nE_m)]} \tag{36}$$

An expression for the current is[4]

$$i = \frac{nFAc^o(DD_mL)^{1/2}}{(t_c)^{1/2}}\left[\frac{C^0_{0,j} - \exp(nE_mC^r_{0,j})}{1 + \exp(nE_m)}\right] \tag{37}$$

This is ratioed to the expected peak current for a thin-layer system[6]

$$i_p = \frac{n^2F^2vAdc^o}{4RT} \tag{38}$$

Equations (37) and (38) can be combined to yield the ratio (at 298 K):

$$\frac{i}{i_p} = \frac{4(0.02567)D_m L}{(V_i - V_f)J_{MAX}} \left[\frac{C^o_{0,j} - \exp(nE_m C^r_{0,j})}{1 + \exp(nE_m)} \right] \tag{39}$$

where J_{MAX} is the number of divisions in the x direction.

In the case of a finite layer, it is easier to define the number of distance intervals. The total experiment time is $t_c = (E_i - E_f)$/(scan rate). Once D and the layer thickness d are defined and the number of distance intervals is J_{MAX}, then $h = d/J_{MAX}$. Since $D_m = 0.45$, we can find k as $D_m(h)^2/D$ and L the number of distance intervals is t_c/k.

Program 2 (page 116) gives the solution for the explicit method. The DO loop labeled 30 comprises the statement of Eq. (14). This is followed by the boundary condition [Eqs. (34) and (36)]. The potential is gradually varied, the current ratio reaches a maximum of 1.00.

It is possible to use a step length of $h/2$ closer to the electrode by changing the lines in Program 2 from:

```
FONEW(1)=FOOLD(1)+DM*(FOSURF-2.D0*FOOLD(1)+FOOLD(2))
FRNEW(1)=FROLD(1)+DM*(FRSURF-2.D0*FROLD(1)+FROLD(2))
W(I)=W(I)*4.D0*0.0256*DM*FLOAT(L)/(FLOAT(JMAX)*(VI-VF))
```

to

```
FONEW(1)= FOOLD(1)+DM*(2.D0*FOSURF-
3.D0*FOOLD(1)+FOOLD(2))
FRNEW(1)=FROLD(1)+DM*(2.D0*FRSURF-
3.D0*FROLD(1)+FROLD(2))
W(I)=W(I)*8.D0*0.0256*DM*FLOAT(L)/(FLOAT(JMAX)*(VI-VF))
```

TABLE 6.1.
Output from Program 2 for the Explicit Finite-Difference Method

	d^a/cm	$N_p{}^a$	E_p /mV	i/ip	$\Delta E_{1/2}$/mV
(a)	1.0×10^{-4}	6	−4.25	0.9176	90.05
(b)	2.5×10^{-4}	6	−19.5	0.7795	107.8
(c)	3.0×10^{-4}	6	−24.0	0.7070	117.7
(d)	4.0×10^{-4}	6	−32.0	0.5687	140.0
(e)	5.0×10^{-4}	6	−37.5	0.4567	168.6
(f)	5.0×10^{-4}	10	−33.2	0.4946	159.2
(g)	5.0×10^{-4}	10	−33.2	0.4946	159.2
(h)	1.0×10^{-4}	10	−3.62	0.8451	90.9

[a]The $D_m = 0.45$, $D = 10^{-9}$ cm²/s, scan rate = 10^{-3} V/s. N_p is the number of points employed; d is the layer thickness; E_p is the potential where the peak current occurs when $E^{o\prime}$ is 0.0V; $\Delta E_{1/2}$ is the peak width at half-height; (g) and (h) are conditions where an $h/2$ spacing was used close to the electrode.

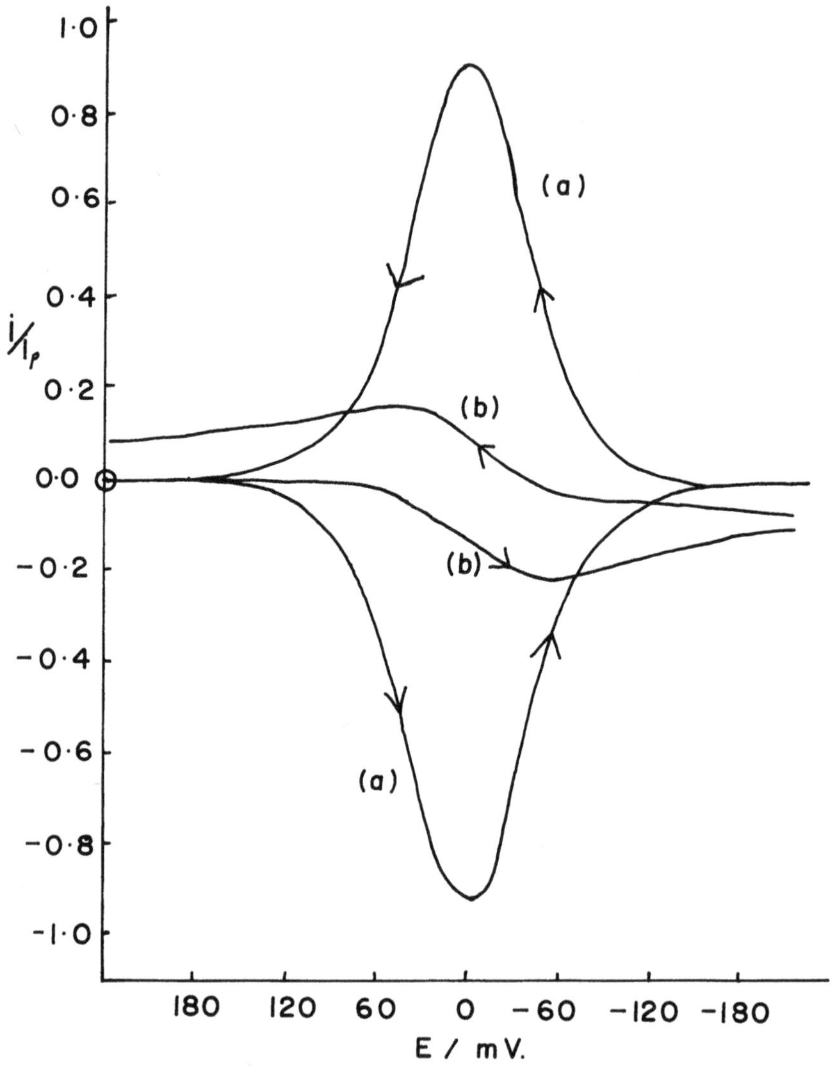

FIGURE 6.7. Cyclic voltammogram generated using Program 2. $D_m = 0.45$, $D = 10^{-10}$ cm^2/s, scan rate = 0.1 V/s, Number of distance intervals = 6. (a) $d = 2.5 \times 10^{-6}$ cm and (b) $d = 9 \times 10^{-6}$ cm.

Figure 6.7 shows the types of cyclic voltammogram for two layer thicknesses of 2.5×10^{-6} cm and 9.0×10^{-6} cm. We see that as layer thickness increases, the peak-to-peak separation increases, which is indicative of bulk solution behavior. Table 6.1 shows the effect of layer thickness on the i/i_p ratio, the potential of peak current E_p, and the peak width at half-height $\Delta E_{1/2}$, which ideally should assume the value 90 mV.

Values for d and D are chosen according to those in reality.[7] As layer thickness decreases [Table 6.1 (a–e)], the peak current ratio i/i_p tends toward 1.00. As the number of points increases [Table 6.1(f)], the i/i_p changes only a little. If $h/2$ rather than h is used for the first distance element, it does not make any difference, as seen in Table 6.1(g, h).

4. Implicit Methods Used for the Cyclic Voltammetry for a Thin-Layer Situation

From the preceding, we know the following:

$$C_{i,j+1} = C_{i,j} + D_m(C_{i+1,j} - 2C_{i,j} + C_{i-1,j}) \tag{10}$$

as seen in Fig. 6.2. The reverse of Eq. (10) is also valid:

$$C_{i,j+1} + D_m(C_{i+1,j+1} - 2C_{i,j+1} + C_{i-1,j+1}) = -C_{i,j} \tag{40}$$

At the point A in Fig. 6.8, the term dC/dt can be approximated as:

$$\frac{dC}{dt} = \frac{C_{i,j+1} - C_{i,j}}{k} \tag{41}$$

At the point A in the C'' term may be an average of Eqs. (10) and (40), i.e.,

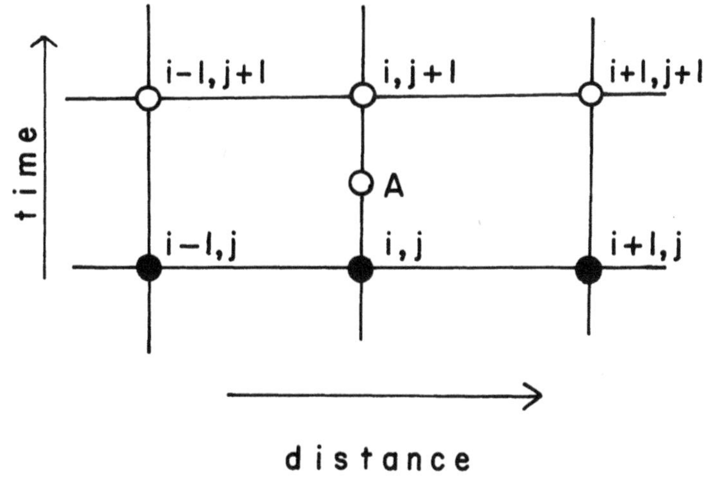

FIGURE 6.8. Distance time scheme for the Crank Nicolson technique.

$$\frac{C_{i,j+1}-C_{i,j}}{k}=\frac{D}{2}\left(\frac{C_{i+1,j+1}+C_{i-1,J+1}-2C_{i,j+1}}{h^2}+\frac{C_{i+1,j}+C_{i-1,j}-2C_{i,j}}{h^2}\right) \quad (42)$$

which becomes

$$(2+2D_m)C_{i,j+1}-D_mC_{i+1,j+1}-D_mC_{i-1,j+1}$$
$$=D_mC_{i+1,j}+D_mC_{i-1,j}-(2D_m-2)C_{ij} \quad (43)$$

Everything on the right-hand side is known, so it is possible to set up a series of algebraic equations to determine concentration on the left-hand side. In the case of the explicit method, we progress forward in time for each distance point; in this case we progress by determining concentrations at all distances simultaneously at each time increment. The resulting series of equations are shown in matrix form as:

$$\begin{bmatrix} (2+2D_m) & -D_m & & & & \\ -D_m & (2+2D_m) & -D_m & & \mathbf{0} & \\ \vdots & \vdots & \vdots & & & \\ & \vdots & \vdots & & & \\ & & \vdots & \vdots & & \\ \mathbf{0} & & \vdots & \vdots & & \\ & & -D_m & (2+2D_m) & -D_m & \\ & & & -D_m & (2+2D_m) & \end{bmatrix}\begin{bmatrix} C_{1,j+1} \\ C_{2,j+1} \\ \vdots \\ \vdots \\ \vdots \\ \vdots \\ C_{N-1,J+1} \\ C_{N,j+1} \end{bmatrix}$$

$$=\begin{bmatrix} (2-2D_m) & D_m & & & & \\ D_m & (2-2D_m) & D_m & & \mathbf{0} & \\ \vdots & \vdots & \vdots & & & \\ & \vdots & \vdots & & & \\ & & \vdots & \vdots & & \\ \mathbf{0} & & \vdots & \vdots & & \\ & & D_m & (2-2D_m) & D_m & \\ & & & D_m & (2-2D_m) & \end{bmatrix}\begin{bmatrix} C_{1,j} \\ C_{2,j} \\ \vdots \\ \vdots \\ \vdots \\ \vdots \\ C_{N-1,j} \\ C_{N,j} \end{bmatrix}=\begin{bmatrix} k_1 \\ k_2 \\ \vdots \\ \vdots \\ \vdots \\ \vdots \\ k_{N-1} \\ k_N \end{bmatrix} \quad (44)$$

Note that in addition to these N points there are also two boundary points at the electrode surface and also at the polymer solution interface. Since these boundary conditions are known at all times, the right-hand side has to be modified as follows

$$
=\begin{bmatrix}
(2+2D_m) & -D_m & & & & \\
-D_m & (2+2D_m) & -D_m & & \mathbf{0} & \\
\vdots & \vdots & \vdots & & & \\
 & \vdots & \vdots & & & \\
 & & \vdots & \vdots & & \\
 & \mathbf{0} & \vdots & \vdots & & \\
 & & -D_m & (2+2D_m) & -D_m & \\
 & & & -D_m & (2+2D_m) &
\end{bmatrix}
\begin{bmatrix}
C_{1,j+1} \\ C_{2,j+1} \\ \vdots \\ \vdots \\ \vdots \\ \vdots \\ C_{N-1,J+1} \\ C_{N,j+1}
\end{bmatrix}
$$

$$
=\begin{bmatrix}
k_1 + D_m(C_{0,j} + C_{0,j+1}) \\ k_2 \\ \vdots \\ \vdots \\ \vdots \\ \vdots \\ \vdots \\ k_{N-1} \\ k_N + D_m(C_{N+1,j} + C_{N+1,j+1})
\end{bmatrix} \tag{45}
$$

In this manner boundary conditions are included. The preceding set of equations are tridiagonal, and they can be solved by using the Thomas algorithm.[1]

Booman and Pence used an implicit method for kinetic complications for bulk solution reactions,[8] and Heinze *et al.*[9] used the Crank–Nicolson method for cyclic voltammetry, Lasia[10] employed it for examining dimerization. A review of the Crank–Nicolson methods appears[11] in Ref. 11. Program 3 (page 117) gives the FORTRAN solution to this problem. To change the program so that the first panel takes a width of $h/2$ rather than h, change the following:

```
FONEW(1)=FOOLD(1)+DM*(FOSURF-2.D0*FOOLD(1)+FOOLD(2))
FRNEW((1)=FROLD(1)+DM*(FRSURF-2.D0*FROLD(1)+FROLD(2))
W(I)=W(I)*4.D0*0.0256*DM*FLOAT(L)/(FLOAT(JMAX)*(VI-VF))
```

to:

```
FONEW(1)  =FOOLD(1)+DM*(2*FOSURF-3.D0*FOOLD(1)+FOOLD(2))
FRNEW(1)  =FROLD(1)+DM*(2*FRSURF-3.D0*FROLD(1)+FROLD(2))
W(I)  =W(I)*8.D0*0.0256*DM*FLOAT(L)/(FLOAT(JMAX)*(VI-VF))
```

The matrix is first defined for the right-hand side, then for the left-hand side. Table 6.2 shows some of the results from Program 3. Table 6.2(a–d) shows the effect of varying layer thickness, these results are similar to those in Table 6.1. Increasing

TABLE 6.2.
Output from Program 3 for the Implicit Finite-Difference Method

	d^a/cm	$N_p{}^a$	E_p/mV	i/ip	$\Delta E_{1/2}$/mV
(a)	1.0×10^{-4}	6	−4.12	0.992	91.0
(b)	2.5×10^{-4}	6	−18.8	0.8432	107.0
(c)	3.0×10^{-4}	6	−23.0	0.766	115.9
(d)	5.0×10^{-4}	6	−34.3	0.5033	162.5
(e)	1.0×10^{-4}	10	−3.68	0.9930	91.4
(f)	3.0×10^{-4}	6	−23	1.532	—
(g)	1.0×10^{-4}	10	−3.68	1.98	—

[a]The $D_m = 0.45$, $D = 10^{-9}$ cm^2/s, scan rate = 10^{-3} V/s; N_p is the number of points employed; d is the layer thickness; Ep is the potential where the peak current occurs when $E^{o\prime}$ is 0.0V; $\Delta E_{1/2}$ is the peak width at half-height; (f) and (g) used $h/2$ spacing close to the electrode, and (e) and (g) had values of 2.45 for D_m.

the value of D_m from 0.45 to 2.45 does not change the result to any extent [Table 6.2(d)]. Decreasing the width of the first panel to $h/2$ makes the ratio of peak heights i/i_p greater than 1. Since this makes the results unreliable, it is better to use equal intervals.

5. Hopscotch Algorithm for Cyclic Voltammetry in a Thin Layer

This method is also based on Eq. (10):

$$C_{i,j+1} = C_{i,j} + D_m(C_{i+1,j} - 2C_{i,j} + C_{i-1,j}) \tag{10}$$

which yields

$$C_{i,j+1} = \left[\frac{C_{i,j} + D_m(C_{i+1,j+1} + C_{i-1,j+1})}{(1 + 2D_m)}\right] \tag{46}$$

Furthermore since:

$$C_{i,j+2} - C_{i,j+1} = C_{i,j+1} - C_{i,j} \tag{47}$$

then Eq. (46) becomes

$$C_{i,j+2} = \frac{2[C_{i,j} + D_m(C_{i+1,j+1} - C_{i-1,j+1})]}{(1 + 2D_m)} - C_{i,j} \tag{48}$$

Equations (46) and (48) can be interpreted as extrapolating further in time to obtain the concentration in the future at a particular position.

From Eq. (48) we obtain

$$C_{i,j+2} = C_{i,j} + \frac{2D_m(C_{i+1,j+1} - 2C_{i,j} + C_{i-1,j+1})}{(1 + 2D_m)} \quad (49)$$

We are interested in the node close to electrode Node 2 when Node 1 is at the electrode surface (see Fig. 6.9). The procedure, as described in Refs. 12–14, follows. Initially $j = 0$.

1. For nodes where $(i + j)$ is odd, $C_{i,j}$ is overwritten using Eq. (38); for the node 2 ($i = 2$), when $(j + 2)$ is odd, store $C_{2,j+1}$ in a parameter $S_{2,j+1}$.
2. When $(j + 2)$ is even, calculate $C_{2,j+1}$ using Eq. (46) and store it in $S_{2,j+1}$.
3. For the rest of the nodes, when $(i + j)$ is even, overwrite $C_{i,j}$ with $C_{i,j+2}$ calculated from Eq. (49).
4. Then increment j by 1 and repeat Steps 1–3.

Program 4 (page 117) shows the solution to the cyclic voltammetric problem for a species in a thin layer, as described previously. Figure 6.9 shows steps in calculating the concentration at each point. Initially $j = 0$ and concentrations in the filled circles are known concentrations. Applying Step 1 to $i = 3$, $C_{3,1}$ is found. Applying Step 2, $C_{2,1}$ is found; applying Step 3 for $i + j$ even (i.e., for $i = 2$ and $i = 4$), $C_{i,j+2}$ is found from Eq. (48). We see that points are calculated in a hopscotch manner. Table 6.3 shows some typical results using Program 4. We see that a considerable number of points are required to achieve a ratio i/ip equal to 1. For

TABLE 6.3.

Output from Program 4 for the Hopscotch Method

	d^a/cm	$N_p{}^a$	E_p /mV	i/ip	$\Delta E_{1/2}$/mV
(a)	1.0×10^{-4}	6	−2.75	0.8037	91.0
(b)	2.5×10^{-4}	6	−15.63	0.7318	100.8
(c)	3.0×10^{-4}	6	−18.57	0.6862	108.8
(d)	5.0×10^{-4}	6	−28.1	0.490	140.6
(e)	2.5×10^{-4}	10	−14.8	0.791	102.0
(f)	5.0×10^{-4}	10	−29.8	0.518	145.1
(g)	1.0×10^{-4}	50	−2.7	0.997	90.4
(h)	2.5×10^{-4}	100	−15.4	0.875	—
(i)	2.5×10^{-4}	100	−15.0	0.867	103.7
(j)	0.8×10^{-4}	10	−2.37	1.016	90.4

[a]The $D_m = 0.45$, $D = 10^{-9}$ cm^2/s, scan rate = 10^{-3} V/s; N_p is the number of points employed; d is the layer thickness; E_p is the potential where the peak current occurs when $E^{0\prime}$ is 0.0V; $\Delta E_{1/2}$ is the peak width at half-height; D_m was taken as 1.45 in (h) and (j).

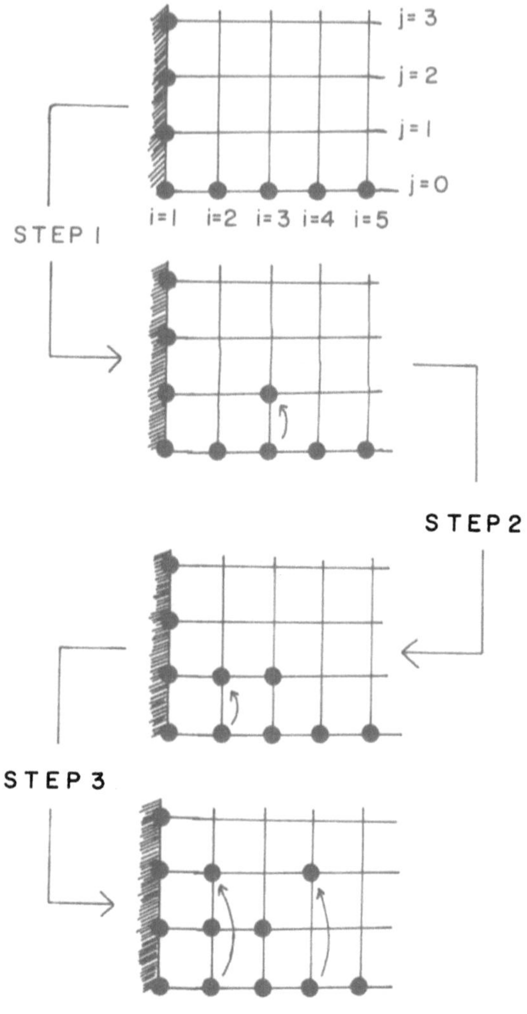

FIGURE 6.9. Using the hopscotch algorithm to evaluate concentrations with time at particular points close to the electrode.

the same number of points in Table 6.1, the value of i/ip is less than 1.00 when $d = 1 \times 10^{-4}$ cm.

From Table 6.3(b,i), we seen that the number of points taken is important: With 100 points, the value of i/ip is close to the value using the Crank–Nicolson method with six points [Table 6.2(b)]. From Table 6.3(j), we note that if D_m is changed to 1.45, the value of i/ip may become greater than 1, which means that it is safer to

D_m as 0.45. The limitation on D_m is not to take the value 2.00;[12] the hopscotch method has been used for chronoamperometry[12] and electrochemistry at microelectrodes.[13,14]

6. Orthogonal Collocation

This method involves solving the diffusion equation:

$$\frac{\partial c}{\partial t} = D\frac{\partial^2 c}{\partial x^2} + f(c) \tag{50}$$

where $f(c)$ represents some kinetic complications in the process. An approximate solution is chosen as:

$$\Phi = \sum_{i=1}^{N} \alpha_i(t)\phi_i(x) \tag{51}$$

to make the residual of the equation equal to zero,[15] i.e.,

$$R = \frac{\partial \Phi}{\partial t} - D\frac{\partial^2 \phi}{\partial x^2} - f(\Phi) = 0.0 \tag{52}$$

In fact, this sets the weighted residual equal to zero:

$$\int_w w_j R(\phi)dx = 0 \tag{53}$$

The choice of the weighting function determines the method.[16] In the collocation method, the weighting function is chosen to be a displaced dirac function:

$$w_j = \delta(x - x_j) \tag{54}$$

which has the very important property that:

$$\int_w w_j R(\phi)dx = R|_{x=xj} \tag{55}$$

so the residual is zero at specified points x_j.[15] The choice of these points can make calculations both convenient and accurate. In orthogonal collocation the points chosen are roots of orthogonal polynomials. Sets of orthogonal polynomials satisfying boundary conditions are chosen, and their roots represent collocation points. An example of polynomials that can be chosen are shifted Jacobi polynomials:

$$\int x^\beta(1-x)^\alpha P_n^{(\alpha,\beta)}(x)P_m^{(\alpha,\beta)}(x)dx = 0 \tag{56}$$

whose roots vary between 0 and 1. However Yen and Chapman[17] chose trial functions whose roots vary between 0 and ∞. Using the notation of Whiting and Carr,[18] the general evolution of the problem follows. Once the system of polynomials is chosen, it essentially limits the problem, so that solution is found only at particular positions within the polymer layer defined by the roots of these polynomials. For example if α is 0.0 and β is 0.0, then for six points concentrations are determined at the dimensionless positions 0.03376, 0.15939, 0.38069: 0.619309, 0.83060, and 0.96623 away from the electrode. The solution takes the general form:

$$C(x_i,t) = \sum_{j=1}^{N+2} b_j(t)\,(x_i)^{j-1} \qquad i = 1,2,3,\ldots, N+2 \tag{57}$$

where there are N+2 collocation points. Notice that the function $C(x_i,t)$ has been divided into two independent parts, one in time and the other in distance. Since positions in distance have been fixed, the problem becomes time-dependent. First and second derivatives of Eq. (57) can be found

$$\left.\frac{dC(x,t)}{dx}\right|_{x=x_i} = \sum_{j=1}^{N+2} \left.\frac{d(x)^{j-1}}{dx}\right|_{x_i} b_j(t) \tag{58}$$

$$\left.\frac{d^2C(x,t)}{dx^2}\right|_{x=x_i} = \sum_{j=1}^{N+2} \left.\frac{d^2(x)^{j-1}}{dx^2}\right|_{x_i} b_j(t) \tag{59}$$

These equations can be evaluated at all N+2 collocation points. Equations (58) and (59) can be expressed in a simpler matrix notation, where the – superscript represents a one-dimensional matrix and the = superscript represents a two-dimensional matrix.

$$\overline{C(t)} = \overline{\overline{Q}}\,\overline{b(t)} \tag{60}$$

where $\overline{\overline{Q}}_{i,j} = x_i^{j-1}$ and

$$\overline{\frac{dC(t)}{dx}} = \overline{\overline{C}}\,\overline{b(t)}; \qquad \overline{\frac{d^2C(t)}{dx^2}} = \overline{\overline{D}}\,\overline{b(t)} \tag{61}$$

where

$$\overline{\overline{C}}_{i,j} = \left.\frac{dx^{j-1}}{dx}\right|_{x=x_i} \qquad \overline{\overline{D}}_{i,j} = \left.\frac{d^2x^{j-1}}{dx^2}\right|_{x=x_i} \tag{62}$$

from Eqs. (60–62):

$$\overline{b(t)} = \overline{\overline{Q}}^{-1}\overline{C(t)} \tag{63}$$

The parameters defined in Eqs. (60–63) can be substituted into Eqs. (58) and (59) to obtain

$$\frac{\overline{dC(t)}}{dx} = \overline{\overline{C}}\,\overline{\overline{Q}}^{-1}\overline{C(t)} = \overline{\overline{A}}\,\overline{C(t)} \tag{64}$$

and

$$\frac{\overline{d^2C(t)}}{dx^2} = \overline{\overline{D}}\,\overline{\overline{Q}}^{-1}\overline{C(t)} = \overline{\overline{B}}\,\overline{C(t)} \tag{65}$$

Once matrices A and B are computed, they are invariant for a given number of collocation points. The final expression to be solved is composed of a series of ordinary differential equations of the form:

$$\left.\frac{\overline{dC(t)}}{dt}\right|_{x=x_i} = \beta\overline{\overline{B}}\,\overline{C(t)} + f[\overline{C(t)}] \tag{66}$$

where β is a dimensionless parameter that is a function of the diffusion coefficient D. This set of ordinary differential equations may be solved by any standard technique.(19,20)

Collocation methods, widely used in electrochemistry,(18,21,22) were first applied by Whiting and Carr.(18) Eddows(23) solved a rotating disk problem using both finite difference and collocation. He stated that the collocation method is more efficient and its efficiency can be optimized by using a change-of-space variable. However when Magno(24) compared collocation to finite difference for an EC_{ir} mechanism, for a large value of k, the orthogonal collocation method was the least valid. This was due to the computation time and the choice of a parameter β, which is related to the diffusion layer thickness. Techniques for β optimization subsequently appeared.(25–27)

7. Thin-Layer Cyclic Voltammetry Using Orthogonal Collocation

Consider the reduction of the species O as before.

$$O + ne^- \rightleftharpoons R \qquad E^{o\prime} \tag{15}$$

The couple is confined within a layer of thickness d. The following diffusion equations must be solved for each half of the couple:

$$\frac{\partial c^o}{\partial t} = D\frac{\partial^2 c^o}{\partial x^2} \tag{67}$$

$$\frac{\partial c^r}{\partial t} = D\frac{\partial^2 c^r}{\partial x^2} \tag{68}$$

The reaction is assumed to be rapid, and as such concentrations of the species at the electrode surface follow the Nernst equation:

$$\frac{c^r(0,t)}{c^o(0,t)} = \exp\left[\frac{nF}{RT}(E^{o\prime} - E)\right] = \theta S_\lambda(t) \tag{69}$$

where the general expression for concentration at a distance x from the electrode is given by $C^0(x,t)$. The notation is from Nicholson and Shain.[28]

$$\theta = \exp\left[\frac{nF(E^{o\prime} - E_i)}{(RT)}\right] \tag{70}$$

and

$$S_\lambda(t)\begin{cases} = \exp[2(m-1)at_\lambda - at] & 2(m-1)t_\lambda \leq t \leq (2m-1)t_\lambda \\ = \exp(at - 2mat_\lambda) & 2(m-1)t_\lambda \leq t \leq 2mt_\lambda \end{cases} \tag{71}$$

where E_i is the initial potential, t_λ is the time when the potential is switched, a is a function of the scan rate, defined as:

$$a = \frac{nFv}{(RT)} \tag{72}$$

where v is the scan rate in Vs^{-1}.

At the electrode surface the amount of O consumed and the amount of R produced are related, so their concentration slopes at the electrode surface are equal in magnitude and opposite in sign:

$$\left.\frac{dc^r}{dx}\right|_{x=0} = -\left.\frac{dc^o}{dx}\right|_{x=0} \tag{73}$$

Furthermore both species are confined within the thin layer, which means

$$\left.\frac{dc^r}{dx}\right|_{x=d} = \left.\frac{dc^o}{dx}\right|_{x=d} = 0 \tag{74}$$

When using orthogonal collocation, it is convenient to make each of the parameters dimensionless as follows:

$$X = \frac{x}{d}$$

$$C = \frac{c}{c^o} \tag{75}$$

$$T = at$$

When these dimensionless parameters are substituted into Eqs. (67) and (68), they become

$$\frac{\partial C^o}{\partial T} = \beta \frac{\partial^2 C^o}{\partial X^2} \tag{76}$$

$$\frac{\partial C^r}{\partial T} = \beta \frac{\partial^2 C^r}{\partial X^2} \tag{77}$$

where $\beta = D/(ad^2)$. The boundary conditions become

$$\frac{C^r(0,t)}{C^o(0,t)} = \theta \, S_\lambda(t) \tag{78}$$

The other two boundary conditions are

$$\left.\frac{dC^r}{dX}\right|_{X=0} = -\left.\frac{dC^o}{dX}\right|_{X=0} \tag{79}$$

and

$$\left.\frac{dC^r}{dX}\right|_{X=1} = \left.\frac{dC^o}{dX}\right|_{X=1} = 0 \tag{80}$$

The following substitutions can be made

$$\left.\frac{dC^o}{dX}\right|_i = \sum_{j=1}^{N+2} A_{i,j}\, C^o(X_j,T) \qquad \left.\frac{d^2C^o}{dX^2}\right|_i = \sum_{j=1}^{N+2} B_{i,j}\, C^o(X_j,T) \tag{81}$$

Substituting these discretized equations into Eq. (79) and expanding, we obtain

$$A_{1,1}C^r(0,t) + A_{1,N+2}C^r(1,T) + \sum_{j=2}^{N+1} A_{1,j}\, C^r(X_j,T)$$

$$= -[A_{1,1}C^o(0,t) + A_{1,N+2}C^o(1,T) + \sum_{j=2}^{N+1} A_{1,j}\, C^o(X_j,T)] \tag{82}$$

Similarly at the boundary we can write for Eq. (80):

$$A_{N+2,1}C^r(0,t) + A_{N+2,N+2}C^r(1,T) + \sum_{j=2}^{N+1} A_{N+2,j}\, C^r(X_j,T) = 0 \tag{83}$$

and

$$A_{N+2,1}C^o(0,t) + A_{N+2,N+2}C^o(1,T) + \sum_{j=2}^{N+1} A_{N+2,j}\, C^o(X_j,T) = 0 \tag{84}$$

From these conditions we obtain expressions for concentrations:

$$C^r(1,T) = -\frac{\left[A_{N+2,1}C^r(0,T) + \sum_{j=2}^{N+1} A_{N+2,j}\, C^r(X_j,T)\right]}{A_{N+2,N+2}} \tag{85}$$

and

$$C^o(1,T) = -\frac{\left[[A_{N+2,1}C^o(0,T) + \sum_{j=2}^{N+1} A_{N+2,j}\, C^o(X_j,T)\right]}{A_{N+2,N+2}} \tag{86}$$

From Eq. (86) we obtain

$$C^o(0,T) = -\frac{\left[\sum_{j=2}^{N+1} D_j C^r(X_j,T) + \sum_{j=2}^{N+1} D_j C^o(X_j,T)\right]}{\{D_1[(1 + \theta S_\lambda(t)]\}} \tag{87}$$

where $D_j = (A_{1,N+2}A_{N+2,j})/A_{N+2,N+2} - A_{1,j}$. The diffusion equation is

$$\frac{\partial C^o}{\partial T} = \beta \frac{\partial^2 C^o}{\partial X^2} \tag{88}$$

$$= \beta[B_{i,1}C^o(0,T) + B_{i,N+2}C^o(1,T) + \sum_{j=2}^{N+1} B_{i,j}C^o(X_j,T)] \tag{89}$$

A similar equation can also be written for R.

Finally the current may be evaluated as:

$$i = nFAD \left.\frac{dc}{dx}\right|_{x=0} \tag{90}$$

$$= \frac{nFAc^o}{d} \sum_{j=1}^{N+2} A_{1,j}\, C^o(X_j,T) \tag{91}$$

TABLE 6.4.
Output from Program 5 for the Orthogonal Collocation Method

	d^a/cm	Ep mV	i/ip	$\Delta E_{1/2}$/mV
(a)	1.0×10^{-4}	3.19	0.9943	90.73
(b)	2.5×10^{-4}	15.69	0.8755	105.4
(c)	4.0×10^{-4}	22.6	0.6735	130.0
(d)	5.0×10^{-4}	26.3	0.5608	147.3

[a]The $D = 10^{-9}$ cm^2/s, scan rate = 10^{-3} V/s; $E_i = -0.300$, $E_\lambda = 0.300$, $T = 298$ K. The number of points employed is 6; d is the layer thickness; E_p is the potential where the peak current occurs when $E^{o\prime}$ is 0.0V; $\Delta E_{1/2}$ is the peak width at half-height.

which in terms of the peak current i_p becomes

$$\frac{i}{i_p} = 4\beta \sum_{j=1}^{N+2} A_{1,j}\, C^o(X_j,T) \tag{92}$$

The structure of the program follows (see Program 5, page 121). The subroutine JCOBI calculates roots and derivatives of the polynomial. The subroutine DFOPR calculates parameters $A_{i,j}$ and $B_{i,j}$ associated with these roots. The subroutine FUN supplies information about the differential equations; F is the vector on the right-hand side of Eq. (89). The subroutine OUT is the output subroutine. The latter two subroutines are supplied to DFOPR from the IMSL library, which solves a system of first-order differential equations with given initial conditions. The β optimization is included in this program; theory behind β is detailed elsewhere.[15,26,27]

Table 6.4 gives the output of Program 5 under certain conditions. We see that for a layer thickness $d = 2.5 \times 10^{-4}$ cm, Program 5 finds a value of i/ip close to that obtained by other techniques [Table 6.4(a)]. Orthogonal collocation is used to simulate rotating ring disk electrodes[29] and cyclic voltammetry for CE and EC mechanisms.[30]* Furthermore orthogonal collocation is used to characterize diffusion at microelectrodes[31] and microelectrode arrays.[32,33]

8. More Advanced Numerical Methods Applied to Electrochemistry

Peaceman and Rachford[34] developed an Alternate Direction Implicit (ADI) method, which retains a tridiagonal matrix and allows two-dimensional equations

*CE represents a mechanism with an electrochemical step with a prior chemical step and EC is an electrochemical step with a subsequent chemical step.

to be solved. Such a method is used in microelectrode studies.[35] To reduce computation time and the preceding methods more amenable to microprocessors, methods were devised using unequal distance intervals, so there is more emphasis on the points closer to the electrode. Joslin and Pletcher[36] were the first to employ such ideas in electrochemistry. Functions such as:[1,36]

$$y = \frac{[1 - \exp(-ax)]}{a} \tag{93}$$

and

$$y = \frac{ax}{(ax + 1)} \tag{94}$$

were employed when a is a constant and y is the transformed distance. Feldberg[37] developed a refinement that allows points in the expanding panels to shift progressively toward the inner boundary closer to the electrode. This is used by Seeber and Stefani,[38] and expanding panel methods are widely used in electrochemistry.[39–41] In recent papers, several authors[42,44] introduced a quasi-explicit algorithm based on the DuFort Frenkel approach that simulates the potential-step technique. The numerical integration algorithm[45] and more recent versions of implicit methods[46] are used for cyclic voltammetry. Numerical methods are also used in spectroelectrochemistry[47] and chronoamperometry and adsorption.[48] In general the choice of method depends on the chemist's computational ability and the ease of using the method rather than its computation efficiency.

9. Digital Methods Applied to Polymer Layers Coated on Electrodes

9.1. Overview

The area of polymer-modified electrodes was traditionally divided into redox species containing layers or heavily conjugated conducting polymer layers. One characteristic of conducting polymer layers is their tendency to have a large double layer charging current component following the faradaic reaction, which is characteristic of a three dimensional conducting phase.[49] More recently the division between these classes of electrode coatings has become less clear.[50] One common property of both layer types is the ion movement requirement once an electrochemical reaction occurs both for such redox species as $[IrCl_6]^{-4}$ in QPVP,[51] or $[Os(bpy)_3]^{2+}$ in Nafion[52] and such conducting polymers as polypyrrole perchlorate.[53] Ion transport may be the rate-determining factor in the polymer redox reaction.

Whether the so-called double-layer charging current is faradaic or not was examined by Hutton,[54] who observed it to be faradaic in the case of polyaniline using ellipsometry. In addition a series of redox potentials, each of which is slightly different for different lengths of molecular weights of polymers was observed by Meerholtz and Heinze.[55] Thus residual unoxidized layers slowly being oxidized may cause the apparent double-layer charging current in the case of conducting polymers. The difference between ions traveling into conducting polymers for the purposes of charge neutrality and double-layer charging is also discussed by Cai and Martin.[56]

Electroinactive coatings, such as hydrolyzed cellulose acetate,[57] are used on electron surfaces to prevent fouling. Transporting material through these layers is characterized using collocation,[58] and analytic solutions to the problem have been found.[59]

9.2. Redox Polymers

Redox species containing polymers are cited as useful in developing new electrode materials.[60,61] Efforts have been made to examine charge transport through these layers using quartz crystal microbalance,[62] luminescence,[63] as well as monitoring current.[64–66] In particular mediation is of interest using such layers as $[Os(bpy)_2(PVP)_{10}Cl]Cl$ for $[Fe(H_2O)_6]^{3+}$ reduction[67] and $[Ru(bpy)_2(PVP)_5Cl]Cl$ for $[Fe(CN)_6]^{4-}$ oxidation.[68] Charge propagation through the layer is not well understood,[69] so the majority of models are for steady-state mediation, such as those of Albery *et al.*[70–72] and Andrieux and Saveant.[73–78] Several valuable parameters for characterizing experimental results appear in these works.

Finite-difference techniques are useful in elucidating time-dependent behavior in such layers, for example the time-dependent rotating disk voltammogram in Fig. 6.10.[68,79] Such nonstationary behavior has also been observed elsewhere.[80–84]

Digital simulation methods, both finite difference,[85,86] finite element,[87,88] and orthogonal collocation,[89] are employed to characterize time dependence in redox-species-containing layers. Originally an effective diffusion coefficient for the redox species was determined when the Nernst–Planck equation was solved for potential step at a redox-species-containing layer. Thus even though migration was taken into account, the current still obeyed the Cottrell equation.[86] This migration occurred when field strength E varied linearly with distance $dE/dx = 10^8\,V\text{cm}^{-2}$. This work followed the paper of Yap,[87] which was a finite-element simulation of the same problem but assumed a constant electric field $dE/dx = 0$. Recently however it was stated that at equilibrium, the potential within the film remains constant, since for any current injected into the layer, there is a subsequent movement of ions to compensate for any difference in potential between the layer and solution.[50]

Orthogonal collocation was applied to the system in Fig. 6.10 consisting of a $[Ru(bpy)_2(PVP)_5Cl]Cl$ layer coated on an electrode[90] used to mediate the oxidation of $[Fe(CN)_6]^{4-}$ in perchlorate electrolyte by the mechanism:

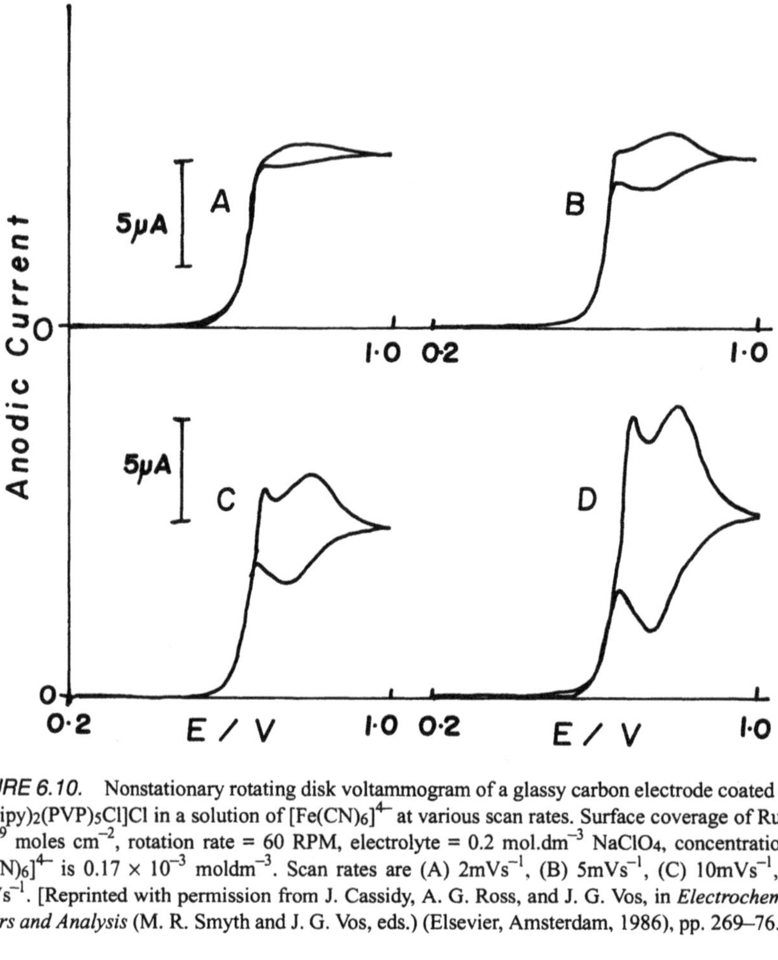

FIGURE 6.10. Nonstationary rotating disk voltammogram of a glassy carbon electrode coated with $[Ru(bipy)_2(PVP)_5Cl]Cl$ in a solution of $[Fe(CN)_6]^{4-}$ at various scan rates. Surface coverage of Ru is 7 × 10^{-9} moles cm^{-2}, rotation rate = 60 RPM, electrolyte = 0.2 mol.dm^{-3} $NaClO_4$, concentration of $[Fe(CN)_6]^{4-}$ is 0.17 × 10^{-3} $moldm^{-3}$. Scan rates are (A) $2mVs^{-1}$, (B) $5mVs^{-1}$, (C) $10mVs^{-1}$, (D) $20mVs^{-1}$. [Reprinted with permission from J. Cassidy, A. G. Ross, and J. G. Vos, in *Electrochemical Sensors and Analysis* (M. R. Smyth and J. G. Vos, eds.) (Elsevier, Amsterdam, 1986), pp. 269–76.]

$$Ru^{2+} \rightarrow Ru^{3+} + e^- \tag{95}$$

$$Ru^{3+} + [Fe(CN)_6]^{4-} \xrightarrow{k_1} Ru^{2+} + [Fe(CN)_6]^{3-} \tag{96}$$

A model was developed to cope with nonstationary behavior observed in Fig. 6.10; see Fig. 6.11 for an example of the resulting simulations,[89] which shows the effect of rotation rate on the current. While results from the simulation exhibit curves similarly shaped to those found in practice,[68,79] there are quite a number of parameters that are not accurately known. Such unknown effects as the layer homogeneity on the rate of electron exchange k_1 and the swollen layer thickness magnitude prevent quantitative comparison between the model and physical results. However certain trends observed in practical results and simulations, such as the decrease in the amount of Ru^{3+} available for reduction by the electrode with increased rotation rate, as seen in Fig. 6.11.

A recent experimental system consisted of an electroactive layer deposited across an array of microelectrodes, each of which was addressable.(91,92) A finite-difference method was used to examine the passage of current from a generating electrode to a sink electrode. In the model (Fig. 6.12),(85) there is a possible distribution of redox centers between the polymer layer and the solution close by. The degree of cross-linking or fluidity of the layer as a function of the oxidation state of the redox species in the layer was also characterized. In the simulation the presence of the microelectrodes was ignored, so the problem was reduced to one dimension, making its solution simpler than situations where nonlinear diffusion

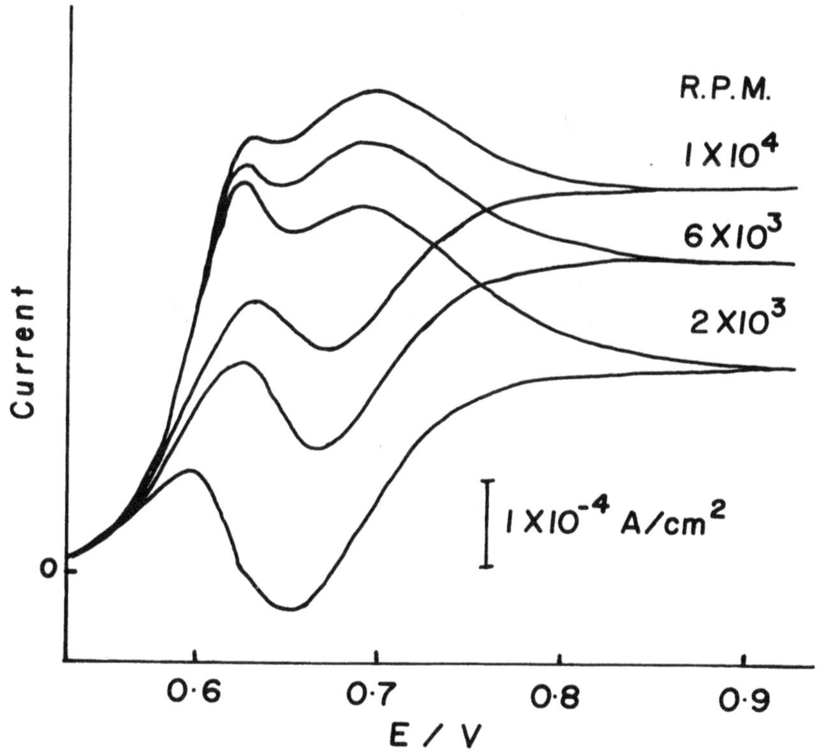

FIGURE 6.11. Simulated rotating disk voltammogram with conditions $E^{o\prime}$ Ru = 0.675V, [A] = 10^{-3} mol.dm^{-3}, surface coverage of Ru = 6×10^{-10} mol cm^{-2}, $d = 6 \times 10^{-5}$ cm; $k_1 = 10^7$mol^{-1}dm^{3}s^{-1}, $D = 10^{-6}$cm^{2}s^{-1}; $D_{Ru} = 10^{-8}$cm^{2}s^{-1}; scan rate = 1 Vs^{-1}. Rotation rates are shown as RPM. [Reprinted with permission from J. F. Cassidy and J. G. Vos, *J. Electroanal. Chem.* **235**, 41 (1987).]

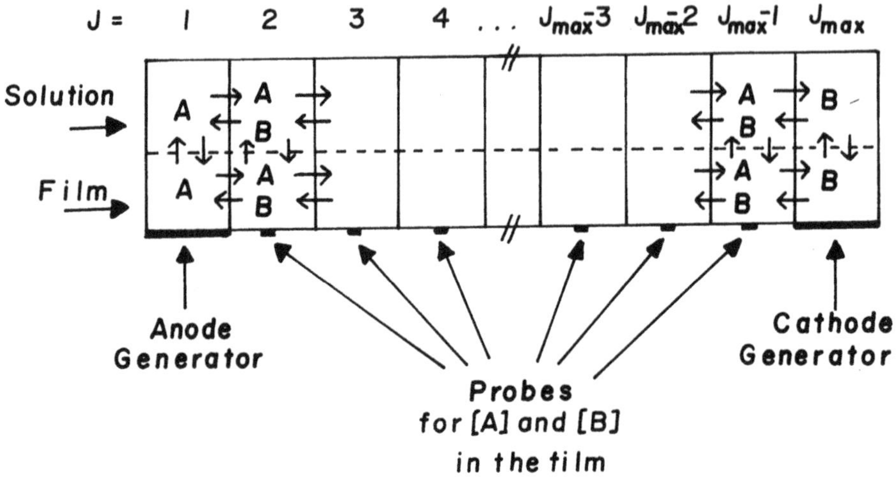

FIGURE 6.12. Discrete model (*not drawn to scale*) of the electrode array solution system, where the reduced form B and the oxidized form A species diffuse into adjacent boxes that may partition between the polymer and solution phases. In this case the array of microelectrodes are not used. [Reprinted with permission from I. Fritsch-Faules and L. R. Faulkner, *J. Electroanal. Chem.* **331**, 997 (1992).]

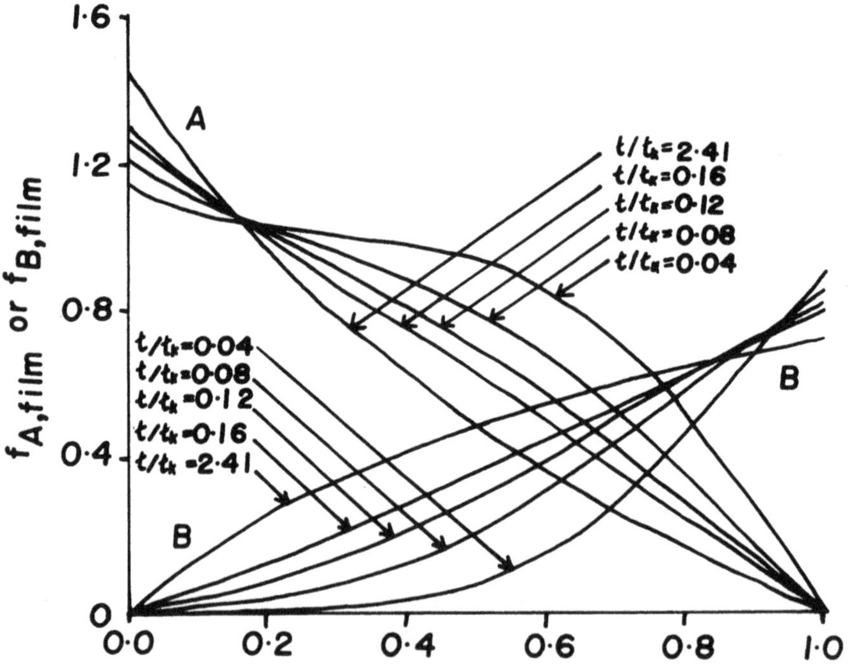

FIGURE 6.13. Transient concentration profiles as a function of time in film that results from a linear increase in fluidity from $x/d = 0$ to $x/d = 1$, where D_m(Box 1) = 0.225 and D_m(Box J_{MAX}) = 0.45. [Reprinted with permission from I. Fritsch-Faules and L. R. Faulkner, *J. Electroanal. Chem.* **331**, 997 (1992).]

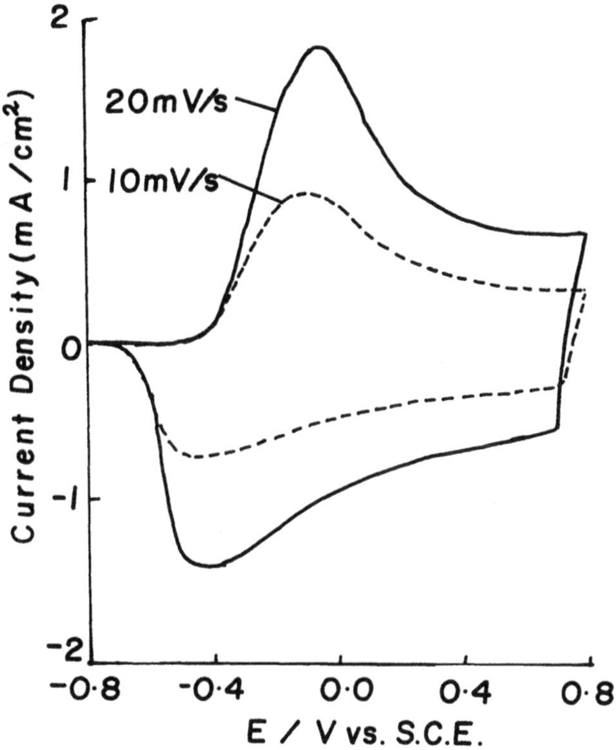

FIGURE 6.14. Experimental cyclic voltammograms for a 1-μm-thick polypyrrole film in 1 mol.dm^{-3} $LiClO_4$ propylene carbonate solution at scan rates of 10 mVs^{-1} (- - -) and 20mVs^{-1}(———). [Reprinted with permission from T. Yeu, V. Nguyen, and R. E. White, *J. Electrochem. Soc.* **138**, 2869 (1991).]

was characterized.[93,94] Ultimately a steady-state current is obtained in this experimental configuration. Figure 6.13[85] shows the effect of fluid mobility, modeled by varying diffusion coefficients on the relaxation of concentration profiles in the film. With different diffusion coefficients, the limiting current varies. The diffusion coefficient is linearly dependent on distance across the array. The diffusion parameter at the reducing electrode D_m (Box J_{MAX}) is 0.45, and the diffusion parameter at the oxidizing electrode D_m (Box 1) = 0.225, t_k is a characteristic time (= d^2/D), where D is the faster diffusion coefficient and d is the layer thickness. This system can be used to model the effects of the oxidation state, for example $[Fe(CN)_6]^{4-}$ in quaternized PVP, whose structure varies with the oxidation state of the $[Fe(CN)_6]^{4-}$/ $[Fe(CN)_6]^{3-}$ couple.[95,96]

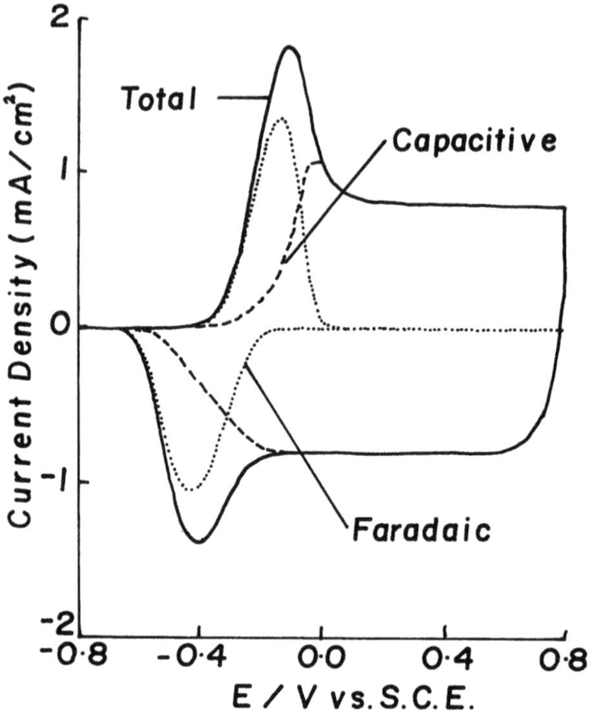

FIGURE 6.15. Partition of a cyclic voltammogram into capacitive and faradaic currents densities at a scan rate of $20 mVs^{-1}$. [Reprinted with permission from T. Yeu, V. Nguyen, and R. E. White, *J. Electrochem. Soc.* **138**, 2869 (1991).]

Finally a more comprehensive model for simulating the Ru^{3+}/Fe^{2+} system was solved using finite-element methods.(88,97) This model takes into account mass transport due to diffusion and migration, electron transfer due to electron hopping, homogeneous chemical reaction in the membrane, heterogeneous reactions, double-layer charging, and Donnan partition equilibrium between the membrane and diffusion layer.

9.3. Conducting Polymers

Ever since Feldberg's original model separating faradaic and capacitance current associated with a conducting polymer that behaves like a porous metal,(49) there have been various attempts to rationalize experimentally the nature of the current at a conducting polymer. Apart from impedance models, there has not been so much effort placed on modeling the system. Experimental efforts to examine processes at conducting-polymer-modified electrodes consisted of voltammetry,(98–100) impedance,(101–103) and quartz crystal microbalance.(104) However the

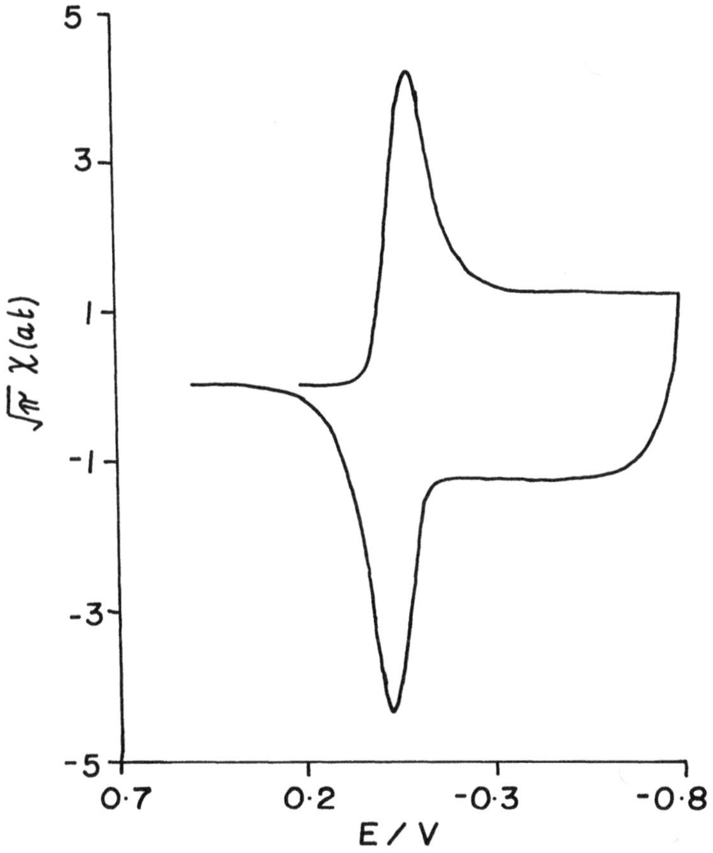

FIGURE 6.16. Simulation of a cyclic voltammogram for finite diffusion in a polymeric film. Ten reversible electron transfer processes at a microscopic redox potential of 0.0 V build up the peak. Further redox steps with equal standard potential differences (50 mV) between each other form the plateau. [Reprinted with permission from J. Heinze, M. Storzbach, and J. Mortensen, *Ber Bunsenges. Physikal. Chemie.* **91**, 960 (1987).]

adsorption and mass transport of gaseous dopants into polypyrrole have been modeled using a finite-difference algorithm of Hindmarsh,[105–107] used specifically for the stiff equations. A typical voltammogram for polypyrrole is shown in Fig. 6.14. Pickup and Osteryoung,[108] who developed a model for chronoamperometry, concluded that the film could behave like a porous metal. This idea was further developed by Yeu *et al.*,[109,110] who used a finite-difference method with Newman's pentadiagonal block matrix equation solver. In this model consisting of a rotating disk electrode coated with a conducting layer, capacitance is an important component, so the cyclic voltammogram can conveniently be partitioned into faradaic and capacitive currents, as seen in Fig. 6.15.

An alternative model[11,111] uses the Crank–Nicholson method to generate a voltammogram that consists of a layer with a series of microscopic formal potentials, most situated at 0.0V and the rest equally spaced 50 mV apart. This also yields a voltammogram (Fig. 6.16) similar to the experimental one (Fig. 6.14). The basis for this is the fact that different oligomers of different chain length possess a range of redox potentials.[55] Thus at least qualitatively, two models may account for the electrochemical behavior of a conducting polymer coated on an electrode.

10. CONCLUSION

Most models are naive in their treatment of an experimental system. Even in the case of diffusion, the effect of a concentration-dependent diffusion coefficient has been treated only as a linear function,[85] though orthogonal collocation[112] has been used to model diffusion under conditions of a concentration-dependent diffusion coefficient.

Taking into account all possible means of electron transport, there are a wide number of parameters required, and many of there are synergistic, so they cannot be evaluated in isolation. Since for more complicated models, there is a range of values that certain parameters may take, so viable conclusions can be reached only with an appropriate experimental configuration and a well-defined model. There are further complications, such as a nucleation-type mechanism in the oxidation of a conducting polymer,[113,114] as well as problems of resistance.[115] Though only cyclic voltammetry for a surface-confined species is covered in this chapter, which forms part of an introduction to the numerical methods of solution, the latter chapters show that more advanced numerical methods can be applied to gain information about mass transport through a redox layer. Furthermore such numerical methods as finite element, as well as random-walk simulations[94] can also be used to elucidate mass transfer.

ACKNOWLEDGMENTS. The author wishes to thank Dr. S. H. Lin, Dr. B. Brown, Prof. J. Heinze, and Dr. A. Lasia for helpful advice in writing these programs. Financial support from EOLAS (grant number ST/91/001/b) is also gratefully acknowledged.

APPENDIX: PROGRAM LISTINGS

A1. Program 1

```
C       PROGRAM 1
C       POTENTIAL STEP FOR INFINITE SOLUTION
C       EXPLICIT FINITE DIFFERENCE
C       WHERE THE DISTANCE SPACINGS ARE ALL EQUAL.
        IMPLICIT REAL*8(A-H,O-W)
        IMPLICIT REAL*4(X-Z)
        DIMENSION FONEW(800),FOOLD(800),FRNEW(800),FROLD(800)
        DIMENSION RAT(800)
        OPEN (UNIT=6,FILE='BK101.DAT',STATUS='NEW')
        DM=0.45
        L=500
C       INITIALISE CONCENTRATIONS
        DO 10 J=1,150
        FOOLD(J)=1.D0
     10 FROLD(J)=0.D0
        DO 20 I=1,L
        XI=FLOAT(I)
        XI=SQRT(XI)
        JMAX=IFIX(4.2*XI)+1
C       DIFFUSIONAL SEGMENT
        DO 30 J=2,JMAX
        FONEW(J)=FOOLD(J)+DM*(FOOLD(J-1)-2*FOOLD(J)+FOOLD(J+1))
        FRNEW(J)=FROLD(J)+DM*(FROLD(J-1)-2*FROLD(J)+FROLD(J+1))
     30 CONTINUE
C       FIRST BOX
        FONEW(1)=FOOLD(1)+DM*(-2.D0*FOOLD(1)+FOOLD(2))
        FRNEW(1)=FROLD(1)+DM*(FOOLD(1)-FROLD(1)+FROLD(2))
C       CALCULATE CURRENT
        W=DSQRT(FLOAT(L)*DM)*FONEW(1)
C       CALCULATE TIME
        T=FLOAT(I)/FLOAT(L)
        CUR=1.D0/DSQRT(3.14159D0*T)
C       CALCULATE THE RATIO OF SIMULATED TO COTTRELL CURRENT.
        RAT(I)=W/CUR
C       RESET THE CONCENTRATION OF THE ARRAY
        DO 40 J=1,JMAX
        FOOLD(J)=FONEW(J)
        FROLD(J)=FRNEW(J)
     40 CONTINUE
     20 CONTINUE
C       OUTPUT DATA
        DO 50 I=1,L
        WRITE(6,100)I,RAT(I)
     50 CONTINUE
    100 FORMAT(I5,F10.5)
        CLOSE(UNIT=6)
        STOP
        END
```

A2. Program 2

Fig. 6.7 (page 92) shows this program's output.

```
C       PROGRAM 2
C       PROGRAM FOR CYCLIC VOLTAMMETRY FOR A THIN LAYER SYSTEM
C       USING AN EXPLICIT METHOD WITH EQUAL INTERVALS.
        IMPLICIT REAL*8(A-H,O-Y)
        IMPLICIT REAL*4(Z)
        DIMENSION FONEW(100000),FOOLD(100000),FROLD(100000)
        DIMENSION FRNEW(100000),W(100000)
        OPEN (UNIT=5,FILE='BK111.DAT',STATUS='OLD')
        OPEN (UNIT=6,FILE='BK112.DAT',STATUS='NEW')
        DM=0.45D0
        READ(5,60)DIFF,DIST,SCAN,NP
     60 FORMAT(3E12.5,I5)
        AH=DIST/NP
        AK=DM*(AH)**2/DIFF
C       SET POTENTIAL LIMITS
        VI=0.3
        VF=-0.3
        EI=VI/0.0256
        EF=VF/0.0256
        TC=(VI-VF)/SCAN
        L=TC/AK
        WRITE(6,61)L,NP
     61 FORMAT(' # TIME DIVISIONS ',I10,' # DISTANCE DIVISIONS',I10)
C       POTENTIAL INCREMENT
        DELTAE=(EF-EI)/FLOAT(L)
C       INITIALISES POTENTIAL AND CONCENTRATION.
        EMOD=EI
        DO 10 J=1,1050
        FOOLD(J)=1.D0
        FROLD(J)=0.D0
     10 CONTINUE
C       THE TOTAL NUMBER OF INCREMENTS IS 2*L
        LT2=2*L
        DO 20 I=1,LT2
        JMAX=NP
C       DIFFUSION SEGMENT
        JMAX1=JMAX-1
        DO 30 J=2,JMAX1
        FONEW(J)=FOOLD(J)+DM*(FOOLD(J-1)-2.D0*FOOLD(J)+FOOLD(J+1))
        FRNEW(J)=FROLD(J)+DM*(FROLD(J-1)-2.D0*FROLD(J)+FROLD(J+1))
     30 CONTINUE
C       CALCULATE THE SURFACE CONCENTRATION
        EMOD=EMOD+DELTAE
        RATIO=DEXP(EMOD)
        FRSURF=(FROLD(1)+FOOLD(1))/(1.D0+RATIO)
        FOSURF=RATIO*FRSURF
C       FIRST BOX
        FONEW(1)=FOOLD(1)+DM*(FOSURF-2*FOOLD(1)+FOOLD(2))
        FRNEW(1)=FROLD(1)+DM*(FRSURF-2*FROLD(1)+FROLD(2))
C       LAST BOX
        FONEW(JMAX)=FOOLD(JMAX)+DM*(FOOLD(JMAX1)-FOOLD(JMAX))
        FRNEW(JMAX)=FROLD(JMAX)+DM*(FROLD(JMAX1)-FROLD(JMAX))
C       CALCULATE CURRENT
        FACT=1.D0/(1.D0+RATIO)
        W(I)=FACT*(FONEW(1)-RATIO*FRNEW(1))
        W(I)=W(I)*4.D0*(0.0256)*DM*FLOAT(L)/(FLOAT(JMAX)*(VI-VF))
```

```
C       RESET CONCENTRATION
        DO 40 J=1,JMAX
        FOOLD(J)=FONEW(J)
        FROLD(J)=FRNEW(J)
   40   CONTINUE
C       TEST FOR SWEEP DIRECTION
        IF(I.EQ.L)DELTAE=-DELTAE
   20   CONTINUE
C       OUTPUT DATA
        DELTAE=-DELTAE
        EMOD=EI
        DO 50 I=1,LT2
        EMOD=EMOD+DELTAE
        V=EMOD*0.0256
        WRITE(6,100)V,EMOD,W(I)
        IF(I.EQ.L) DELTAE=-DELTAE
   50   CONTINUE
  100   FORMAT(3F10.5)
        CLOSE(UNIT=5)
        CLOSE(UNIT=6)
        STOP
        END
```

A3. Program 3

Table 6.2 (page 96) shows partial output of this program.

```
C       PROGRAM 3
C       PROGRAM FOR THIN LAYER CYCLIC VOLTAMMETRY USING IMPLICIT METHOD
C       THERE ARE NP INTERNAL POINTS AND TWO BOUNDARY POINTS.
        IMPLICIT REAL*8(A-H,O-Z)
        DIMENSION W(20),RMAT(400),G(20)
        DIMENSION FONEW(20),FOOLD(20),FROLD(20),FRNEW(20),WZ(22000)
        DIMENSION RKO(20),RKR(20),RNEW(20),RK(20)
        OPEN (UNIT=5,FILE='BK111.DAT',STATUS='OLD')
        OPEN (UNIT=6,FILE='BK71.DAT',STATUS='NEW')
        DM=2.45
C       SET POTENTIAL LIMITS
        VI=0.3
        VF=-0.3
        EI=VI/0.0256
        EF=VF/0.0256
        READ(5,60)DIFF,DIST,SCAN,NP
   60   FORMAT(3E12.5,I5)
        AH=DIST/NP
        AK=DM*(AH)**2/DIFF
        TC=(VI-VF)/SCAN
        L=TC/AK
C       POTENTIAL INCREMENT
        DELTAE=(EF-EI)/FLOAT(L)
        N=NP
        IF=1
        NX=N-1
        WRITE(6,61)L,N
   61   FORMAT(' # TIME INTERVALS = ',I10,' # DISTANCE INTERVALS',I10)
```

```
C      INITIALISES POTENTIAL AND CONCENTRATION.
       EMOD=EI
       DO 10 J=1,N
       FOOLD(J)=1.D0
       FROLD(J)=0.D0
   10  CONTINUE
C      THE TOTAL NUMBER OF INCREMENTS IS 2*L
       LT2=2*L
       DO 20 I=1,LT2
C      DIFFUSION SEGMENT
C      SET UP THE RIGHT HAND SIDE MATRIX
       DO 16 LL=1,N
       DO 16 K=1,N
   16  RMAT((LL-1)*N+K)=0.D0
       DO 17 LL=1,N
   17  RMAT((LL-1)*N+LL)=2.D0*(1.D0-DM)
       DO 14 LL=1,NX
       RMAT((LL-1)*N+LL+1)=DM
   14  RMAT(LL*N+LL)=DM
C      CALCULATE THE RHS CONSTANT MATRIX FOR THOMAS ALGORITHM
       DO 18 LL=1,N
       RKO(LL)=0.D0
       RKR(LL)=0.D0
   18  CONTINUE
       DO 19 K=1,N
       DO 19 LL=1,N
       RKO(K)=RMAT((LL-1)*N+K)*FOOLD(LL)+RKO(K)
   19  RKR(K)=RMAT((LL-1)*N+K)*FROLD(LL)+RKR(K)
C      RECALCULATE THE MATRIX FOR THE LHS OF THE SYSTEM
       DO 30 LL=1,N
   30  RMAT((LL-1)*N+LL)=2.D0*(1.D0+DM)
       DO 21 LL=1,NX
       RMAT((LL-1)*N+LL+1)=-DM
   21  RMAT(LL*N+LL)=-DM
C      CALCULATE THE OLD AND NEW SURFACE CONCENTRATIONS
       RATIO=DEXP(EMOD)
       DRSURF=(FOOLD(1)+FROLD(1))/(1.D0+RATIO)
       DOSURF=RATIO*DRSURF
       EMOD=EMOD+DELTAE
       RATIO=DEXP(EMOD)
       FRSURF=(FROLD(1)+FOOLD(1))/(1.D0+RATIO)
       FOSURF=RATIO*FRSURF
       FONEW(1)=FOOLD(1)+DM*(FOSURF-2*FOOLD(1)+FOOLD(2))
       FRNEW(1)=FROLD(1)+DM*(FRSURF-2*FROLD(1)+FROLD(2))
       FONEW(N)=FOOLD(N)+DM*(FOOLD(NX)-FOOLD(N))
       FRNEW(N)=FROLD(N)+DM*(FROLD(NX)-FROLD(N))
C      FIXING UP THE BOUNDARY CONDITIONS WITHIN THE MATRICES
       RKO(1)=RKO(1)+DM*(DOSURF+FOSURF)
       RKR(1)=RKR(1)+DM*(DRSURF+FRSURF)
       RKO(N)=RKO(N)+DM*(FOOLD(N)+FONEW(N))
       RKR(N)=RKR(N)+DM*(FROLD(N)+FRNEW(N))
       CALL THOMAS(FRNEW,RKR,RMAT,N)
       CALL THOMAS(FONEW,RKO,RMAT,N)
C      CALCULATE CURRENT
       FACT=1.D0/(1.D0+RATIO)
       WZ(I)=FACT*(FONEW(1)-RATIO*FRNEW(1))
       WZ(I)=WZ(I)*4.D0*0.0256*DM*FLOAT(L)/(FLOAT(N)*(VI-VF))
C      RESET CONCENTRATION
       DO 40 J=1,N
       FOOLD(J)=FONEW(J)
       FROLD(J)=FRNEW(J)
   40  CONTINUE
```

```
C     TEST FOR SWEEP DIRECTION
      IF(I.EQ.L)DELTAE=-DELTAE
   20 CONTINUE
C     OUTPUT DATA
      DELTAE=-DELTAE
      EMOD=EI
      DO 50 I=1,LT2
      EMOD=EMOD+DELTAE
      V=EMOD*0.0256
      WRITE(6,100)V,WZ(I)
      IF(I.EQ.L) DELTAE=-DELTAE
   50 CONTINUE
  100 FORMAT('    POTENTIAL = 'F10.5,' W =',F10.6)
      CLOSE(UNIT=5)
      CLOSE(UNIT=6)
      STOP
      END
C     SUBROUTINE FOR SOLVING A TRIDIAGONAL MATRIX SYSTEM.
      SUBROUTINE THOMAS(RNEW,RK,RMAT,N)
      IMPLICIT REAL*8(A-H,O-Z)
      DIMENSION W(20),RMAT(400),G(20)
      DIMENSION RNEW(20),RK(20)
      W(1)=RMAT(N+1)/RMAT(1)
      G(1)=RK(1)/RMAT(1)
      DO 10 I=2,N
      W(I)=RMAT(I*N+I)/(RMAT((I-1)*N+I)-RMAT((I-2)*N+I)*W(I-1))
   10 G(I)=(RK(I)-RMAT((I-2)*N+I)*G(I-1))/(RMAT((I-1)*N+I)-
     1RMAT((I-2)*N+I)*W(I-1))
      RNEW(N)=G(N)
      N11=N-1
      DO 20 I=1,N11
      J=N-I
   20 RNEW(J)=G(J)-W(J)*RNEW(J+1)
      RETURN
      END
```

A4. Program 4

Table 6.3 (page 97) shows typical output of this program.

```
C     PROGRAM 4
C     HOPSCOTCH METHOD APPLIED TO THE CYCLIC VOLTAMMETRY
C     IN A THIN LAYER.
      IMPLICIT REAL*8(A-H,O-Z)
      DIMENSION FONEW(215),FRNEW(215),W(500000)
      OPEN(UNIT=5,FILE='BK111.DAT',STATUS='OLD')
      OPEN(UNIT=6,FILE='BK81.DAT',STATUS='NEW')
      READ(5,60)DIFF,DIST,SCAN,NP,NF
```

```
C     EVERY NFth POINT IS OUTPUT
   60 FORMAT(3E12.5,2I5)
      DM=1.45
      AH=DIST/NP
      AK=DM*(AH)**2/DIFF
      VI=0.3
      JC=0
      VF=-0.3
      TC=(VI-VF)/SCAN
      L=TC/AK
      LT2=2*L
      Q=1.D0/(1.D0+2*DM)
      Q2=2.D0*DM*Q
      JMAX=NP
      WRITE(6,61)L,NP
   61 FORMAT(' # TIME INTERVALS',I10,'#DISTANCE INTERVALS',I10)
C     K IS THE TIME PROGRESS COUNTER
      K=0
C     INITIAL CONDITIONS
      DO 10 J=1,JMAX
      FONEW(J)=1.D0
      FRNEW(J)=0.D0
   10 CONTINUE
      JMAX1=JMAX-1
      EI=VI/0.02569
      EF=VF/0.02569
C     POTENTIAL INCREMENT
      DELTAE=(EF-EI)/FLOAT(L)
C     INITIALISES POTENTIAL AND CONCENTRATION
      EMOD=EI
C     STARTUP FOR ODD NODES
      DO 20 J=3,JMAX
      IF(MOD(J,2).EQ.0) GO TO 20
      FONEW(J)=FONEW(J)+DM*(FONEW(J-1)-2*FONEW(J)+FONEW(J+1))
      FRNEW(J)=FRNEW(J)+DM*(FRNEW(J-1)-2*FRNEW(J)+FRNEW(J+1))
   20 CONTINUE
C     START OF ITERATION LOOP.
   23 K=K+1
      EMOD=EMOD+DELTAE
      RATIO=DEXP(EMOD)
      FRNEW(1)=(FRNEW(2)+FONEW(2))/(1.D0+RATIO)
      FONEW(1)=RATIO*FRNEW(1)
      FONEW(JMAX)=FONEW(JMAX)+DM*(FONEW(JMAX1)-FONEW(JMAX))
      FRNEW(JMAX)=FRNEW(JMAX)+DM*(FRNEW(JMAX1)-FRNEW(JMAX))
C     EXPANDING DIFFUSION LAYER.
      KM1=K-1
C     FOR NODE 2
      MEV=2+KM1
      IF(MOD(MEV,2).EQ.1) GO TO 25
C     MEV IS EVEN CALCULATE NODE 2 THIS WAY( EQ 19)
C     S IS FOR O AND  SS IS FOR R
      S=(FONEW(2)+DM*(FONEW(3)+FONEW(1)))*Q
      SS=(FRNEW(2)+DM*(FRNEW(3)+FRNEW(1)))*Q
      GO TO 26
C     MEV IS ODD NODE 2 ALREADY CALCULATED.
   25 S=FONEW(2)
```

```
      SS=FRNEW(2)
   26 CONTINUE
C     DO HOPSCOTCH FOR MEV EVEN FOR THE REST OF NODES
      DO 29 JJ=2,JMAX1
      MEV=JJ+KM1
      IF(MOD(MEV,2).EQ.1) GO TO 29
      FRNEW(JJ)=FRNEW(JJ)+Q2*(FRNEW(JJ+1)-2.DO*FRNEW(JJ)+FRNEW(JJ-1))
      FONEW(JJ)=FONEW(JJ)+Q2*(FONEW(JJ+1)-2.DO*FONEW(JJ)+FONEW(JJ-1))
   29 CONTINUE
C     FARADAIC CONVERSION AND CURRENT FLOW CALCULATION.
      FACT=1.DO/(1.DO+RATIO)
      W(K+1)=FACT*(S-RATIO*SS)
      W(K+1)=W(K+1)*4*0.02569*DM*FLOAT(L)/(FLOAT(JMAX)*(VI-VF))
      JC=JC+1
      IF(JC.EQ.NF) GO TO 51
      GO TO 52
   51 JC=0
      V=EMOD*0.02569
      WRITE(6,100)V,W(K+1)
C     RETURN FOR NEXT ITERATION IF K<L
   52 IF(K.EQ.L) DELTAE=-DELTAE
      IF(K.LT.LT2) GO TO 23
C     OUTPUT DATA
      DELTAE=-DELTAE
      EMOD=EI
      DO 50 I=1,LT2
      EMOD=EMOD+DELTAE
      V=EMOD*0.02569
      WRITE(6,100)V,W(I)
      IF(I.EQ.L)DELTAE=-DELTAE
   50 CONTINUE
  100 FORMAT(4F10.5)
      CLOSE(UNIT=5)
      CLOSE(UNIT=6)
      STOP
      END
```

A5. Program 5

Table 6.4 (page 105) shows typical output of this program.

```
C     PROGRAM 5
C     A CYCLIC VOLTAMMETRY PROGRAM FOR  A THIN LAYER OF SOLUTION
C     ORTHOGONAL COLLOCATION
      IMPLICIT REAL*8(A-H,O-Z)
      COMMON AZ,BETA,SCAN,TEMP,THETA,SLAMBD,E0,ESWIT,ESTART
      COMMON XLAM,DELTAE,AM1,CONC,CO0,CO1,CR0,CR1
      COMMON DD(100),DIFF,E
      COMMON AA(100),BB(100),POT(10000),CUR(10000),RR(20)
      COMMON N,N2,N3,NSCAN,NACT,NFL,NFLIP,NMR,NELECT,NFX,NPT
      DIMENSION ROO(20),VECT(20),DIF1(20),DIF2(20),DIF3(20)
      DIMENSION Y(100),F(100),AUX(16,100),PRMT(5),DERY(100)
      EXTERNAL FUN,OUT
      OPEN(UNIT=5,FILE='BK51.DAT',STATUS='OLD')
      OPEN(UNIT=6,FILE='BK52.DAT',STATUS='NEW')
      OPEN(UNIT=7,FILE='BK53.DAT',STATUS='NEW')
      N=6
      NO=1
```

```
C     NMR IS A COUNTER FOR THE OUTPUT OF THE CONC. PROFILES.
      NMR=0
      N1=1
C     EVERY NPTth CURRENT IS OUT /NFX #COUNTER
C     EVERY NPUth PROFILE IS OUT /NFY #COUNTER
      AL=0.D0
      BE=0.D0
      N2=N+2
      NFX=0
      ND=15
      N3=2*N
      CALL JCOBI(ND,N,N0,N1,AL,BE,DIF1,DIF2,DIF3,ROO)
      DO 18 I=1,N2
   18 RR(I)=ROO(I)
      DO 4 I=1,N2
      CALL DFOPR(ND,N,N0,N1,I,1,DIF1,DIF2,DIF3,ROO,VECT)
      DO 4 J=1,N2
    4 AA((J-1)*N2+I)=VECT(J)
      DO 7 I=1,N2
      CALL DFOPR(ND,N,N0,N1,I,2,DIF1,DIF2,DIF3,ROO,VECT)
      DO 7 J=1,N2
    7 BB((J-1)*N2+I)=VECT(J)
      DO 27 I=1,N2
   27 DD(I)=AA(N2*(N2-1)+1)*AA(I*N2)/AA(N2*N2)-AA((I-1)*N2+1)
      READ(5,20)ESTART,ESWIT,DELTAE,E0
      READ(5,22)CONC,SCAN,TEMP,DIFF,DIST
      READ(5,21)NSCAN,NELECT,NOPT,NPT
C     NOPT = 0
C     NOPT = 1 OPTIMISATION OF STEP LENGTH OCCURS
   20 FORMAT(4E10.4)
   22 FORMAT(5E10.4)
   21 FORMAT(4I5)
      AZ=NELECT*96487.*SCAN/(8.314*TEMP)
      THETA=DEXP(96487.D0*(E0-ESTART)/(8.314*TEMP))
      BETA=DIFF/(AZ*DIST**2)
      NPU=4*NPT
      PRMT(1)=ESTART
      PRMT(2)=NSCAN*DABS(ESTART-ESWIT)*96487.D0/(8.314D0*TEMP)*2.D0
      PRMT(4)=1.0D-5
      DO 9 I=1,N
      Y(I)=0.D0
      Y(I+N)=1.D0
    9 CONTINUE
      DO 10 I=1,N3
   10 F(I)=1.D0/FLOAT(N3)
      DELTAT=DELTAE*96487.D0/(8.314D0*TEMP)
C     BETA OPTIMIZATION
      J=0
      IF(NOPT.EQ.0)GO TO 174
   72 DO 73 I=1,N
      Z=BB(I*N2+I+1)+BB(N2*(N2-1)+I+1)*AA(N2*I)/AA(N2*N2)
      H1=Z*BETA
      H1=0.65/DABS(H1)
      IF(H1.LT.DELTAT) GO TO 79
      H2=Z+DD(I+1)*(-BB(I*N2+1)+BB(N2*(N2-1)+I+1)*AA(N2)/
     1AA(N2*N2))/(2*DD(1))
      H2=BETA*H2
      H2=0.65D0/DABS(H2)
      IF(H2.LT.DELTAT) GO TO 79
      H3=BB(I*N2+I+1)
      Z=-BB(I+1)*DD(I+1)/DD(1)+BB(N2*(N2-1)+I+1)*AA(N2)*DD(I+1)/
     1(AA(N2*N2)*DD(1))
```

```
      H3=BETA*(H3+Z)
      H3=0.65/DABS(H3)
      IF(H3.LT.DELTAT) GO TO 79
      H4=BETA*(BB(I*N2+I+1)+Z/2)
      H4=0.65/DABS(H4)
      IF(H4.LT.DELTAT) GO TO 79
   73 CONTINUE
      IF(J.EQ.0)GO TO 198
      GO TO 153
   79 DELTAT=DELTAT*0.5
      J=J+1
      IF(J.EQ.1000)GO TO 64
      GO TO 72
   64 WRITE(7,65)
   65 FORMAT('ERROR!')
      GO TO 199
  198 DELTAT=DELTAT*2.0
      GO TO 72
  199 DELTAT=0.001
  153 NACT=1
  174 BETA=DIFF/(AZ*(DIST)**2)
      WRITE(7,80)BETA,DELTAT
   80 FORMAT('       BETA =    ',E12.6,'   DELTAT = ',E12.6)
      NFL=1
      NFLIP=0
      XLAM=96487.DO*DABS(ESTART-ESWIT)/(8.314DO*TEMP)
      PRMT(3)=DELTAT
      CALL DHPCG(PRMT,Y,F,N3,IHLF,FUN,OUT,AUX)
      CLOSE(UNIT=5)
      CLOSE(UNIT=7)
      CLOSE(UNIT=6)
      STOP
      END
C     SUBROUTINE TO CALCULATE THE VALUES OF F(I)
      SUBROUTINE FUN(X,Y,F)
      IMPLICIT REAL*8(A-H,O-Z)
      COMMON AZ,BETA,SCAN,TEMP,THETA,SLAMBD,E0,ESWIT,ESTART
      COMMON XLAM,DELTAE,AM1,CONC,CO0,CO1,CR0,CR1
      COMMON DD(100),DIFF,E
      COMMON AA(100),BB(100),POT(10000),CUR(10000),RR(20)
      COMMON N,N2,N3,NSCAN,NACT,NFL,NFLIP,NMR,NELECT,NFX,NPT
      DIMENSION Y(100),F(100),PRMT(5)
      AM=NSCAN
      TEND=AM*XLAM*2.DO
      AM1=NACT
      TEND1=AM1*XLAM*2.DO
      XXX=NFL*XLAM
      IF(X.GE.XXX)NFLIP=FLIP+1
      IF(X.GE.XXX)NFL=NFL+1
      IF(NFLIP.EQ.2)NFLIP=0
      IF(X.GE.TEND1)GO TO 207
      GO TO 208
  207 NACT=NACT+1
      AM1=AM1+1.
  208 IF(NFLIP.EQ.1)GO TO 77
      SLAMBD=DEXP(2*(AM1-1)*XLAM-X)
      E=ESTART+(-2*(AM1-1)*XLAM+X)*(ESWIT-ESTART)/XLAM
      GO TO 81
   77 SLAMBD=DEXP(X-2*AM1*XLAM)
      E=2.DO*ESWIT-(-2*(AM1-1)*XLAM+X)*(ESWIT-ESTART)/XLAM
      E=E-ESTART
```

```
   81 CO0=0.D0
      DO 34 J=1,N
   34 CO0=CO0+DD(J+1)*Y(N+J)+DD(J+1)*Y(J)
      CO0=-CO0/((1.D0+SLAMBD*THETA)*DD(1))
      CR0=THETA*SLAMBD*CO0
      CO1=0.D0
      DO 35 J=1,N
   35 CO1=CO1+AA(N2*(J+1))*Y(J)
      CO1=CO1+AA(N2)*CO0
      CO1=-CO1/AA(N2*N2)
      CR1=0.D0
      DO 36 J=1,N
   36 CR1=CR1+AA(N2*(J+1))*Y(J+N)
      CR1=CR1+AA(N2)*CR0
      CR1=-CR1/AA(N2*N2)
      DO 1 I=1,N
      TDDER1=0.D0
      TDDER2=0.D0
      DO 2 J=1,N
      TDDER1=TDDER1+BB(J*N2+I+1)*Y(J)
      TDDER2=TDDER2+BB(J*N2+I+1)*Y(J+N)
    2 CONTINUE
      F(I)=BETA*(BB(I+1)*CO0+BB(N2*(N2-1)+I+1)*CO1+TDDER1)
      F(I+N)=BETA*(BB(I+1)*CR0+BB(N2*(N2-1)+I+1)*CR1+TDDER2)
    1 CONTINUE
      RETURN
      END
C     OUTPUT SUBROUTINE
      SUBROUTINE OUT(X,Y,F,IHLF,NDIM,PRMT)
      IMPLICIT REAL*8(A-H,O-Z)
      COMMON AZ,BETA,SCAN,TEMP,THETA,SLAMBD,E0,ESWIT,ESTART
      COMMON XLAM,DELTAE,AM1,CONC,CO0,CO1,CR0,CR1
      COMMON DD(100),DIFF,E
      COMMON AA(100),BB(100),POT(10000),CUR(10000),RR(20)
      COMMON N,N2,N3,NSCAN,NACT,NFL,NFLIP,NMR,NELECT,NFX,NPT
      DIMENSION Y(100),F(100),PRMT(5)
      SUM=0.0
      DO 88 J=1,N
      SUM=SUM+AA(J*N2+1)*Y(J+N)
   88 CONTINUE
      SUM=SUM+AA(1)*CR0+AA(N2*(N2-1)+1)*CR1
      SUM=4.D0*SUM*BETA
      NFX=NFX+1
      IF(NFX.EQ.NPT)NMR=NMR+1
      IF(NFX.EQ.NPT)POT(NMR)=E
      IF(NFX.EQ.NPT)CUR(NMR)=SUM
      IF(NFX.EQ.NPT)WRITE(6,4)POT(NMR),CUR(NMR)
      IF(NFX.EQ.NPT)NFX=0
    4 FORMAT(F12.7,3X,F12.7)
      RETURN
      END
      SUBROUTINE JCOBI(ND,N,N0,N1,AL,BE,DIF1,DIF2,DIF3,ROOT)
      IMPLICIT REAL*8(A-H,O-Z)
      DIMENSION DIF1(ND),DIF2(ND),DIF3(ND),ROOT(ND)
      AB=AL+BE
      AD=BE-AL
      AP=BE*AL
      DIF1(1)=(AD/(AB+2)+1)/2
      DIF2(1)=0.0
      IF(N.LT.2)GO TO 15
```

```
      DO 10 I=2,N
      Z1=I-1
      Z=AB+2*Z1
      DIF1(I)=(AB*AD/Z/(Z+2)+1)/2
      IF(I.NE.2)GO TO 11
      DIF2(I)=(AB+AP+Z1)/Z/Z/(Z+1)
      GO TO 10
   11 Z=Z*Z
      Y=Z1*(AB+Z1)
      Y=Y*(AP+Y)
      DIF2(I)=Y/Z/(Z-1)
   10 CONTINUE
   15 X=0.0
      DO 20 I=1,N
   25 XD=0.
      XN=1.0
      XD1=0.0
      XN1=0.0
      DO 30 J=1,N
      XP=(DIF1(J)-X)*XN-DIF2(J)*XD
      XP1=(DIF1(J)-X)*XN1-DIF2(J)*XD1-XN
      N1=1
      XD=XN
      XD1=XN1
      XN=XP
   30 XN1=XP1
      ZC=1.0
      Z=XN/XN1
      IF(I.EQ.1) GO TO 21
      DO 22 J=2,I
   22 ZC=ZC-Z/(X-ROOT(J-1))
   21 Z=Z/ZC
      X=X-Z
      IF(DABS(Z).GT.1.D-09) GO TO 25
      ROOT(I)=X
      X=X+0.0001
   20 CONTINUE
      NT=N+N1+NO
      IF(NO.EQ.0)GO TO 35
      DO 31 I=1,N
      J=N+1-I
   31 ROOT(J+1)=ROOT(J)
      ROOT(1)=0.0
   35 IF(N1.EQ.1)ROOT(NT)=1
      DO 40 I=1,NT
      X=ROOT(I)
      DIF1(I)=1.0
      DIF2(I)=0.
      DIF3(I)=0.0
      DO 40 J=1,NT
      IF(J.EQ.I) GO TO 40
      Y=X-ROOT(J)
      DIF3(I)=Y*DIF3(I)+3*DIF2(I)
      DIF2(I)=Y*DIF2(I)+2*DIF1(I)
      DIF1(I)=Y*DIF1(I)
   40 CONTINUE
```

```
      RETURN
      END
      SUBROUTINE DFOPR(ND,N,NO,N1,I,ID,DIF1,DIF2,DIF3,ROOT,VECT)
      IMPLICIT REAL*8(A-H,O-Z)
      DIMENSION DIF1(ND),DIF2(ND),DIF3(ND),ROOT(ND),VECT(ND)
      NT=N+N1+NO
      IF(ID.EQ.3) GO TO 10
      DO 20 J=1,NT
      IF(J.NE.I)GO TO 21
      IF(ID.NE.1) GO TO 5
      VECT(I)=DIF2(I)/DIF1(I)/2
      GO TO 20
    5 VECT(I)=DIF3(I)/DIF1(I)/3
      GO TO 20
   21 Y=ROOT(I)-ROOT(J)
      VECT(J)=DIF1(I)/DIF1(J)/Y
      IF(ID.EQ.2)VECT(J)=VECT(J)*(DIF2(I)/DIF1(I)-2/Y)
   20 CONTINUE
      GO TO 50
   10 Y=0.0
      DO 25 J=1,NT
      X=ROOT(J)
      AX=X*(1-X)
      IF(NO.EQ.0)AX=AX/X/X
      IF(N1.EQ.0)AX=AX/(1-X)/(1-X)
      VECT(J)=AX/DIF1(J)**2
   25 Y=Y+VECT(J)
      DO 60 J=1,NT
   60 VECT(J)=VECT(J)/Y
   50 RETURN
      END
      SUBROUTINE DHPCG(PRMT,Y,DERY,NDIM,IHLF,FUN,OUT,AUX)
      IMPLICIT REAL*8(A-H,O-Z)
      DIMENSION PRMT(5),Y(200),DERY(200),AUX(16,200)
      N=1
      IHLF=0
      X=PRMT(1)
      H=PRMT(3)
      PRMT(5)=0.DO
      DO 1 I=1,NDIM
      AUX(16,I)=0.DO
      AUX(15,I)=DERY(I)
    1 AUX(1,I)=Y(I)
      IF(H*(PRMT(2)-X))3,2,4
    2 IHLF=12
      GO TO 4
    3 IHLF=13
    4 CALL FUN(X,Y,DERY)
      CALL OUT(X,Y,DERY,IHLF,NDIM,PRMT)
      IF (PRMT(5))6,5,6
    5 IF(IHLF)7,7,6
    6 RETURN
    7 DO 8 I=1,NDIM
    8 AUX(8,I)=DERY(I)
      ISW=1
      GO TO 100
    9 X=X+H
      DO 10 I=1,NDIM
   10 AUX(2,I)=Y(I)
```

```
11 IHLF=IHLF+1
   X=X-H
   DO 12 I=1,NDIM
12 AUX(4,I)=AUX(2,I)
   H=0.5D0*H
   N=1
   ISW=2
   GO TO 100
13 X=X+H
   CALL FUN(X,Y,DERY)
   N=2
   DO 14 I=1,NDIM
   AUX(2,I)=Y(I)
14 AUX(9,I)=DERY(I)
   ISW=3
   GO TO 100
15 DELT=0.D0
   DO 16 I=1,NDIM
16 DELT=DELT+AUX(15,I)*DABS(Y(I)-AUX(4,I))
   DELT=0.066666666666666666D0*DELT
   IF(DELT-PRMT(4))19,19,17
17 IF(IHLF-10)11,18,18
18 IHLF=11
   X=X+H
   GO TO 4
19 X=X+H
   CALL FUN(X,Y,DERY)
   DO 20 I=1,NDIM
   AUX(3,I)=Y(I)
20 AUX(10,I)=DERY(I)
   N=3
   ISW=4
   GO TO 100
21 N=1
   X=X+H
   CALL FUN(X,Y,DERY)
   X=PRMT(1)
   DO 22 I=1,NDIM
   AUX(11,I)=DERY(I)
22 Y(I)=AUX(1,I)+H*(0.375D0*AUX(8,I)+
  10.7916666666666667D0*AUX(9,I)-
  2.2083333333333333D0*AUX(10,I)+
  3.0416666666666667D0*DERY(I))
23 X=X+H
   N=N+1
   CALL FUN(X,Y,DERY)
   CALL OUT(X,Y,DERY,IHLF,NDIM,PRMT)
   IF(PRMT(5))6,24,6
24 IF(N-4)25,200,200
25 DO 26 I=1,NDIM
   AUX(N,I)=Y(I)
26 AUX(N+7,I)=DERY(I)
   IF(N-3)27,29,200
27 DO 28 I=1,NDIM
   DELT=AUX(9,I)*2.D0
   DELT=DELT+DELT
28 Y(I)=AUX(1,I)+0.33333333333333333D0*H*
  1(AUX(8,I)+DELT+AUX(10,I))
   GO TO 23
```

```
 29 DO 30 I=1,NDIM
    DELT=AUX(9,I)+AUX(10,I)
    DELT=3.D0*DELT
 30 Y(I)=AUX(1,I)+0.375D0*H*(AUX(8,I)+DELT+AUX(11,I))
    GO TO 23
100 DO 101 I=1,NDIM
    NNK=N+7
    Z=H*AUX(NNK,I)
    AUX(5,I)=Z
101 Y(I)=AUX(N,I)+0.4D0*Z
    Z=X+0.4D0*H
    CALL FUN(Z,Y,DERY)
    DO 102 I=1,NDIM
    Z=H*DERY(I)
    AUX(6,I)=Z
102 Y(I)=AUX(N,I)+0.29697760924775360D0*AUX(5,I)+
   10.15875964497103583D0*Z
    Z=X+0.45573725421878943D0*H
    CALL FUN(Z,Y,DERY)
    DO 103 I=1,NDIM
    Z=H*DERY(I)
    AUX(7,I)=Z
103 Y(I)=AUX(N,I)+0.21810038822592047D0*
   1AUX(5,I)-3.05096514889293O8D0*
   2AUX(6,I)+3.8328647604670103D0*Z
    Z=X+H
    CALL FUN(Z,Y,DERY)
    DO 104 I=1,NDIM
104 Y(I)=AUX(N,I)+0.17476028226269037D0*AUX(5,I)-
   10.55148066287873294D0*AUX(6,I)+1.20553559939665235D0*
   2AUX(7,I)+0.17118478121951903O*H*DERY(I)
    GO TO(9,13,15,21),ISW
200 ISTEP=3
201 IF(N-8)204,202,204
202 DO 203 N=2,7
    DO 203 I=1,NDIM
    AUX(N-1,I)=AUX(N,I)
203 AUX(N+6,I)=AUX(N+7,I)
    N=7
204 N=N+1
    DO 205 I=1,NDIM
    AUX(N-1,I)=Y(I)
205 AUX(N+6,I)=DERY(I)
    X=X+H
206 ISTEP=ISTEP+1
    DO 207 I=1,NDIM
    DELT=AUX(N-4,I)+1.333333333333D0*H
   1*(AUX(N+6,I)+AUX(N+6,I)-AUX(N+5,I)
   1+AUX(N+4,I)+AUX(N+4,I))
    Y(I)=DELT-0.925619*AUX(16,I)
207 AUX(16,I)=DELT
    CALL FUN(X,Y,DERY)
    DO 208 I=1,NDIM
    DELT=0.125D0*(9.D0*AUX(N-1,I)-AUX(N-3,I)+3.D0*H*(DERY(I)+
   1AUX(N+6,I)+AUX(N+6,I)-AUX(N+5,I)))
    AUX(16,I)=AUX(16,I)-DELT
208 Y(I)=DELT+0.0743801652892562D0*AUX(16,I)
    DELT=0.D0
    DO 209 I=1,NDIM
```

```
209 DELT=DELT+AUX(15,I)*DABS(AUX(16,I))
    IF(DELT-PRMT(4))210,222,222
210 CALL FUN(X,Y,DERY)
    CALL OUT(X,Y,DERY,IHLF,NDIM,PRMT)
    IF(PRMT(5))212,211,212
211 IF(IHLF-11)213,212,212
212 RETURN
213 IF(H*(X-PRMT(2)))214,212,212
214 IF(DABS(X-PRMT(2))-0.1*DABS(H))212,215,215
215 IF(DELT-0.02D0*PRMT(4))216,216,201
216 IF(IHLF)201,201,217
217 IF(N-7)201,218,218
218 IF(ISTEP-4)201,219,219
219 IMOD=ISTEP/2
    IF(ISTEP-2*IMOD)201,220,201
220 H=H*2.D0
    IHLF=IHLF-1
    ISTEP=0
    DO 221 I=1,NDIM
    AUX(N-1,I)=AUX(N-2,I)
    AUX(N-2,I)=AUX(N-4,I)
    AUX(N-3,I)=AUX(N-6,I)
    AUX(N+6,I)=AUX(N+5,I)
    AUX(N+5,I)=AUX(N+3,I)
    AUX(N+4,I)=AUX(N+1,I)
    DELT=AUX(N+6,I)+AUX(N+5,I)
    DELT=3.D0*DELT
221 AUX(16,I)=8.962962962962963D0*
   1(Y(I)-AUX(N-3,I))-3.361111111111111D0*
   2H*(DERY(I)+DELT+AUX(N+4,I))
    GO TO 201
222 IHLF=IHLF+1
    IF(IHLF-10)223,223,210
223 H=0.5D0*H
    ISTEP=0
    DO 224 I=1,NDIM
    Y(I)=0.00390625D0*(80.D0*AUX(N-1,I)+135.D0*AUX(N-2,I)+
   140.D0*AUX(N-3,I)+AUX(N-4,I))-0.1171875D0*(AUX(N+6,I)-6.D0*
   2AUX(N+5,I)-AUX(N+4,I))*H
    AUX(N-4,I)=0.00390625D0*(12.D0*AUX(N-1,I)+135.D0*
   1AUX(N-2,I)+108.D0*AUX(N-3,I)+AUX(N-4,I))-0.0234375D0*
   2(AUX(N+6,I)+18.D0*AUX(N+5,I)-9.D0*AUX(N+4,I))*H
    AUX(N-3,I)=AUX(N-2,I)
224 AUX(N+4,I)=AUX(N+5,I)
    X=X-H
    DELT=X-H-H
    CALL FUN(DELT,Y,DERY)
    DO 225 I=1,NDIM
    AUX(N-2,I)=Y(I)
    AUX(N+5,I)=DERY(I)
225 Y(I)=AUX(N-4,I)
    DELT=DELT-H-H
    CALL FUN(DELT,Y,DERY)
    DO 226 I=1,NDIM
    DELT=AUX(N+5,I)+AUX(N+4,I)
    DELT=DELT*3.D0
    AUX(16,I)=8.962962962962963D0*(AUX(N-1,I)-Y(I))-
   13.361111111111111D0*H*(AUX(N+6,I)+DELT+DERY(I))
226 AUX(N+3,I)=DERY(I)
    GO TO 206
    END
```

REFERENCES

1. D. Britz, *Digital Simulation in Electrochemistry* (Springer Verlag, Berlin, 1981).
2. S. W. Feldberg, in *Electroanalytical Chemistry*, vol. 3 (A. J. Bard, ed.) (Marcel Dekker, New York, 1969), pp. 199–249.
3. J. R. Sandifer and R. P. Buck, *J. Electroanal. Chem.* **49**, 161 (1974).
4. The Southampton Electrochemical Group, *Instrumental Methods in Electrochemistry* (Ellis Horwood, Chichester, England 1985).
5. M. K. Hanafey, R. L. Scott, T. H. Ridgway, and C. N. Reilley, *Anal. Chem.* **50**, 116 (1978).
6. A. J. Bard and L. R. Faulkner, *Electrochemical Methods* (Wiley, New York, 1980).
7. J. Cassidy, W. Breen, A. McGee, J. G. Vos, and M. E. G. Lyons, *Electroanalysis* **4**, 751 (1992).
8. G. L. Booman and D. T. Pence, *Anal. Chem.* **37**, 1366 (1965).
9. J. Heinze, M. Storzbach, and J. Mortensen, *J. Electroanal. Chem.* **165**, 61 (1984).
10. A. Lasia, *J. Electroanal. Chem.* **146**, 397 (1983).
11. M. Storzbach and J. Heinze, *J. Electroanal. Chem.* **346**, 1 (1993).
12. D. Shoup and A. Szabo, *J. Electroanal. Chem.* **160**, 1 (1984).
13. D. Shoup and A. Szabo, *J. Electroanal. Chem.* **160**, 18 (1984).
14. D. Shoup and A. Szabo, *J. Electroanal. Chem.* **160**, 27 (1984).
15. S. Pons, in *Electroanalytical Chemistry*, vol. 13 (A. J. Bard, ed.) (Marcel Dekker, New York, 1984), pp. 115–90.
16. B. A. Finlayson, *Method of Weighted Residuals and Variational Principles* (Academic Press, New York, 1972).
17. S. Yen and T. Chapman, *J. Electroanal. Chem.* **135**, 305 (1982).
18. L. F. Whiting and P. W. Carr, *J. Electroanal. Chem.* **81**, 1 (1977).
19. G. E. Forsythe, M. A. Malcolm, and L. B. Moler, *Computer Methods for Mathematical Computation* (Prentice-Hall, Englewood Cliffs, NJ, 1977).
20. C. F. Gerald, *Applied Numerical Analysis*, 2nd ed. (Addison Wesley, Reading, MA, 1978).
21. B. S. Pons and P. P. Schmidt, *Electrochim. Acta* **25**, 987 (1980).
22. S. Pons, B. Speiser, and J. F. McAleer, *Electrochim. Acta* **27**, 1177 (1982).
23. M. J. Eddows, *J. Electroanal. Chem.* **159**, 1 (1983).
24. F. Magno, G. Bontempelli, and D. Perosa, *Anal. Chim. Acta* **147**, 65 (1983).
25. B. Speiser and A. Rieker, *J. Electroanal. Chem.* **102**, 1 (1979).
26. B. Speiser, *J. Electroanal. Chem.* **110**, 69 (1980).
27. B. Speiser, S. Pons, and A. Rieker, *Electrochim. Acta* **27**, 1171 (1982).
28. R. S. Nicholson and I. Shain, *Anal. Chem.* **36**, 706 (1964).
29. R. S. Parikh and K. C. Liddell, *J. Electrochem. Soc.* **136**, 679 (1989).
30. P. Hertl and B. Speiser, *J. Electroanal. Chem.* **217**, 225 (1987).
31. J. F. Cassidy, S. Pons, A. S. Hinman, and B. Speiser, *Can. J. Chem.* **62**, 716 (1984).
32. J. Cassidy, J. Ghoroghchian, F. Sarfarasi, and S. Pons, *Can. J. Chem.* **63**, 3577 (1985).
33. J. Cassidy, J. Ghoroghchian, F. Sarfarasi, J. J. Smith, and S. Pons, *Electrochim. Acta* **32**, 629 (1986).
34. D. W. Peaceman and H. H. Rachford, *J. Soc. Indust. Appl. Math.* **3**, 28 (1955).
35. J. Heinze, *J. Electroanal. Chem.* **124**, 73 (1981).
36. T. Joslin and D. Pletcher, *J. Electroanal. Chem.* **49**, 171 (1974).
37. S. Feldberg, *J. Electroanal. Chem.* **127**, 1 (1981) [error corrected in *J. Electroanal. Chem.* **129**, 388 (1981)].
38. R. Seeber and S. Stefani, *Anal. Chem.* **53**, 1011 (1981).
39. J. Margarit and M. Levy, *J. Electroanal. Chem.* **49**, 369 (1974).
40. K. B. Prater and A. J. Bard, *J. Electrochem. Soc.* **117**, 207 (1970).
41. T. Kuwana and N. Winograd, in *Electroanalytical Chemistry*, vol. 7 (A. J. Bard, ed.) (Marcel Dekker, New York, 1974), pp. 1–39.

42. M. Marques da Silva, L. A. Avaca, and E. R. Gonzalez, *J. Electroanal. Chem.* **250**, 457 (1988).
43. M. Marques da Silva, L. A. Avaca, and E. R. Gonzalez, *J. Electroanal. Chem.* **269**, 1 (1989).
44. S. W. Feldberg, *J. Electroanal. Chem.* **290**, 49 (1990).
45. M. Rudolph, *J. Electroanal. Chem.* **292**, 1 (1990).
46. A Lasia, *J. Electroanal. Chem.* **191**, 185 (1985).
47. W. Wanzhi, X. Qingji, and Y. Shouzhao, *J. Electroanal. Chem.* **328**, 9 (1992).
48. D. Britz, J. Heinze, J. Mortensen, and M. Storzbach, *J. Electroanal. Chem.* **240**, 27 (1988).
49. S. W. Feldberg, *J. Am. Chem. Soc.* **106**, 4671 (1984).
50. K. Doblhofer, *J. Electroanal. Chem.* **331**, 1015 (1992).
51. N. Oyama, T. Shimomura, K. Shigehara, and F. C. Anson, *J. Electroanal. Chem.* **112**, 271 (1980).
52. R. Naegeli, J. Redepenning, and F. C. Anson, *J. Phys. Chem.* **90**, 6227 (1986).
53. G. L. Duffitt and P. G. Pickup, *J. Chem. Soc. Faraday Trans.* **88**, 1417 (1992).
54. R. S. Hutton, M. Kalaji, and L. M. Peter, *J. Electroanal. Chem.* **270**, 429 (1989).
55. K. Meerholz and J. Heinze, *Angew. Chem. Int. Engl. Ed.* **29**, 692 (1990).
56. Z. Cai and C. R. Martin, *J. Electroanal. Chem.* **300**, 35 (1991).
57. J. Wang and P. Tuzhi, *J. Electroanal. Soc.* **134**, 586 (1987).
58. J. Cassidy, W. Breen, and M. E. G. Lyons, *Electroanalysis* **3**, 293 (1991).
59. W. Breen, J. F. Cassidy, and M. E. G. Lyons, *Anal. Chem.* **63**, 2263 (1991).
60. R. W. Murray, in *Electroanalytical Chemistry*, vol. 13 (A. J. Bard, ed.) (Marcel Dekker, New York, 1984), pp. 191–260.
61. R. J. Forster and J. G. Vos, in *Analytical Voltammetry* (M. R. Smyth, J. G. Vos, eds.) (Elsevier, Amsterdam, 1992), pp. 465–518.
62. M. D. Ward, *J. Phys. Chem.* **92**, 2049 (1988).
63. S. M. Oh and L. R. Faulkner, *J. Am. Chem. Soc.* **111**, 5613 (1989).
64. J. Facci and R. W. Murray, *J. Phys. Chem.* **85**, 2870 (1981).
65. N. Oyama, T. Ohsaka, and T. Ushirogouchi, *J.Phys. Chem.* **88**, 5274 (1984).
66. N. Oyama and F. C. Anson, *Anal. Chem.* **52**, 1192 (1980).
67. R. J. Forster and J. G. Vos, *J. Chem. Soc. Faraday Trans.* **87**, 1863 (1991).
68. J. F. Cassidy and J. G. Vos, *J. Electroanal. Soc.* **135**, 863 (1988).
69. M. E. G. Lyons, H. G. Fay, T. McCabe, J. Corish, J. G. Vos, and A. J. Kelly, *J. Chem. Soc., Faraday Trans.* **86**, 2905 (1990).
70. W. J. Albery, M. G. Boutelle, and A. R. Hillman, *J. Electroanal. Chem.* **180**, 99 (1985).
71. W. J. Albery and A. R. Hillman, *J. Electroanal. Chem.* **170**, 27 (1984).
72. W. J. Albery, P. N. Bartlett, and D. H. Craston, *J. Electroanal. Chem.* **194**, 223 (1985).
73. C. P. Andrieux and J. M. Saveant, *J. Electroanal. Chem.* **111**, 377 (1980).
74. C. P. Andrieux, J. M. Dumas-Bouchiat, and J. M. Saveant, *J. Electroanal. Chem.* **131**, 1 (1982).
75. C. P. Andrieux, J. M. Dumas-Bouchiat, and J. M. Saveant, *J. Electroanal. Chem.* **169**, 9 (1984).
76. C. P. Andrieux and J. M. Saveant, *J. Electroanal. Chem.* **171**, 65 (1984).
77. C. P. Andrieux and J. M. Saveant, *J. Electroanal. Chem.* **142**, 1 (1982).
78. C. P. Andrieux and J. M. Saveant, *J. Electroanal. Chem.* **134**, 163 (1982).
79. J. F. Cassidy, A. G. Ross, and J. G. Vos, in *Electrochemistry, Sensors and Analysis* (M. R. Smyth and J. G. Vos, eds.) (Elsevier, Amsterdam, 1986), pp. 269–76.
80. M. Sharp, D. D. Montgomery, and F. C. Anson, *J. Electroanal. Chem.* **194**, 247 (1985).
81. C. Lin and A. B. Bocarsly, *J. Electroanal. Chem.* **300**, 325 (1991).
82. M. Sharp, *J. Electroanal. Chem.* **230**, 109 (1987).
83. E. T. Turner-Jones, *J. Electrochem. Soc.* **132**, 245C (1985).
84. K. Niwa and K. Doblhofer, *Electrochim. Acta* **31**, 549 (1986).
85. I. Fritsch-Faules and L. R. Faulkner, *J. Electroanal. Chem.* **331**, 997 (1992).
86. R. Lange and K. Doblhofer, *J. Electroanal. Chem.* **237**, 13 (1987).
87. W. P. Yap and R. A. Durst, *J. Electroanal. Chem.* **216**, 11 (1987).

88. E. Deiss, O. Haas, and C. Daul, *J. Electroanal. Chem.* **337**, 299 (1992).
89. J. F. Cassidy and J. G. Vos, *J. Electroanal. Chem.* **235**, 41 (1987).
90. J. F. Cassidy and J. G. Vos, *J. Electroanal. Chem.* **218**, 341 (1987).
91. I. Fritsch-Faules and L. R. Faulkner, *Anal. Chem.* **64**, 1118 (1992).
92. I. Fritsch-Faules and L. R. Faulkner, *Anal. Chem.* **64**, 1127 (1992).
93. M. R. Deakin, R, M. Wightman, and C. A. Amatore, *J. Electroanal. Chem.* **215**, 49 (1986).
94. S. Licht, V. Cammarata, and M. S. Wrighton, *J. Phys. Chem.* **94**, 6133 (1990).
95. R. Lange and K. Doblhofer, *J. Electroanal. Chem.* **216**, 241 (1987).
96. K. Doblhofer, H. Braun, and R. Lange, *J. Electroanal. Chem.* **206**, 93 (1986).
97. C. Daul and O. Haas, in *Electrochemistry, Sensors and Analysis* (M. R. Smyth and J. G. Vos, eds.) (Elsevier, Amsterdam, 1986), pp. 277–84.
98. R. M. Penner, L. S. VanDyke, and C. R. Martin, *J. Phys. Chem.* **92**, 5274 (1988).
99. D. J. Walton, C. E. Hall, and A. Chyla, *Analyst* **117**, 1305 (1992).
100. H. Mao and P. G. Pickup, *J. Am. Chem. Soc.* **112**, 1776 (1990).
101. H. Mao and P. G. Pickup, *J. Phys. Chem.* **93**, 6480 (1989).
102. J. Tanguy, N. Mermilliod, and M. Hoclet, *J. Electrochem. Soc.* **134**, 795 (1987).
103. X. Ren and P. G. Pickup, *J. Electrochem. Soc.* **139**, 2097 (1992).
104. C. K. Baker, Y-J. Qiu, and J. R. Reynolds, *J. Phys. Chem.* **95**, 4446 (1991).
105. H. Reiss and D. Kim, *J. Phys. Chem.* **90**, 1973 (1986).
106. D. Kim, H. Reiss, and H. M. Rabeony, *J. Phys. Chem.* **92**, 2673 (1988).
107. W. D. Murphy, H. M. Rabeony, and H. Reiss, *J. Phys. Chem.* **92**, 7007 (1988).
108. P. G. Pickup and R. A. Osteryoung, *J. Electroanal. Chem.* **195**, 271 (1985).
109. T. Yeu, T. V. Nguyen, and R. E. White, *J. Electrochem. Soc.* **135**, 1971 (1988).
110. T. Yeu, K-M. Yin, J. Carbajal, and R.E. White, *J. Electrochem. Soc.* **138**, 2869 (1991).
111. J. Heinze, M. Storzbach, and J. Mortensen, *Ber. Bunsenges. Physikal. Chemie* **91**, 960 (1987).
112. S. H. Lin, *J. Chem. Soc., Faraday Trans.* **88**, 1573 (1992).
113. K. Doblhofer and R. Lange, *J. Electroanal. Chem.* **229**, 239 (1987).
114. L. M. Abrantes, J. C. Mesquita, M. Kalaji, and L. M. Peter, *J. Electroanal. Chem.* **307**, 275 (1991).
115. S. Gottesfeld, A. Redondo, I. Rubinstein, and S. W. Feldberg, *J. Electroanal. Chem.* **265**, 15 (1989).

7

In Situ Ellipsometry and FTIR Spectroscopy Applied to Electroactive Polymer-Modified Electrodes

Simon J. Higgins, Paul A. Christensen, and Andrew Hamnett

1. INTRODUCTION

Characterizing films on electrode surfaces has always been a challenging problem in electrochemistry. Recent developments have lent new impetus to efforts to adapt classical spectroscopic techniques to make them suitable for studies of these films. Such studies have found considerable contemporary importance in the chemical modification of electrodes(1) and in the discovery of electronically conducting heterocyclic polymers.(2) Two techniques have especially played key roles in this area: ellipsometry and *in situ* Fourier transform infrared (FTIR), the first providing a measure of film thickness and overall composition and the second, data on molecular composition. Most importantly both methods can be used to study films *in situ* during electrochemical growth, cycling, and degradation, thus providing information of direct relevance to that obtained by other electrochemical techniques.

Ellipsometry was originally developed specifically for the optical characterization of thin films. In principal therefore it is an ideal technique for charac-

---

Simon J. Higgins • Department of Chemistry, Donnan Robert Robinson Laboratories, University of Liverpool, Box 147, Liverpool, L69 3BX, United Kingdom. *Paul A. Christensen and Andrew Hamnett* • Chemistry Department, Bedson Building, University of Newcastle upon Tyne, Newcastle upon Tyne, NE1 7RU, United Kingdom.

Electroactive Polymer Electrochemistry, Part 2: Methods and Applications, edited by Michael E. G. Lyons. Plenum Press, New York, 1996.

terizing the optical properties of electroactive polymer-modified electrodes. Because it is a reflectance technique, it is straightforward to perform the experiment *in situ* in the electrochemical cell, allowing measurements to be performed as a function of potential. Additionally ellipsometry is very sensitive, so submonolayer amounts of material can be detected. A related further advantage is its high degree of surface selectivity. This is important since, in an *in situ* experiment, where it is often difficult to distinguish between a given species (for example, solvent or ions) in the surface film and the huge excess of the same species in solution.

Clearly ellipsometry has much to recommend it for studies in modified electrodes; however until recently its application to electrochemical problems was restricted by several severe problems. Firstly it is not mathematically straightforward to extract optical data from parameters measured by ellipsometers.(3) Secondly collecting each data point used to involve tedious and slow manual determination. In the last 15 years however, two developments have overcome these problems: the development of inexpensive microcomputers and automation of the data collection process. The only drawback now remaining is the capital cost of suitable equipment, and even this is declining rapidly as new and simpler types of ellipsometer become available.

In situ FTIR(4) also had to overcome serious difficulties in its application to electrochemical problems. Unlike ellipsometry, where the wavelengths used are in a region of low solvent absorbance, IR is strongly absorbed by most familiar organic solvents and most particularly by water. This leads inevitably either to thin-layer cells or the development of internal reflection techniques. The former has the advantage of simplicity in interpreting spectral data, but it severely limits the type of electrochemistry that can be carried out. The latter requires not only a suitable high-refractive index substrate, such as Ge or Si, but also an adherent very thin layer of metal as the electrode. Technically this is difficult to fabricate so that the metal layer is continuous, and a substantial lateral resistance is inevitable.

In this chapter, we first review the two techniques in some detail to illustrate their difficulties and pitfalls, then we consider some examples of their recent application to electrode films in some detail. Only the briefest explanation consistent with this aim is made of experimental aspects, and the technique itself, since these topics are covered in much more detail elsewhere.(3–6)

2. EXPERIMENTAL APPROACHES

2.1. Basis of Ellipsometry

An ellipsometer measures the polarization state of light. Applied to electrode surfaces, monochromatic plane-polarized light is incident on the working electrode. The state of polarization of the reflected light is different: it is in general *elliptically* polarized, and the change in polarization depends on the optical properties (that is, the refractive index) of the surface. The ellipse is characterized by two parameters: the azimuthal angle between its major axis and the plane of incidence, and the ratio of the minor to major axes. If our electrode is covered with a film, the state of polarization of the reflected beam also depends on the refractive index *and thickness* of the film.

In principle ellipsometric measurements are a way of determining film thickness *in situ*, as a function of redox state, and this was an early driving force behind attempts to apply ellipsometry to electroactive polymers.[7] The thickness of an electroactive polymer film is a highly significant parameter. As a function of the redox state of the material, thickness can be expected to vary with, among other parameters, polymer tertiary structure and solvent and ion populations. Nonoptical methods of determining film thickness are rather unsatisfactory, since they are invariably made *ex situ*.

Because both electrode and electroactive polymer materials are nearly always optically absorbing over the usual wavelength range used for ellipsometry (400–900 nm), their refractive indices are complex quantities $N_j = n_j - ik_j$, where n_j is the real and k_j the imaginary component of the refractive index. The quantity k is also known as the extinction coefficient, and it is related to the absorption coefficient α by $\alpha = 4\pi k/\lambda$. Therefore assuming all three media (electrolyte, electroactive polymer film, and electrode material) are homogeneous and the electrolyte is transparent, we have six unknowns ($n_{electrolyte}$, $N_{polymer}$, $N_{electrode}$, and the film thickness L). Ellipsometry measures only two parameters, the two angles that characterize the ellipse of polarization (either azimuth α and ellipticity ε or more commonly the mathematically related parameters Δ and Ψ. Fortunately we can measure $n_{electrolyte}$ independently using refractometry, and $N_{electrode}$ can be determined unequivocally by ellipsometry before the film is grown. Therefore in the presence of the film, we have three unknowns ($N_{polymer}$ and L) but only two experimental parameters. This is the key remaining difficulty in applying ellipsometry to any problem. Several approaches have been used in attempts to surmount it.[6] When reviewing the literature, we must consider the advantages and limitations of these approaches, so a brief summary of the main methods follows.

2.2. Pure Ellipsometry

With many older ellipsometers, the user is restricted to measurement of Δ and Ψ (hereafter referred to as *two-parameter ellipsometry*), so the usual approach is to take measurements at different stages of film growth, plot Δ and Ψ for the growing film, then fit to a model Δ–Ψ curve assuming constant $N_{polymer}$. Many papers in the literature rely on this philosophy. However as we show, while a good fit of experimental data to a theoretical Δ–Ψ curve for a film of constant N and increasing L can be taken as *evidence* of homogeneous film growth at constant

N_{polymer}, it cannot be taken as *proof*, since some other untested growth mechanism may also fit the observed Δ–Ψ evolution.

2.3. Ellipsometry and Intensity Measurements

With automatic rotating analyzer ellipsometers, in addition to Δ and Ψ, we can measure the ratio of reflected to incident light intensities. Although this parameter depends on the same quantities as Δ and Ψ, its functional dependence is different, so it can be treated as a third, independent parameter. The advantage with this method (referred to hereafter as a three-parameter ellipsometry) is that for each individual data point, we can solve uniquely for N_{polymer} and L using Δ, Ψ, and the intensity measurement I. This calculation is of course made under the assumption of film homogeneity, but deviations from homogeneity become apparent on considering the evolution of N_{polymer} and L with time.

The technique requires careful control of the experimental setup, since while accuracy in measuring Δ and Ψ is not strongly affected by small fluctuations in incident light intensity, measurement of absolute intensity clearly is. We can in principle allow for fluctuations by using a split-beam arrangement, where both incident and reflected light intensity are monitored. However if for example soluble but optically absorbing products are formed during film growth, the total light absorbed is the sum of absorption due to the film and that due to solution species. Problems can also arise if the growing film becomes macroscopically rough and begins to scatter light. Even with these provisos, the technique has proved very useful, and examples of its application follow.

2.4. Spectroscopic Ellipsometry

By measuring Δ and Ψ for the same film at different wavelengths, a unique solution can usually be found to the ellipsometric problem providing the functional form of N can be approximated accurately with fewer parameters than the number of wavelengths measured. In practice it is normal to approximate an absorption band in the visible or near ultraviolet (UV) by a Lorentzian line shape, requiring three parameters. The real part of the refractive index can also be approximated for a wide variety of films where the absorption band maximum is not accessible by an elementary Kramers–Kronig-type formula or even more straightforwardly as an expansion in even inverse powers of the frequency ω. If measurements are taken at, say, 30 wavelengths, and the physically reasonable assumption is made that thickness L is independent of the wavelength of the probe beam, then we normally have to determine less than 10 parameters, a process that can be carried out using statistical algorithms based on a nonlinear least-squares fit approach. Until recently this approach suffered from the disadvantage that a complete spectroscopic run took several minutes or even longer, but the recent development of rapid ellipsometers with no moving parts, such as the stokesmeter, has led to deployment of a new generation of instruments that are both extremely rapid and capable of genuine spectroscopic measurement.

2.5. In Situ FTIR

While it was long recognized that it would be valuable to apply IR spectroscopy *in situ* to the study of the electrode–electrolyte interface, realization of the technique was prevented by two main problems:

The extremely strong absorption of IR light by all common solvents and most especially water.

The need to pick out the weak absorption of near-electrode or adsorbed species from the strong absorptions of the solvent etc., and the noise.

The first problem is relatively easy to solve by employing a thin layer of electrolyte, and it has been approached in three ways, according to the mode and angle of incidence of the IR light on the electrode.

2.5.1. Transmittance

A thin-layer cell is fabricated based on two optically transparent windows. The thickness of the electrolyte layer is dictated by the thickness of the optically transparent working electrode, usually a gold minigrid, indium-doped tin-oxide-coated glass slide or a reticulated vitreous carbon slice. This approach is not surface-sensitive, and it is effectively restricted to nonaqueous solvents, because the minimum electrolyte thicknesses that can be achieved cause severe attenuation in the IR beam.

2.5.2. Total Internal Reflectance or Attenuated Total Reflectance

This approach, employed along with transmittance in the earliest *in situ* studies, exploits the phenomenon of critical-angle internal reflectance. If light is incident at an interface with a second medium that has a *lower* (real) refractive index than the first, then for angles of incidence in excess of $\sin^{-1}(n_1/n_2)$, a propagating refracted ray cannot exist in the second phase, so total internal reflection occurs. To realize this experiment in the IR, a prism of Ge or Si is used as the first phase, and the IR beam is propagated into the prism at an appropriate angle, so that it suffers one or more reflections at the inner surface of the crystal. At each point of incidence, an evanescent or nonpropagating wave is set up and directed into the medium of lower refractive index adjacent to the prism.

If this medium is transparent to IR, e.g., air, then the penetration depth of the evanescent wave is *ca.* 0.1λ, where λ is the wavelength of the probing IR beam. This penetration is reduced if the medium adjacent to the crystal is absorbing, and energy can be absorbed from the evanescent wave.[4] Thus in an electrochemical cell, only a very thin layer adjacent to the attenuated total reflection (ATR) prism is sampled by the IR beam. In early experiments the crystal itself was employed as the working electrode. The advent of more sensitive detection methods, more powerful IR sources, and in particular, Fourier Transform instruments, allowed the ATR prism to be coated with a thin layer of metal to serve as the working electrode. This obviously attenuates the evanescent wave, so a fine balance must be struck

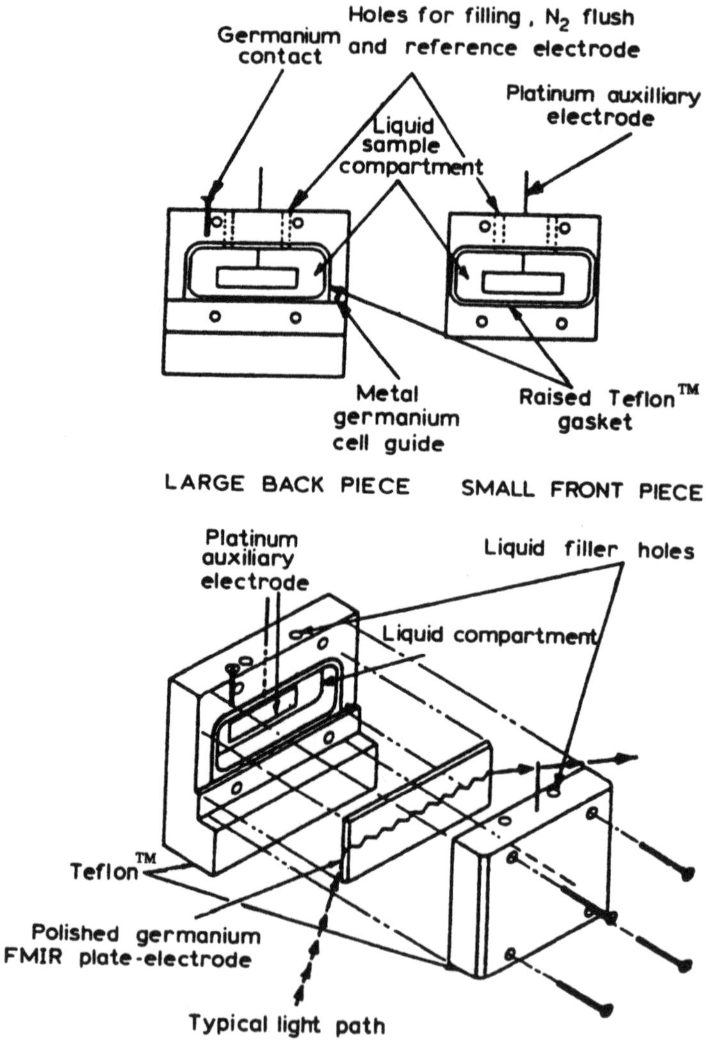

FIGURE 7.1. Experimental arrangement of an internal reflectance or ATR spectroelectrochemical cell. After H. B. Mark, Jr. and B. S. Pons, *Anal. Chem.* **38**, 119 (1966).

between the layer being too thick (and completely damping the evanescent wave) and too thin, so that the metal layer is resistive.

It was quickly realized that homogeneity of the film would greatly improve if the ATR crystal was first "wetted" with a layer of metal, such as chromium. Figure 7.1 shows schematic representation of an electrochemical ATR cell.

2.5.3. External Reflectance

This approach is by far the most commonly adopted, probably because it is the simplest. A thin layer of electrolyte is formed by pressing a polished, reflecting electrode against the IR transparent window of the spectroelectrochemical cell (see Fig. 7.2, the thickness of the electrolyte layer is usually $\gtrsim$ *ca.* 1 μm). The IR beam thus passes twice through the electrolyte layer; a high angle of incidence at the electrode increases absorption by any surface layer. A balance must then be struck between maximizing the angle of incidence of the IR beam on the cell window (and hence at the reflective working electrode) and minimizing the amount of light reflected only from the front and rear surfaces of the window.

Both internal and external reflectance approaches have advantages and disadvantages. The major drawback of the external reflectance technique is that the thin layer imposes severe restrictions on the diffusion of species to and from the working

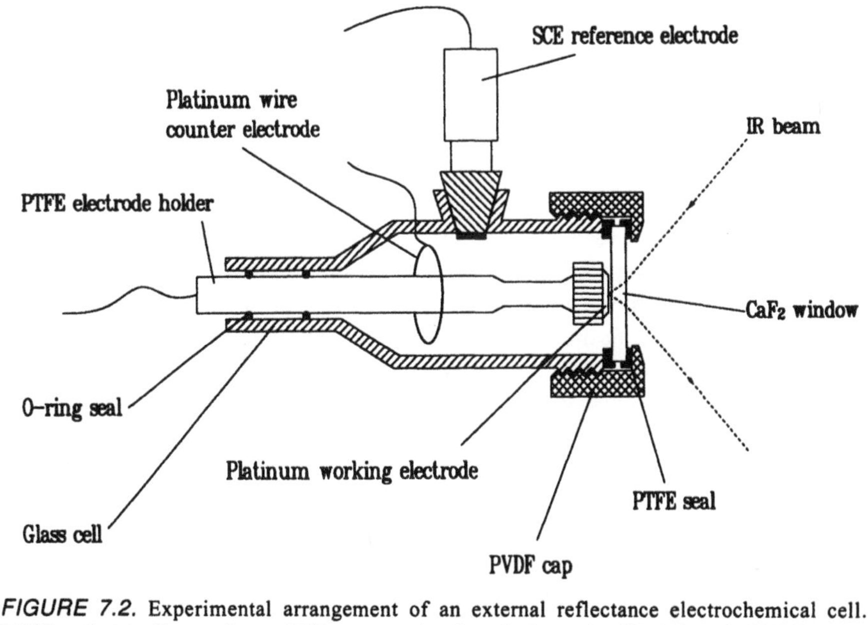

FIGURE 7.2. Experimental arrangement of an external reflectance electrochemical cell. PTFE=polytetrafluoroethene; SCE=saturated calomel electrode; IR=infrared; PVDF=polyvinylidenedifluoride.

electrode. This leads to a severe ohmic drop across the electrode surface at high current and precludes the use of the technique to study the *growth* of a conducting polymer, since films formed with the electrode against the window often form merely as an unrepresentative annulus around the outside of the electrode. In addition if the film is too thick, the oxidized form absorbs all the IR light, and artefactual spectra result.

In the ATR approach, the thickness of the polymer and/or electrolyte adjacent to the crystal that can be probed is determined by the penetration depth of the evanescent wave. This in turn is determined by the optical properties of the crystal–polymer–electrolyte interface and any metal films on the ATR element. Hence the solution layer thickness is not restricted, and diffusion is no longer a problem. Consequently the ATR approach complements the external reflectance method, since it is ideally suited to studying the growth of conducting polymers, as well as their behavior on electrochemical switching. However if the polymer is too thick, the evanescent wave probes only the thin layer adjacent to the electrode, which may not be representative of the behavior of the polymer as a whole. A more serious complication of ATR is attaining precisely determined electrode coatings on the ATR prism. The cost of prisms commonly employed is such that the user cannot afford to make too many mistakes. As previously mentioned ATR and transmittance were the first methods employed in *in situ* IR spectroscopic studies of the electrode–electrolyte interface. However sensitivity was a major problem with these early studies, and *in situ* IR–ATR became reasonably well-exploited only with the advent of FTIR spectrometers. A general renaissance in IR spectroscopy resulted due to the power and sensitivity of these instruments.

Turning now to the question of sensitivity, there are two primary means whereby the signal-to-noise ratio (S/N) of an optical signal can be improved for a system employing dispersive optics. The first method is to increase the detection time at each wavelength; the second is to employ phase-sensitive detection. In the latter case, the optical response of the electrode can be modulated either by modulating the electrode potential or the polarization of the incident light, detecting in either case only the modulation in reflection that is in phase with, and at the same frequency as, the driving modulation. The frequency of *potential* modulation is normally rather low, *ca.* 10 Hz, and the primary purpose is to alter the coverage reversibly of any surface species. However the operative word here is reversible, and this restricts the technique, called electrochemically modulated infrared spectroscopy (EMIRS),[(8)] to systems that show *fast reversible kinetics*. Unfortunately electrochemical switching of conducting polymers is neither reversible in the preceding sense, nor is it sufficiently rapid in general for EMIRS to be applicable.

The second means of increasing the S/N ratio, which has also been employed with FTIR spectrometers, involves modulating the polarization state of the light, a technique termed polarization modulation infrared reflection–absorption spectroscopy (PM-IRRAS).[(9)] This technique relies on a subsidiary selection rule, valid when the underlying substrate is very strongly absorbing. Under these conditions the amplitude of the electric vector of the beam polarized *perpendicular to the plane of reflectance* (s-polarized) or parallel to the electrode surface becomes very small. It follows that s-polarized light is blind to any adsorbed species, and by measuring the difference in reflected intensity of s- and p-polarized light, only absorption by surface adsorbates is detected.

In vacuum studies this is indeed the case, but for studies involving liquid–solid interfaces, the different intensity of the two polarizations extends throughout the thin layer, with the result that differences due to adsorbed species tend to be swamped by unbalanced solvent absorptions. The application of this technique to polymer films suffers from the additional disadvantage that the relative amplitudes of electric vectors for the two polarizations are a strong function of the distance from the electrode surface, making quantitative interpretation very difficult.

In fact the substantial thickness of most polymer films on electrode surfaces does not demand the use of modulation techniques to enhance sensitivity, particularly if an FTIR spectrometer is used, since the very high stability and throughput of this type of instrument, combined with rapid data collection and the ability to sum and average spectra allow us to record difference spectra for such films with extremely simple equipment. The use of difference spectra allows us to subtract contributions from the solvent, as well as remove effects due to source and detector response envelopes, cell windows, etc. The normal mode of operation is to measure the spectrum at some potential E_R, then step to a potential E_n to remeasure the spectrum. The normalized difference spectrum is then $(\Delta R/R) = (S_n - S_R)/S_R$, and the S/N ratio increases as the squareroot of the number of scans. As a result of this data manipulation, peaks pointing upward in the spectra correspond to species *lost* on moving from E_R to E_n, whereas peaks pointing downward to negative values of $(\Delta R/R)$ correspond to *gains* on moving to E_n.

3. ELLIPSOMETRY AND FTIR OF ELECTROACTIVE POLYMERS

3.1. Polyaniline and Related Systems

The story of ellipsometric studies of polyaniline (PA) shows that increasing sophistication in the technique has led to successively more valuable information from experimental data. In early ellipsometric studies of PA film growth, and subsequent potential cycling, either single λ (633 nm)[7,10,12] or photoelastic modulation (650–350 nm)[12–14] ellipsometry (where Δ and Ψ only were measured) were employed (i.e., two-parameter ellipsometry). With one exception,[10] conclusions from these studies are remarkably consistent: Ellipsometric data collected both for the oxidized and reduced forms of the film could be fitted to a model involving a

TABLE 7.1
Polyaniline Optical Properties Determined Ellipsometrically

Ref. No.	λ (/nm)	Oxidized Form (n	k)	Reduced Form (n	k)
6[a]	600	1.346	?	1.61	0
8[b]	600	1.17	0.20	1.56	0.03
8[b]	550	1.34	0.09	1.62	0.01
8[b]	450	1.43	0.12	1.65	0.05
9[c]	633	1.578	0.407	1.638	0.019
9[d]	633	1.537	0.350	1.705	0.046
10[e]	633	1.125	0.354	1.705	0

[a]0.1-M aniline, 1.0-M H_2SO_4 film grown by cycling, −0.2–+0.8 V (SCE).
[b]1.0-M aniline, 2.0-M HCl galvanostatic growth. Readings for oxidized film taken at +0.52 V (Ag quasi-ref) for reduced film at 0.0 V.
[c]0.1-M aniline, 0.1-M H_2SO_4 film grown galvanostatically.
[d]0.1-M aniline, 0.1-M $HClO_4$ film grown galvanostatically.
[e]0.1-M aniline, 0.1-M H_2SO_4, or 0.1-M $HClO_4$, film grown galvanostatically.

single homogeneous film of constant n and k and increasing thickness, up to a thickness of *ca.* 150–200 nm. Optical data for the films are summarized in Table 7.1. From the plots in Refs. 7, 10, 13, 14, and 19, it appears that the precision of the fits varies, perhaps according to the exact deposition conditions used, with growth by cycling in HCl[(12)] apparently giving the best fit; however this is difficult to assess, since no mathematical information on the goodness of fit is given. For films grown thicker than 200 nm, the outer layer apparently roughens, for data no longer fit this simple model. Other techniques[(7)] suggest that PA grows as a smooth, dense film until this thickness, after which it grows in a rough, fibrillar form.

A fascinating and important observation is that during oxidation of the polymer, n fell to remarkably low values, particularly at longer λ (Table 7.1). Gottesfeld *et al.* in their work on PA[(13–15)] first pointed out that this is in accord with the predicted behavior of n as a function of λ around a strong chromophore that behaves as a Lorentz oscillator (Fig. 7.3). Immediately at low energy of the absorption peak, (maximum in k), n rises, while at high energy, n falls.

Optical properties of electroactive polymers as measured ellipsometrically depend greatly on conditions used to grow the film. Greef *et al.* point out significant optical differences between *reduced* PA films grown on different materials[(12)] (for PA, on Pt, n = 1.55–1.58, whereas on Au, n could attain values as high as to 1.7). Explanations in terms of the presence of larger amounts of solvents within reduced films with lower n values[(10)] (n for aqueous electrolytes is typically *ca.* 1.34) are not valid. Just as a very low value of n can be traced to a Lorentzian oscillator at

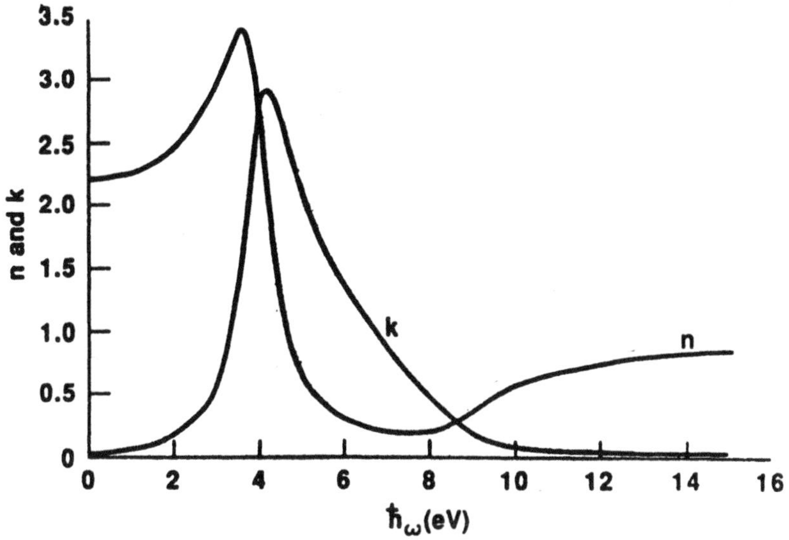

FIGURE 7.3. Variation of the real (n) and imaginary (k) components of the complex refractive index with energy from a Lorentz oscillator. Specific values of the model parameters are $\hbar\omega_0 = 4$ eV, where ω_0 is the oscillator resonance frequency: $\hbar\Gamma = 1$ eV, where Γ represents oscillator damping and provides an energy loss mechanism; $4\pi Ne_0^2/m_e = 60$, where N is the volume density of the electronic oscillators, e_0 the electronic charge, and m_e the electronic mass.

lower energy, an apparently high value of n can simply mean an intense Lorentzian oscillator at higher energies (Fig. 7.3).

Limitations of two-parameter ellipsometry are well-illustrated in subsequent work by Gottesfeld *et al.* who report combined ellipsometric and intensity measurements (three-parameter ellipsometry) on the galvanostatic growth of PA films.[15] Results show that in fact the film grows by a mechanism so that nucleation at limited points on the electrode occurs, producing a thick film with refractive index close to that of the electrolyte (since it does in fact largely consist of electrolyte). Subsequently as the film grows, it also increases in density (that is, the volume fraction of PA in the film seen by the ellipsometer increases at the expense of electrolyte), as the initial nucleation sites grow laterally and begin to overlap. This observation is consistent with results from investigations by other workers on the growth of heterocyclic-conducting polymers. Clearly whatever electrochemical method is used to grow the film, the necessity for nucleation to occur cannot be obviated, so this mechanism probably applies to all films previously described.

In a fascinating demonstration of morphology control by electrode modification, observed using simultaneous ellipsometry and electrochemical quartz crystal microbalance (EQCM) measurements, Gottesfeld *et al.* have recently shown that the preceding growth mechanism for PA on Au electrodes can be greatly altered by

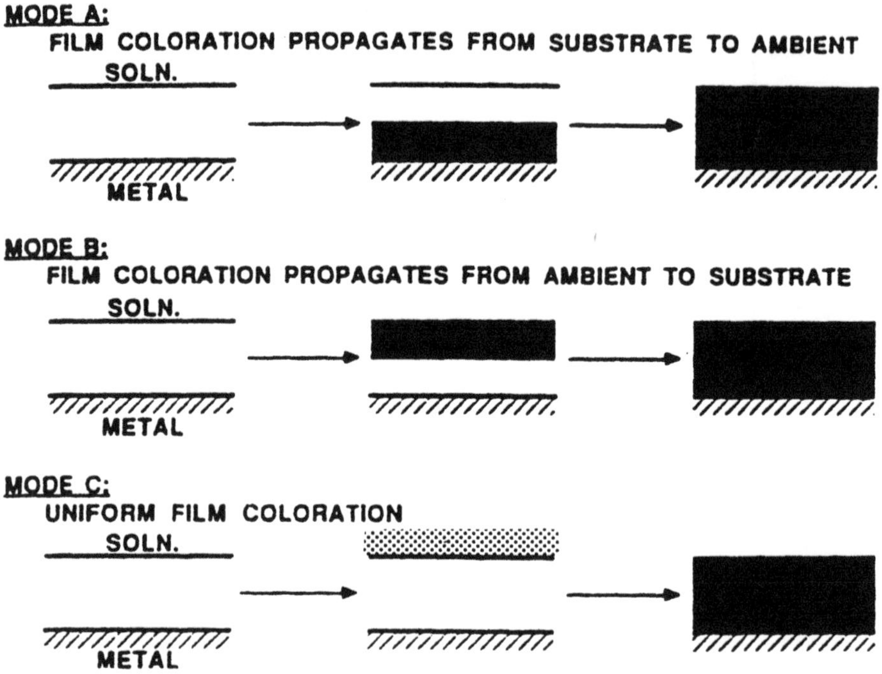

FIGURE 7.4. Three possible modes of electrochemical film conversion that can be differentiated by ellipsometric measurements. Model A corresponds to a process of charge injection limited by electron transport; Model B to a rate limited by insertion of ions from solution, and Model C to a case of high levels of both ionic and electronic transport rates within the film.

precoating the electrode with a monolayer of 4-aminothiophenol.[16] Film grown with the modified electrode is much denser in early stages of growth, both optically (as judged by ellipsometry) and physically (as judged by EQCM measurements). Presumably nucleation is facilitated by the presence of aniline moieties already firmly attached to the electrode by the strong Au-S bond.

The use of two-parameter ellipsometry to examine film redox conversion is exemplified by the results of Gottesfeld *et al.* for PA.[14] They compared ellipsometric results obtained during film cycling with theoretical Δ–Ψ plots made assuming three different models of conversion (Fig. 7.4, Models 1–3). In principle which model is most closely followed provides information about which process is rate-determining in film conversion if changes in film thickness are assumed not to be significant. If electron mobility through the film is greater than ion mobility, Model 1 should apply; if ion mobility is greater than electron mobility, Model 2 should apply. If both ion and electron mobilities are high compared to the sampling rate, a uniform change throughout the film, i.e., Model 3, should apply; on oxidation the ellipsometric behavior most closely resembled Model 3. This was unexpected: Given the electronically insulating nature of the reduced form, we would have expected Model 2 to be obeyed. Often electrochemical data obtained in conjunction with *in situ* techniques can prove useful, and this is true here. In the voltammogram, an irreversible prepeak is seen before the major peak due to oxidation of the polymer. This may be due to nucleation of the oxidized form of the polymer from the electrode through the insulating form to the electrolyte, which may require only a small portion of the total charge. Following this, uniform conversion of the rest of the film can occur. It is reported that the quality of the fit ruled out the possibility of thickness changes larger than some 10–15% during the redox process. However such thickness changes may be significant. To determine thickness changes directly, three-parameter or spectroscopic measurements are required; this has not yet been reported for PA films.

3.2. Polypyrrole and Related Polymers

Like polyaniline, polypyrrole has been the subject of several studies using different types of ellipsometer. Bard *et al.* used two-parameter single-wavelength ellipsometry to examine the growth (by potential cycling in KCl electrolyte) and potential switching of polypyrrole.(17) The key feature of this work is an assessment of the mechanism of film conversion, the philosophy being closely similar to that of Gottesfeld *et al.* previously described. Of the three models shown in Fig. 7.4, Model 1 best fits the data for both oxidation and reduction processes, but there were large deviations, particularly in the later stages of conversion. A better fit was obtained using a model where conversion happens by a Nernstian diffusion process, although again from the solution inward. This was modeled using a multilayer approach, where each successive layer consists of different proportions of fully oxidized and fully reduced material.

Using three-parameter ellipsometry, extensive work has been done on the growth and subsequent electrochemistry of polypyrrole films. In a preliminary communication, the potentiostatic growth of polypyrrole from neutral aqueous electrolytes was examined.(18) The most striking result here was a clear demonstration that the evolution of n, k, and L with time (for a film grown using KNO_3 electrolyte) is consistent with a growth mechanism involving a three-stage process. First immediately on switching the potential to initiate film formation, a film starts to grow, with values of n and k close to those of the electrolyte (*ca.* 1.39 and 0.06, *cf.* 1.335 and 0.00). Film thickness increases linearly with time. After a time that varies with experimental conditions, thickness stops changing significantly while n and k rise sharply. Eventually optical properties begin to level off, and again the film grows linearly with time, although at a different rate. Typical data are shown in Fig. 7.5. The growth of films containing different counterions of electrocatalytic

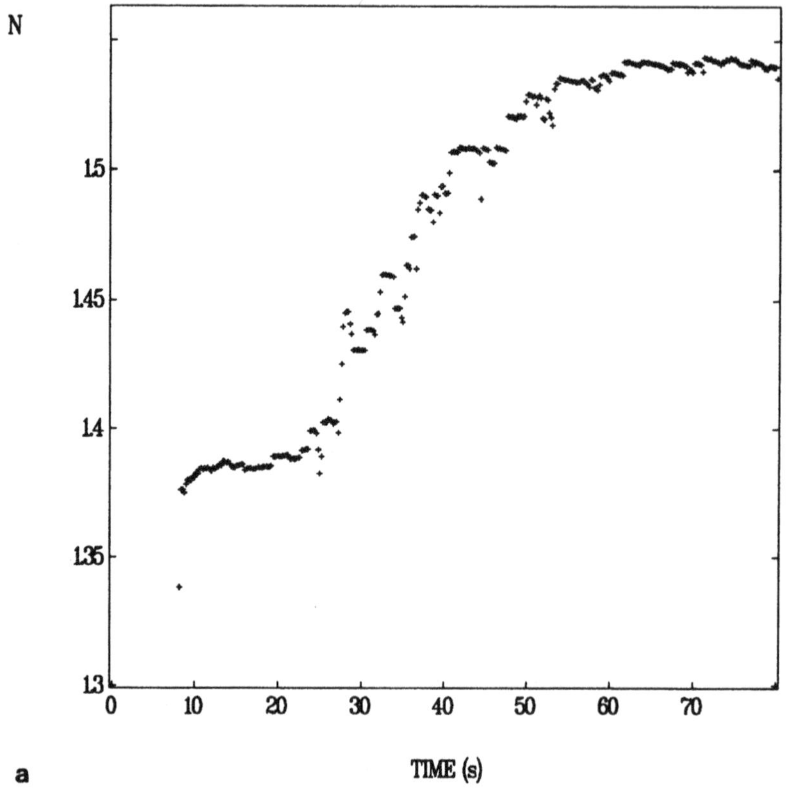

FIGURE 7.5. Evolution of the optical properties (a,b) and film thickness(c) as a function of time for a polypyrrole film grown potentiostatically on a Pt sputter-deposited film electrode at +0.60 V (SCE) using an aqueous solution of pyrrole (0.025 M) and KNO_3 (0.5 M). (From Ref. 19).

interest ($[Fe(CN)_6]^{3-}$, 2,6-anthroquinone disulphonate) was also examined. Results here were not so clearly defined because of absorption of light due to solution species ($[Fe(CN)_6]^{3-}$) and the evolution of film surface roughness, but they were consistent with the same general process.

In later work effects of varying film growth conditions on the growth mechanism, and the final optical properties of the films, were examined in more detail.[19] Films were produced under conditions most commonly encountered in the literature (neutral aqueous, or acidic aqueous, solutions; acetonitrile containing varying amounts of water), and effects of monomer concentration and growth potential were also tested; results are summarized in Table 7.2. The most striking fact is that the magnitude of n for the film as grown is directly related to growth conditions. The more optimal these are (as assessed from literature relating to film conductivity and mechanical properties), the lower is n at 633 nm. For example in acetonitrile solution, employing low concentrations of monomer and high growth potentials,

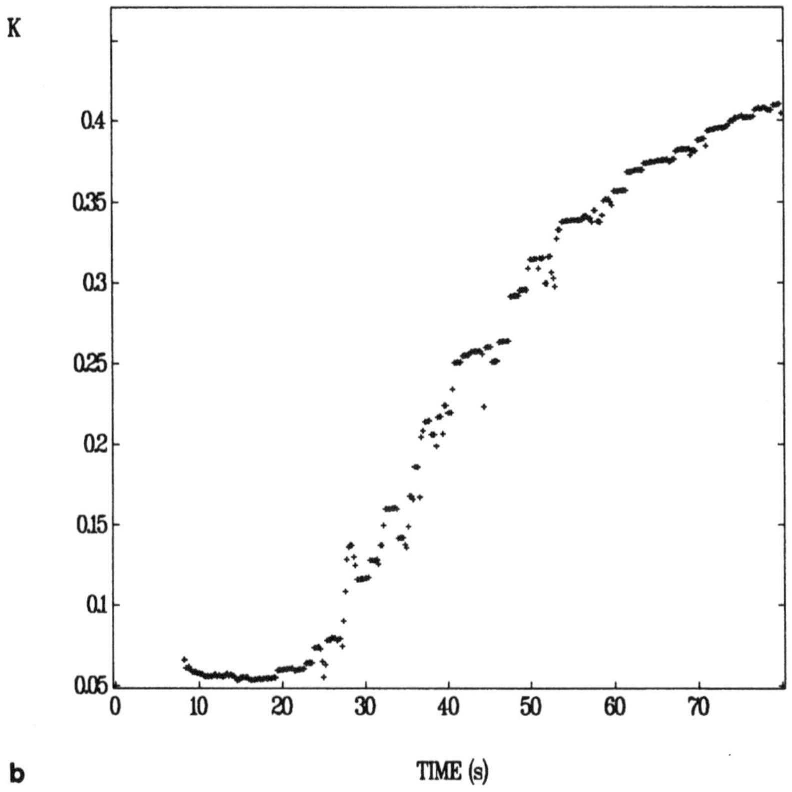

FIGURE 7.5. (*Continued*)

films with high values of k were made, but n was higher than for films made using high monomer concentrations and less extreme potentials, which are known to produce better, more conductive films. In aqueous solution acidic conditions are known to produce better films than neutral conditions, and this, too, is reflected in n values in this study. Interestingly if more concentrated pyrrole solutions (0.1 M) are employed in neutral media, the film's optical properties approach values seen for growth in acid media.[20] This is because protons are produced during polymerization:

$$n\,C_4H_5N - (2+x)ne^- \rightarrow [(C_4H_3N)_n]^{x+} + 2n\,H^+$$

which leads to a suitably acidic pH near the electrode. The growth mechanism for all films made, assessed by examining the evolution of n, k, and L as before, was similar to that expounded in the preliminary communication. Additionally a further test of the validity of this result was made by performing growth experiments at different wavelengths; Optical properties differed, but the evolution of the thickness with time was very similar.

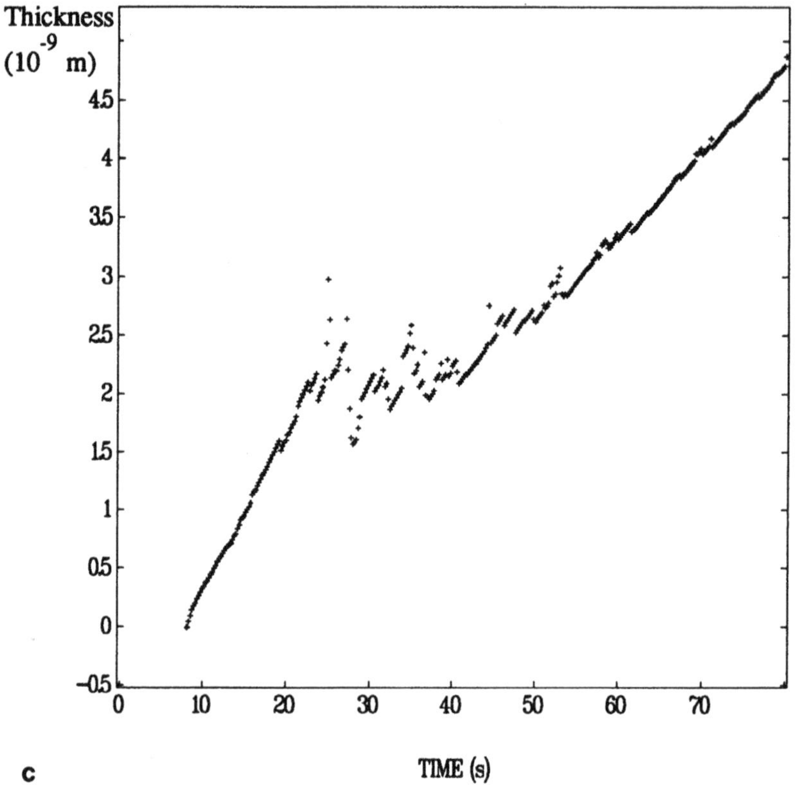

FIGURE 7.5. (Continued)

On reducing the films to neutral, insulating form, optical changes were seen that were entirely consistent with the Lorentz oscillator model previously applied to polyaniline.[19] These are summarized in Table 7.2. In particular *n* rises to values abnormally high for an organic polymer (1.6–1.8), and for the optically superior films, it rises with increasing light energy. This is consistent with the presence of a strong oscillator at higher energy than that of the probing beam. Since the band gap of polypyrrole is approximately 3.6–3.8 eV,[21] absorption is presumably the band-to-band transition.

It is very significant that optical properties of the film as grown determine its subsequent behavior on potential cycling; the lower the *n* value at the end of growth, the lower the *n* value at the anodic limit during potential cycling.[19] In other words the optical quality of the film (and if we relate charge carrier absorptions to physical properties, film conductivity, etc.) depends on conditions used to grow it, and not noticeably on subsequent electrochemical experiments.

TABLE 7.2
Variation in Polypyrrole Optical Properties As a Function of Growth Conditions

Notes[a]	Anode (/V vs. SCE)	Oxidized Form *n*	Oxidized Form *k*	Oxidized Form *L* (/nm)	Cathode (/V vs. SCE)	Reduced Form *n*	Reduced Form *k*	Reduced Form *L* (/nm)
0.2-M py, CH_3CN/ +0.4-M Et_4NBF_4	+0.4	1.46	0.35	50.8	–0.6	1.53	0.248	52.2
0.025-M py, 0.2-M $NaBF_4$, H_2O:+0.65 V	+0.35	1.54	0.19	37.8	–0.55	1.61	0.11	42.4
As above but +0.7 V	+0.35	1.60	0.31	64.0	–0.55	1.77	0.20	69.5
As above but $NaClO_4$ electrolyte	+0.35	1.58	0.31	47.0	–0.55	1.74	0.12	49.4
As above but pH2 ($HClO_4$ added)	+0.35	1.50	0.30	72.0	–0.55	1.70	0.16	81.0
As above +0.1 M $K_3[Fe(CN)_6]$	+0.40	1.48	0.29	62.0	–0.60	1.75	0.10	79.0

[a]All measurements at 633 nm. Film grown in medium indicated in Notes, then transferred to appropriate fresh electrolyte and held at the potential shown until no further ellipsometric changes occurred prior to measurement. Anodic form of film was measured first, followed immediately by cathodic form.

Because these experiments were performed using three-parameter ellipsometry, it was possible to measure film thickness, as well as *n* and *k*, without *a priori* assumptions about the film. Again solutions were checked by ensuring that the film thickness/potential relationship was the same, within experimental error, for successive experiments conducted at different wavelengths. In fact the quality of fits during cycling experiments is significantly better than during growth, because no interference from soluble, absorbing products is encountered. Behavior of film thickness during potential cycling is extremely interesting and at first sight rather unexpected. The conventional view of the oxidation of these polymers has been that charge-balancing anions with some solvent enter the film on oxidation. An increase in film thickness may therefore have been expected. The observed substantial decrease must therefore be connected with structural changes in the polymer on oxidation, possibly due to electrostrictive processes. Their complexity is illustrated by inflexion in the rate of thickness change seen at *ca.* –0.2 V (oxidation) and –0.3 V (reduction) in this example. (However it is noteworthy that this behavior varies with the exact conditions used originally to grow the film; see Ref. 17.)

Recently rapid-scanning spectroscopic ellipsometry, using optical multichannel detection, has allowed real-time spectroscopic ellipsometric investigation of modified electrodes. The first conducting polymer to benefit from such a study is polypyrrole.[21,22] Advantages of this technique are firstly that the dielectric function of the polymer is determined directly over quite a wide energy range (1.5–3.5 eV) and secondly that statistical methods, using knowledge of the dielectric function of the polymer, can then be used to assess the mechanism of growth in more detail and with greater confidence than when single wavelength measurements are used. The best fit to the data during growth was to a model involving the initial roughening of the working (Au) electrode with monomer adsorption, followed by filling the voids thus generated with polypyrrole and the formation of low-density polypyrrole on top. Subsequently this densifies and simultaneously grows from the electrode surface. These conclusions are strikingly similar to those obtained using single-wavelength measurements (Pt-working electrodes).[18,19]

As applied in this work, there are some drawbacks to the technique, however, that will probably be overcome as the technique progresses. For example, although data collection is rapid, for good resolution growth conditions employed [0.005 M pyrrole, 0.62 V (SCE) potentiostatic deposition] had to be such that film formed very slowly, and data (translating from the ε_1 and ε_2 data in Ref. 21 to $n = 1.58$, $k = 0.29$ at 633 nm, for example) indicate that the film does not have optimum optical properties. Also the method of determining film thickness (intensity measurements are not possible) was to minimize electrode-related artifacts in the dielectric function. This requires electrode material to have distinct optical features in the spectral range of interest. Gold meets this requirement, having a d-band to Fermi-level transition at 2.5 eV, but the more commonly employed platinum does not.

Early optical and epr studies, carried out *ex-situ* (see for example Refs. 23 and 24), were somewhat contradictory in terms of the identity of carriers responsible for the electronic conduction in polypyrrole. However two plausible conduction mechanisms were quickly identified that were distinguished by the magnitude of residence time τ for the carrier at a specific molecular site. If τ is small compared to the vibrational time scale, the carrier–lattice interaction is small and the conduction metallike, whereas values of τ larger than the vibrational time scale permit the lattice to distort around the carrier, trapping it and reducing its mobility substantially. The carrier and its accompanying local distortion then propagate together as an entity, termed a polaron, and such a species is always associated with *strong* carrier–lattice interactions.

In fact the situation is even more involved in practice, since removing a single charge from the polymer chain gives rise to a radical cation, to which the term polaron is normally applied, whereas removing a second charge may give rise to an associated dicationic species termed a bipolaron. The formation of the latter was unexpected, since it was initially believed that positive charges on the chain would repel each other, but *ex situ*, and later *in situ* epr studies[25–29] showed that the epr-active polaron species decreased in concentration at higher potentials, being steadily replaced by carriers that were epr-silent as the level of oxidation increased. It was also clear from optical studies *ex situ*[23,24] that optical properties of the film were also changing at higher potentials: Both polarons and bipolarons were giving rise to electronic states intermediate in energy between the highest occupied and lowest unoccupied π-levels of the neutral polymer. Extended Hückel calculations were able to give a semiquantitative account of these levels as arising from the carrier-induced lattice distortions. These calculations also revealed the reason for the surprising stability of the bipolaron, which arose because the presence of a single positive charge gives rise to a change in the double-bond alternation to a less stable form. The presence of a second positive charge restores the more stable bond alternation based on preserving the aromaticity of the heterocyclic ring. The length of the bipolaron is then a compromise between like-charge repulsion energy and aromatic stabilization.

The initial investigation of conducting polymers by IR revealed the growth of a number of very strong bands, termed infrared active vibrations (IRAV) bands, in the ring-stretching region. A fundamental insight into the origin of these bands was provided by a simple experiment on neutral polyacetylene and subsequently on other heterocyclic polymers, particularly polythiophene. The polymer was irradiated with suprabandgap light while the IR spectrum was simultaneously monitored.[30–33] Bands similar to those in the oxidized form were found, and calculations showed that the π-structure in the lowest unoccupied π-levels was associated with an alternating π-structure similar to that expected on charge formation. It was realized that the high intensity of IRAV bands was caused by the strong coupling of the charge carrier to particular vibrational modes: excitation of these modes facilitated carrier movement and hence gave rise to a very large dipole moment change.

Direct evidence that both polarons and bipolarons formed during electrochemical oxidation of polypyrrole *in situ* was first obtained through a joint ellipsometry and IR study on polypyrrole growth on a platinum electrode in aqueous perchlorate solutions.[20] The ellipsometric study showed behavior essentially identical to that just discussed and demonstrated that the film did indeed grow by nucleation and consolidation. The average length of the polymer chains nucleating on the surface was found to be ~125Å. The authors also established that the ellipsometric thickness of their neutral films substantially exceeded that expected on the basis of pure polymer with a calculated water content of 37% by volume.

A detailed analysis of ellipsometric data during cyclic voltammetry was carried out to characterize the film and relate its behavior to the evolution of the FTIR spectra. The cyclic voltammogram of the polypyrrole grown as described is shown in Fig. 7.6; the first sweep in particular shows a marked peak in the positive-going scan at –0.2 V. This potential undoubtedly marks a definite transition; above this potential, for example, the charge versus potential plot for the film becomes linear. The variation of n, k, and L for this film derived from ellipsometric data is shown in Fig. 7.7; it is evident that there is little change in L until –0.2 V is reached on the anodic scan, but L falls rapidly above this potential, in agreement with other studies, before rising slowly at much higher potentials.

This decrease in L was postulated to be derived from the expulsion of solvent and protons from the film—the latter being the kinetically driven ionic compensation mechanism as the film backbone becomes positively charged. Each proton can be expected to transport at least four water molecules with it from the film. On the

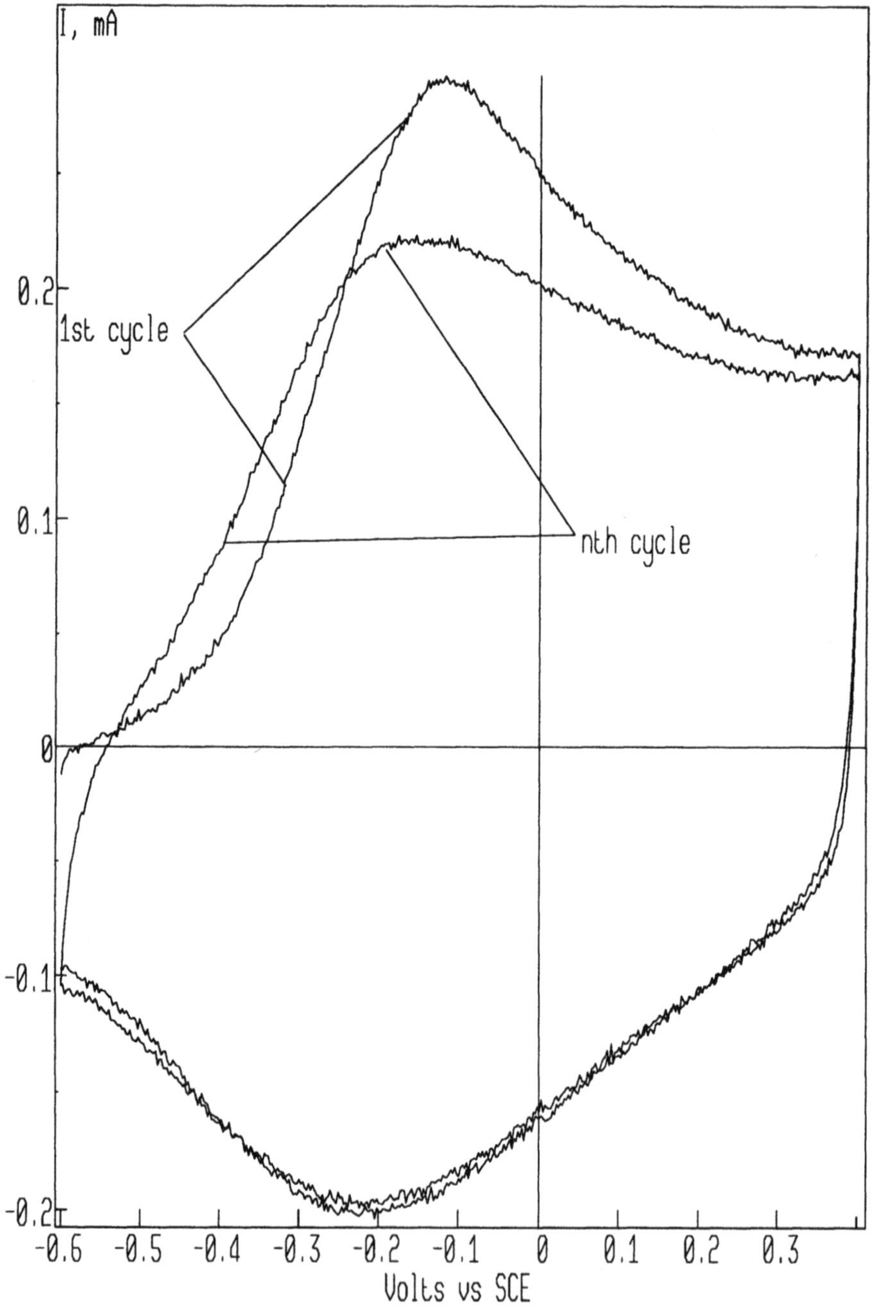

FIGURE 7.6. Typical cyclic voltammograms of a polypyrrole film on a platinum foil electrode in N_2-saturated 1M $NaClO_4$. Voltammograms were collected immediately after holding the film at −0.6 V for 5 minutes and cycling for 5 minutes; scan rate: 100 mV/s, film thickness: 84nm. (From Ref. 20)

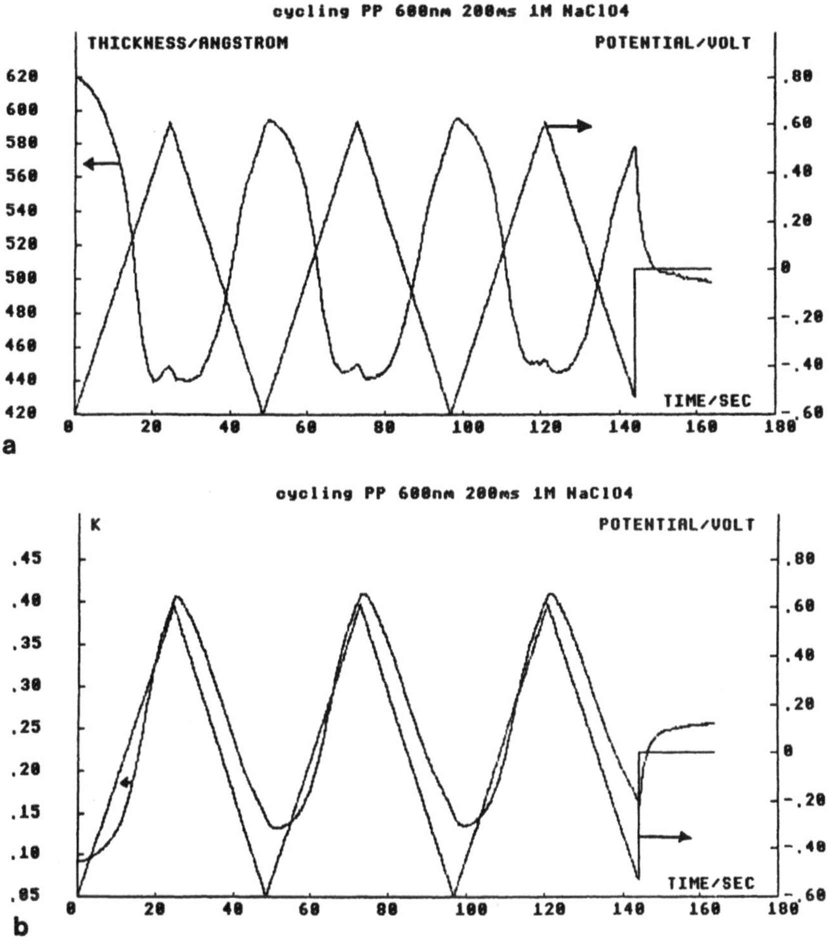

FIGURE 7.7. Processed ellipsometric data obtained during electrochemical cycling of a polypyrrole film under the same experimental conditions as in Fig. 7.6, (from –0.6 V to +0.6 V at 50 mV/s). Samples were collected every 0.2 ms. The working electrode was a Pt film sputtered onto a glass slide; (a) n, (b) k, and (c) L. The potential ramp is also shown for reference.

basis of the water content previously calculated and the known oxidation of the film from electrochemical data, this corresponds to the loss of at least 4×10^{-8} mol water, which leads to a decrease in thickness very close to that actually observed. This postulate of proton ejection as the kinetically driven process is strongly supported by Rajeshwar *et al.*,[34] who grew a polyyrrole film on a gold minigrid optically transparent electrode in a UV–Vis spectrometer. On cycling this film in the presence of a suitable pH indicator in solution, color changes were observed that could be explained only by proton ejection from the film and their subsequent reincorporation during the cathodic scan. Further evidence came from careful

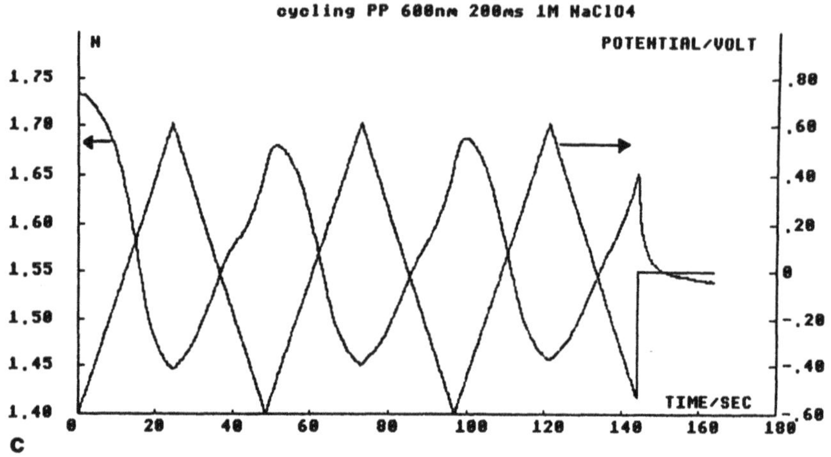

FIGURE 7.7. (*Continued*)

analysis of IR data in the region from 1600–3500 cm^{-1}, which were consistent with the formation of OH^- ions within the film.

It is clear from Fig. 7.7 that *k* shows little change at 633 nm until the potential is increased –0.2V, where it increases monotonically. However it is evident from the behavior of *L* over several cycles that there is considerable hysteresis in the thickness; this can be interpreted in terms of relatively slow morphological changes. This phenomenon was first noted by Heinze *et al.*,[35] who suggested that a film, such as polypyrrole, may exist in two conformational forms, *A* and *B*. The *A* form was thought to be highly irregular in morphology and the stable form in the neutral film, but form *B* shows a more planar configuration, essentially stabilized by a Zener exchange mechanism, which allows carriers to move more freely. Expulsion of water is accompanied by a transition from *A* to *B* forms, but on rereduction, not all the polymer film is reconverted into the *A* form, accounting for the hysteresis in thickness. It is evident that if this hypothesis is correct, the film can be more easily oxidized on subsequent cycles, since any residual form *B* has the correct conformation to accommodate carriers.

Ellipsometric data are clearly in agreement with the suggestion by Heinze *et al.* that a critical concentration of carriers is needed before any conversion of *A* into *B* takes place. The strongly peaked initial cyclic voltammogram can also be explained as follows: The initial conversion of pure *A* into *B* is strongly cooperative in nature, but residual *B* formed in the neutral film during subsequent dynamic scans permits film conversion to take place over a wider potential range and gives rise to a more symmetrical cyclic voltammogram.

Figures 7.8(a) and (b) show *in situ* FTIR spectra of a 40-nm polypyrrole film on Pt collected at successively higher potentials and referenced to the spectrum of

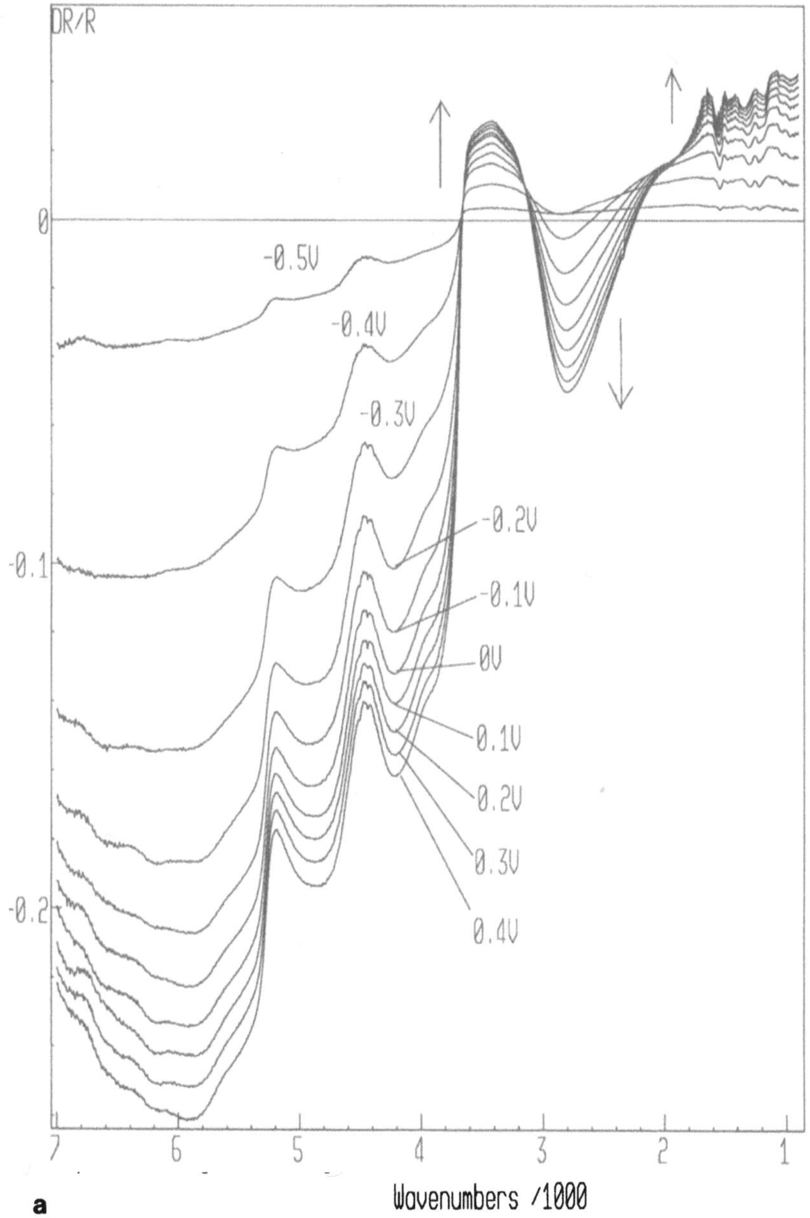

a

FIGURE 7.8. Normalized spectra obtained from the film in Fig. 7.6. Spectra (8-cm^{-1} resolution, 100 coadded and averaged scans requiring *ca.* 1 minute per spectrum) were collected at the potentials shown and normalized to the reference spectrum collected at the base potential of −0.6 V. (a) Full spectral range and (b) 900–1640 cm^{-1}; for clarity, spectra taken at 0.4 V and −0.5 V are omitted.

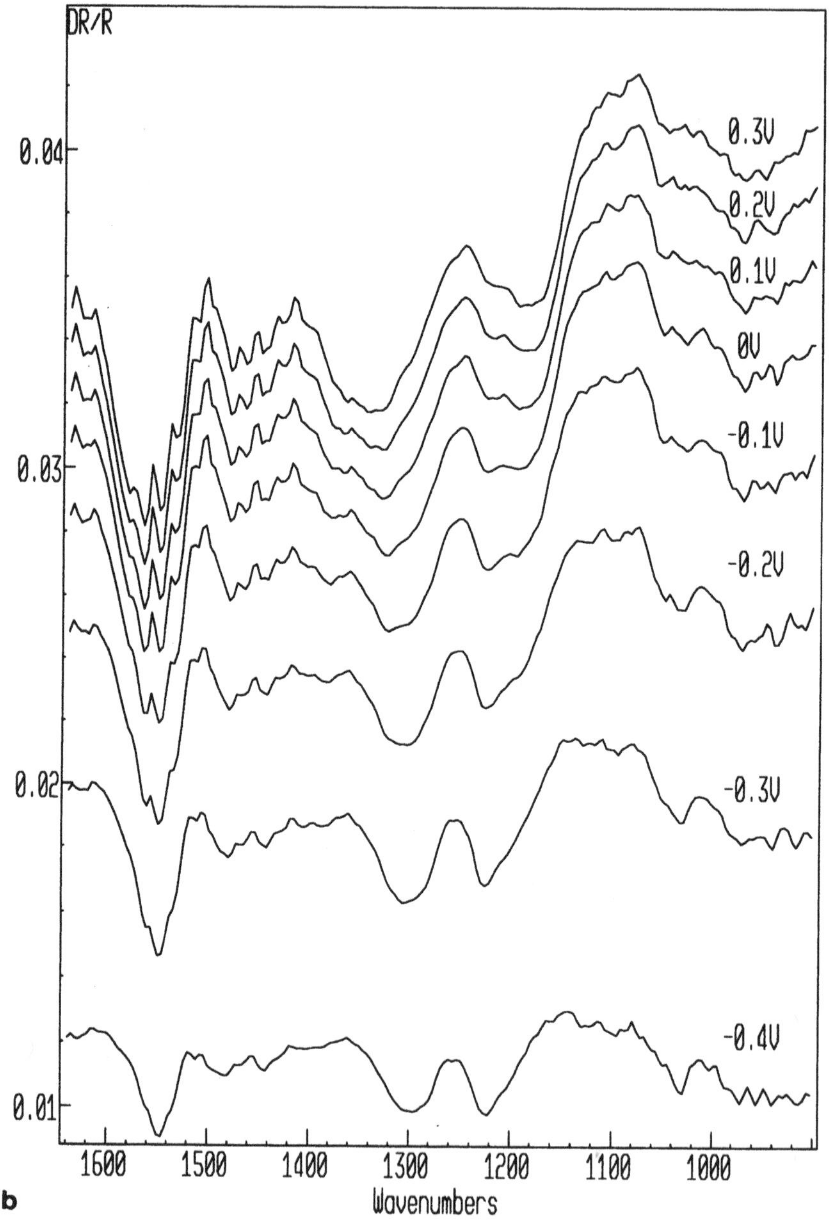

b

FIGURE 7.8. *(Continued)*

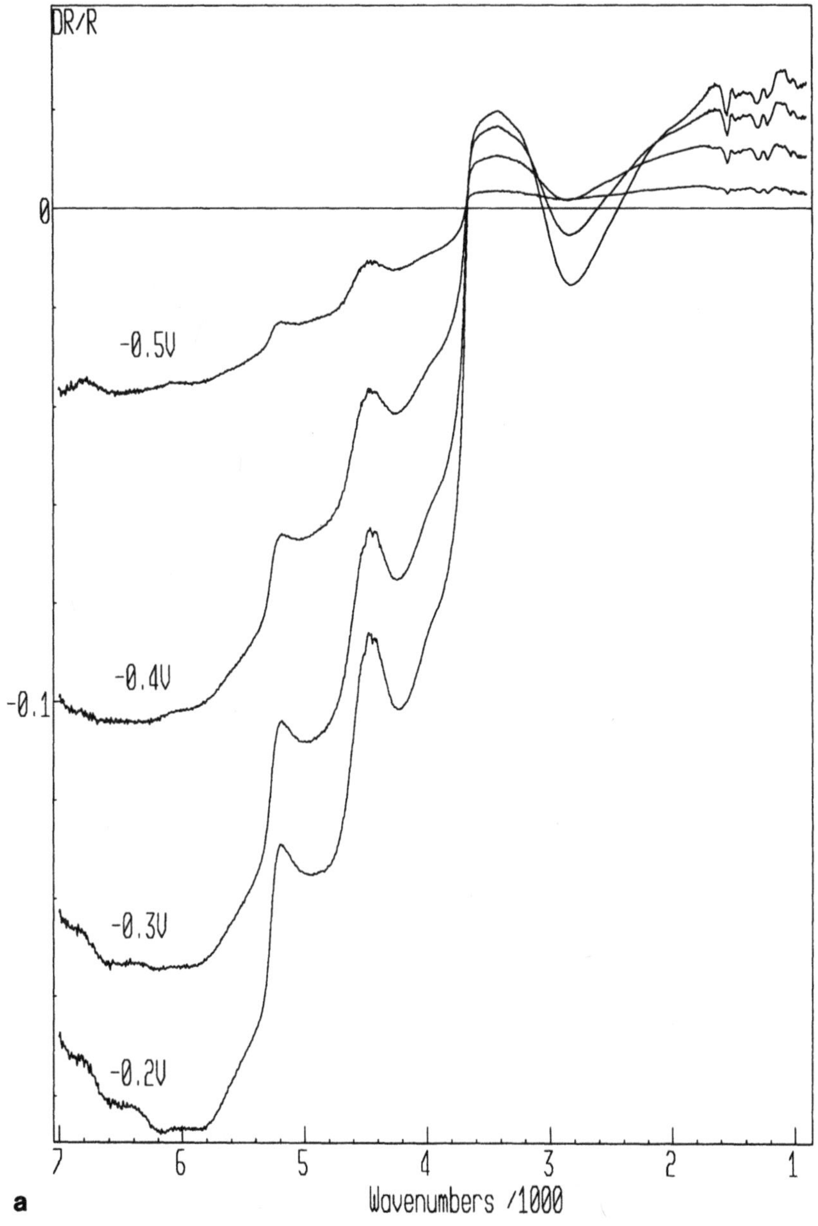

a

FIGURE 7.9. (a,b) Spectra in Fig. 7.8 collected at potentials from –0.6 to –0.2 V, normalized to that taken at –0.6 V. (a) Full-spectral range and (b) 900–1640 cm^{-1}. (c,d) Spectra in Fig. 7.8 collected at potentials from –0.2 Vto 0.4 V, normalized to that taken at –0.2 V. (c) Full-spectral range and (d) 900–1640 cm^{-1}.

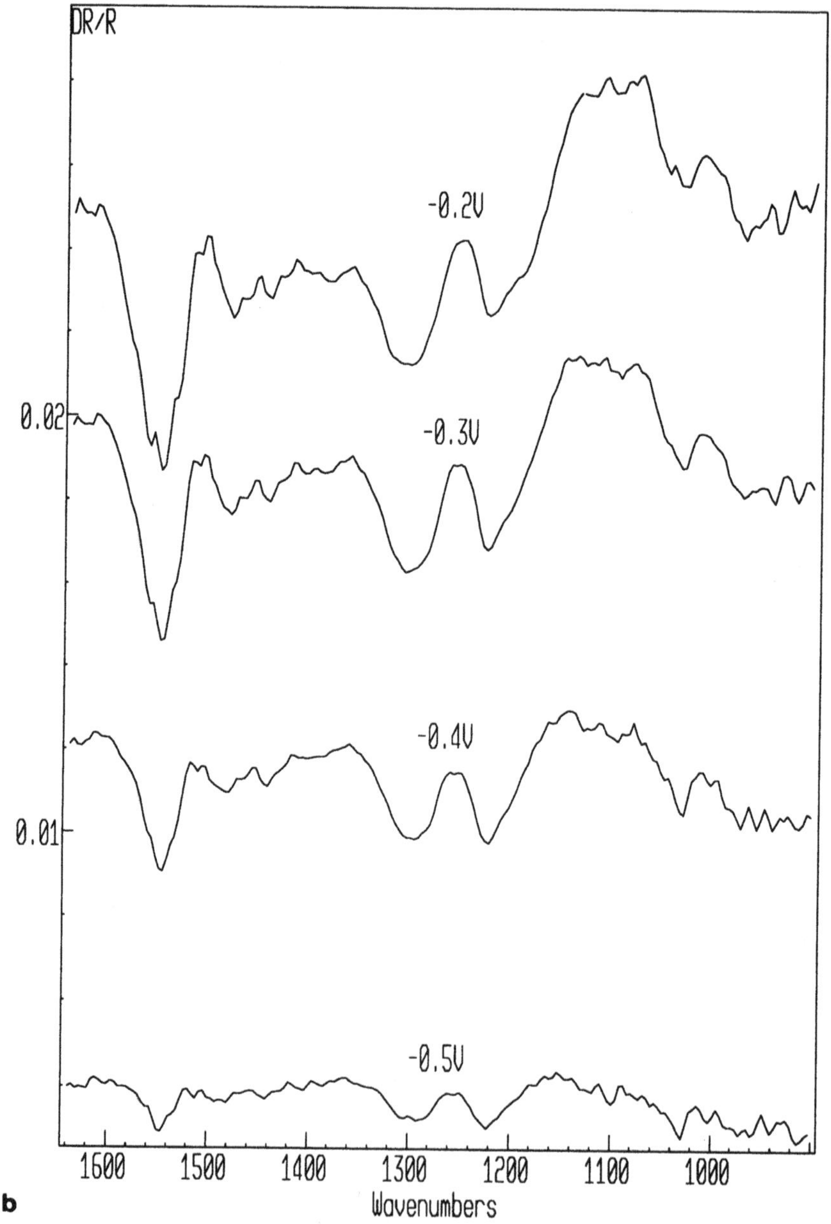

b

FIGURE 7.9. (*Continued*)

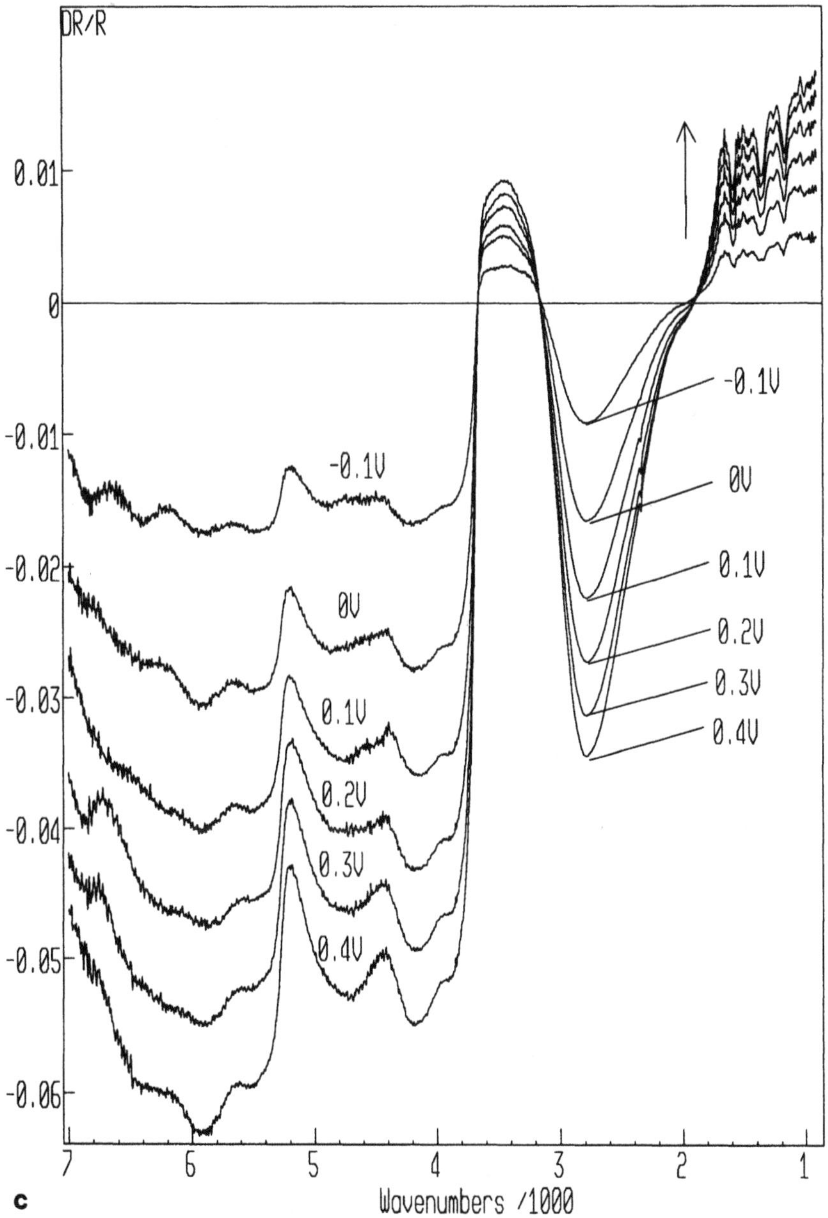

c

FIGURE 7.9. (*Continued*)

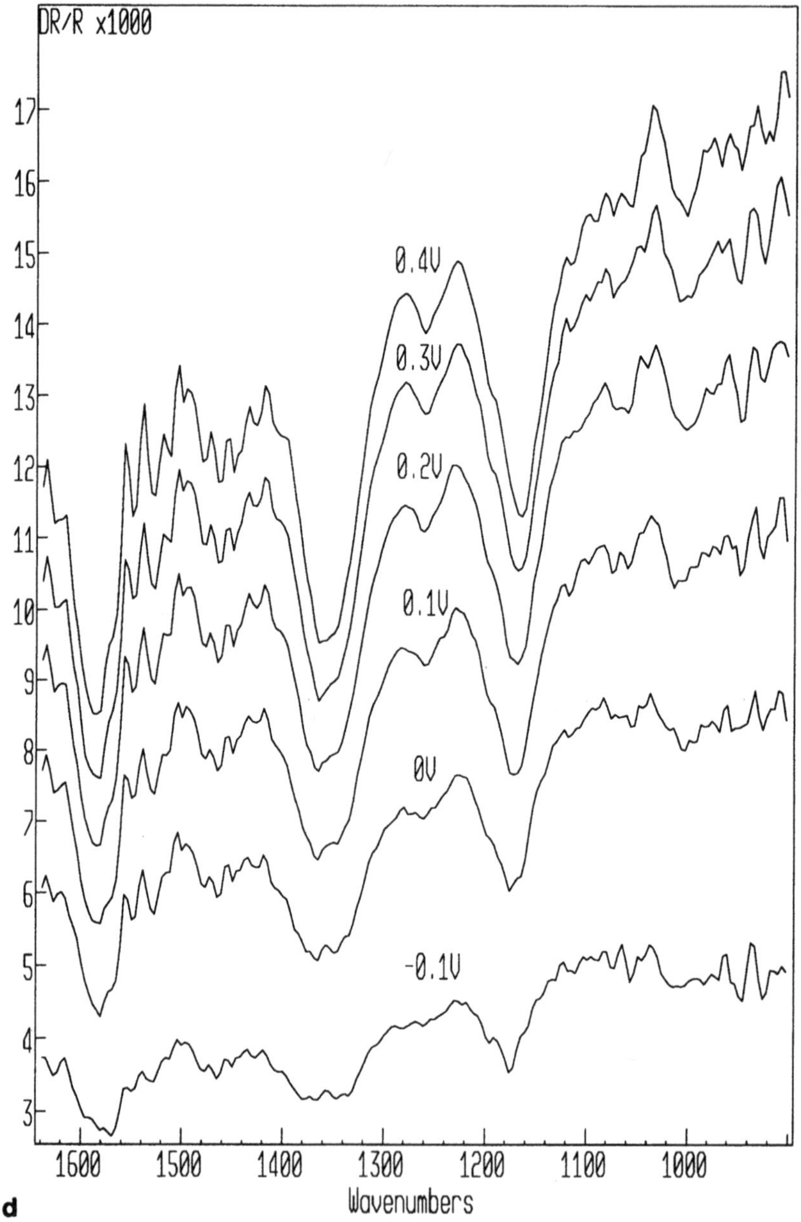

d

FIGURE 7.9. (*Continued*)

a neutral film at –0.6 V. It is clear from the figures that at higher potentials, an intense absorption band is visible, extending into the near IR and arising from a low-energy *electronic* transition. This band increases particularly rapidly up to –0.2 V, above which the rate of increase is slower, which again indicates a branch point at this potential. This change in the rate of increase of absorbance with potential is not accompanied however by an equal change in the rate of increase of charge. This again suggests a change in carrier *type* above –0.2 V; data are best understood by assuming that one type of carrier, clearly identified by electron paramagnetic resonance (epr) work as a single charge *polaron*, forms at lower potentials, while a second type of carrier with a similar absorption coefficient, a *bipolaron*, forms above –0.2 V. The structure apparent in Fig. 7.8 at 3500, 4500, and 5200 cm^{-1} is not associated with the film but arises from artefacts due to intense solvent and window absorption, which gives rise to single-beam breakthrough, discussed in Ref. 36.

More subtle changes in the optical properties of the film can give rise to other marked optical effects. One example is the general increase ($\Delta R/R$) values below 2000 cm^{-1} in more oxidized films. This can be traced to the fact that the refractive index of the film shows a marked rise in this region as a result of the growth of the electronic absorption band in the near IR. This substantial change in n can be explained by the same Kramers–Kronig relationship responsible for the *decrease* of n in the visible region. Essentially the effect is to increase the reflectivity of the film–electrolyte interface, effectively hiding absorptions within the film due to solvent and polymer absorptions.

At lower frequencies than the electronic band, we find the true IRAV bands, identical in origin to those found in *ex situ* studies. Up to –0.2 V, these bands are sharp, but above this potential, they broaden and appear less well-defined. To emphasize this difference, spectra at higher potentials were all referenced to the spectrum of the partially oxidized film at –0.2 V. This brings out more clearly changes that take place in this second potential region; results are shown in Fig. 7.9. It is evident from this figure that a new set of IRAV bands appears above –0.2

TABLE 7.3
IRAV Bands

Type I (polarons)/cm^{-1}	Type II (bipolarons)/cm^{-1}
1196 (sh)	1165[a]
1227	1258
1308	1358
1558	1588

[a]Potential-dependent band.

V, and both this and the set seen below –0.2 V are listed in Table 7.3. Again we anticipate that the set below –0.2 V can be associated with polaron carriers, whereas the set above –0.2 V can be ascribed to *bipolarons*. The sharpness of both sets of peaks suggests that both carriers probably extend over well-defined lengths, both of which are expected to be small compared to the overall coherence length of the polymer chain (i.e., the length over which the carrier can move freely before having to jump to a neighboring chain or propagate through a chain defect). An additional point can be made: the highest frequency IRAV band is associated with a combination of intraring and interring modes. Its frequency offers a rough measure of the physical extension of the carrier, increasing in frequency as the conjugation length of the carrier decreases. On this basis bipolarons are smaller than polarons.

It is clear from Figs. 7.8 and 7.9 that polarons are generated up to –0.2 V, then attain an approximately constant concentration, since no polaron loss features can be discerned in the spectra in Fig. 7.10. Above –0.2 V only bipolaron generation can be seen; in fact the first bipolarons are seen as low as –0.4 V, but their concentration becomes significant only above –0.2 V.

This concept of the initial formation of polarons and the subsequent formation of bipolarons that coexist with polarons already formed initially emerged from *in situ* epr work, where correlations of spin density and total film charge had also led to the postulate by Genoud and coworkers of a definite length for both types of carriers.[25–27] This model was extended by Christensen and Hamnett[20] who assumed (1) polarons of monomer conjugation length l_p are longer than bipolarons, of length l_b; (2) there is a Gaussian distribution of polymer coherence lengths with a mean of 34 monomer units, as derived from ellipsometric measurements; (3) a single charge is placed on each polymer chain before any second charges are accommodated; and (4) a second charge placed on a polymer chain leads to the formation of a second polaron unless combined carrier lengths exceed the chain coherence length, in which case a shorter *bipolaron* is formed. The best fit obtained by quantitatively fitting this model to IR data suggested that $l_p \sim 12$ monomer units and $l_b \sim 3$ monomer units; the latter is close to that postulated by many other workers. This model does predict a rather flat maximum in the polaron concentration as a function of potential. The fact that no decrease in polaron concentration can be discerned may reflect experimental difficulties inherent in this type of data manipulation.

If conducting polymers are to find applications, it is necessary to understand factors determining their stability to switching and their overoxidation. A study using both ellipsometry and *in situ* FTIR spectroscopy has been made of what happens to polypyrrole films on overoxidation in aqueous solution.[20] As the potential is taken into the region where irreversible oxidation is seen in voltammetry, the value of *n* increases and *k* decreases (600-nm wavelength), as expected if the film loses conjugation. The charge carrier population decreases, so the

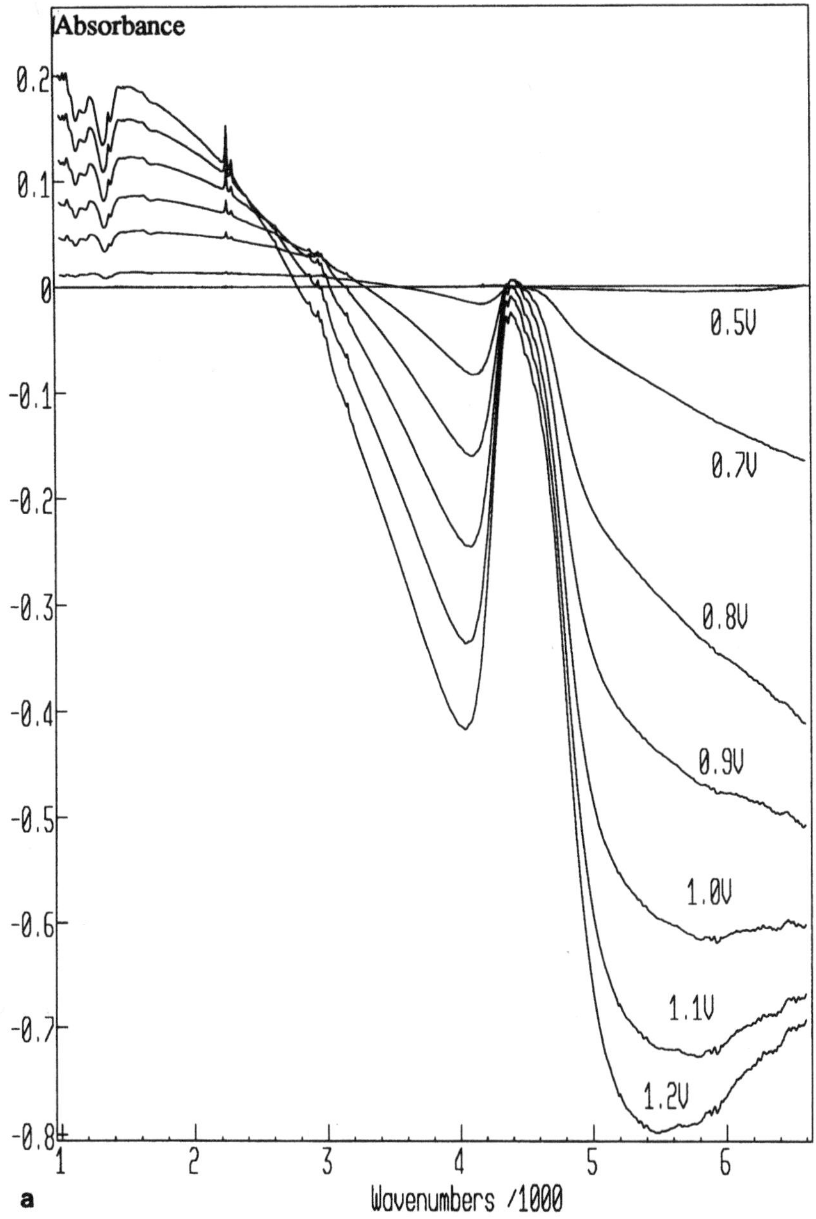

a

FIGURE 7.10. (a), (b) Spectra (8-cm^{-1} resolution, 100 coadded and averaged scans requiring *ca.* 1 minute per spectrum) taken during oxidation of a 106-nm-thick polythiophene film on a Pt electrode in acetonitrile/0.1 M tetrabutylammonium tetrafluoroborate. Spectra were collected at potentials shown, then ratioed to the reference spectrum taken at 0 V. Resultant normalized spectra were then logged to give plots of absorbance versus wave number. (a) Full-spectral range; (b) IRAV region. (c), (d) Plots of $1/\xi_I$ versus $1/\xi_E$; (c) Calculated from spectra in (a) and (b); (d) Reduction process. (From Ref. 50)

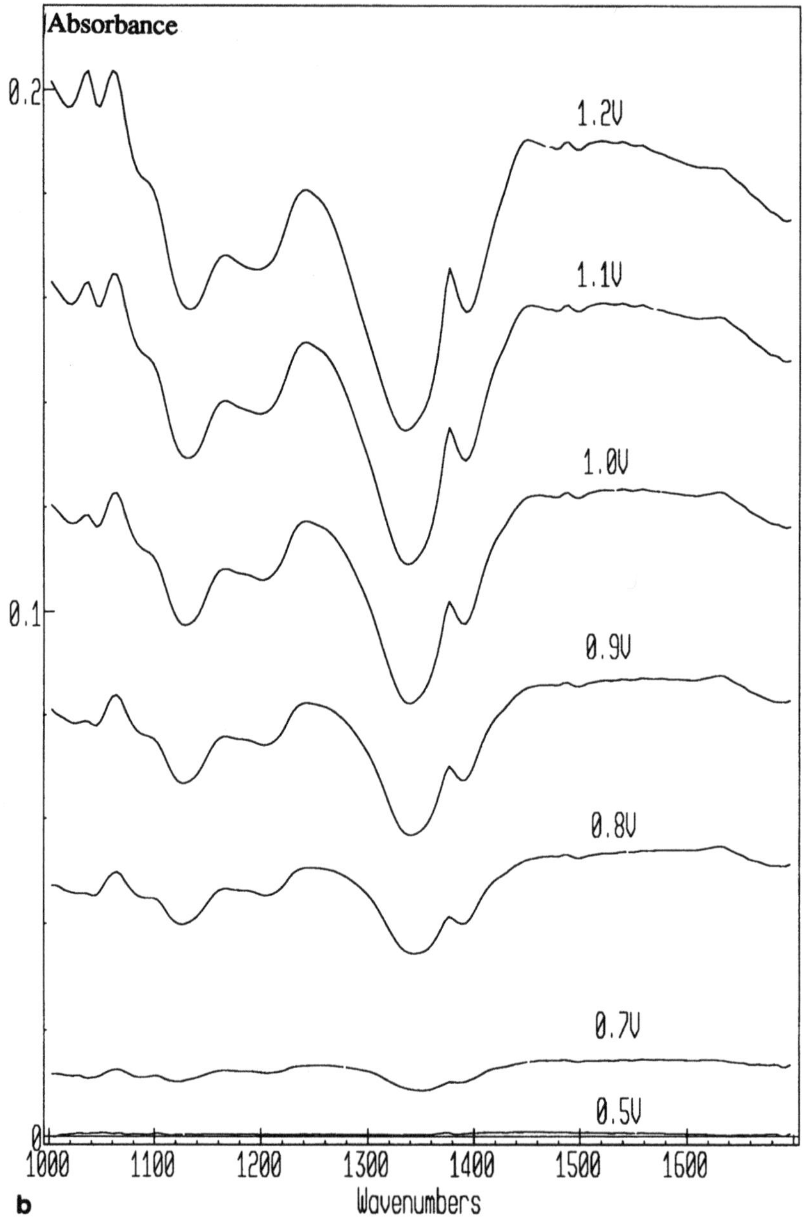

FIGURE 7.10. (Continued)

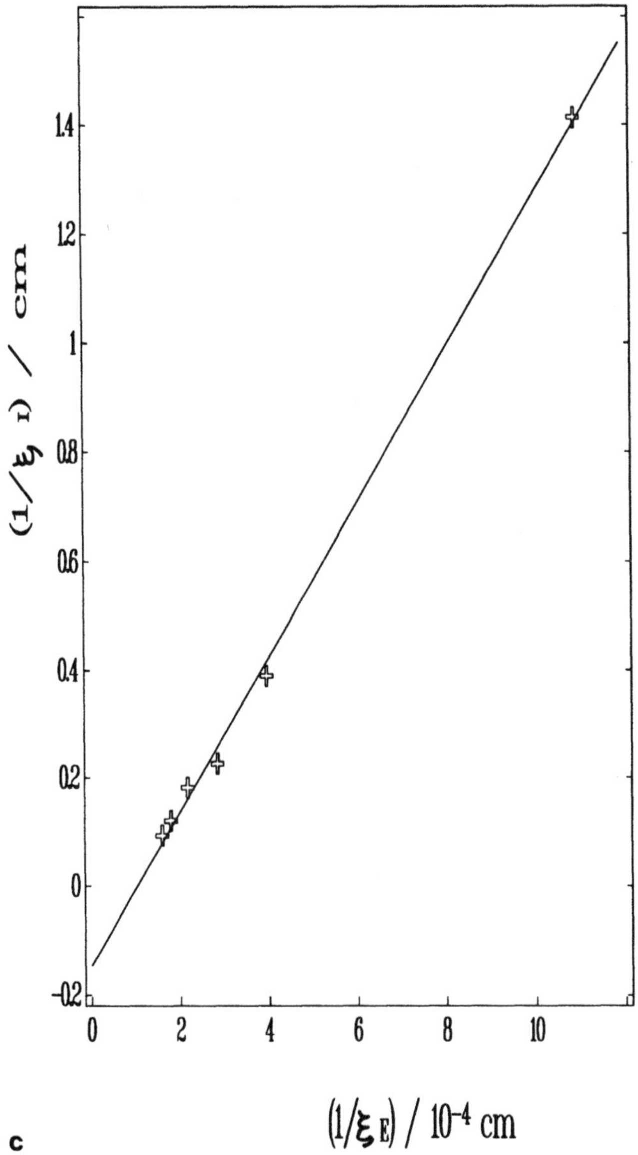

FIGURE 7.10. *(Continued)*

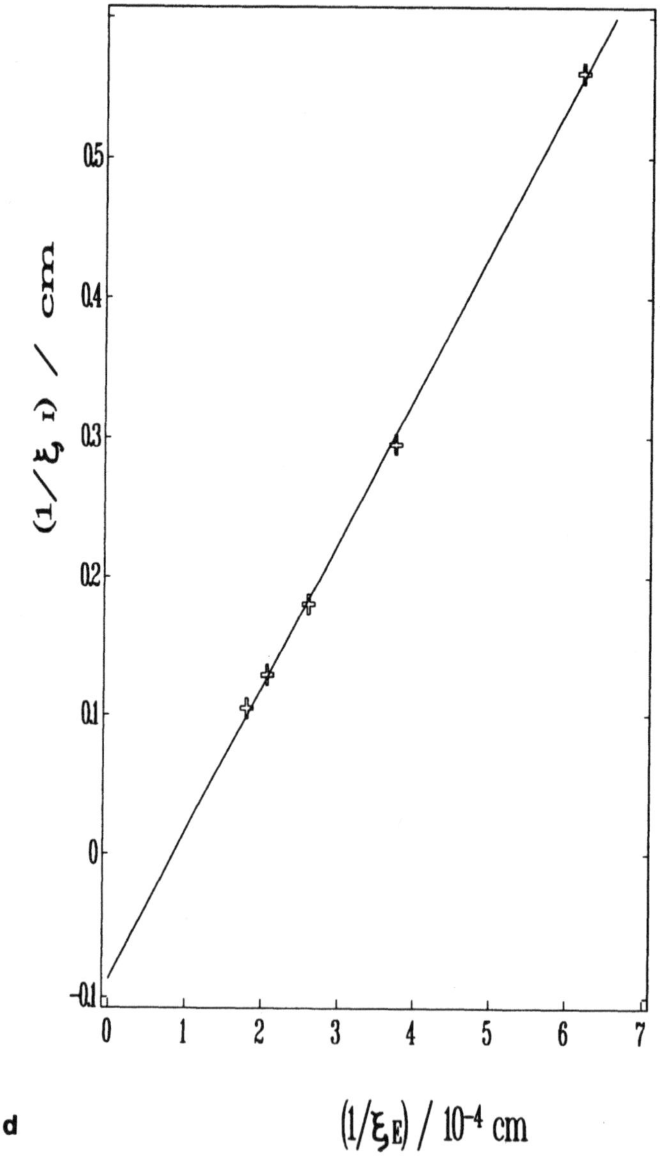

FIGURE 7.10. (*Continued*)

intensity of the mid-IR bands falls. Film thickness increases, and infrared spectroscopic evidence shows the formation of CO_2 and carbonyl groups.

We conclude this section by reviewing the state of our understanding of polypyrrole gained as a result of ellipsometric and IR experiments. First of all these experiments were conducted *in situ*, and they reflect the dynamic electrochemistry taking place. Secondly they show a clear change in the evolution of the film near to −0.2 V, associated with a change in the character of the carriers. Thirdly they establish that the film initially shrinks as the potential is increased—quite an unexpected result and undoubtedly due to a combination of the kinetically driven expulsion of protons as the most facile ionic compensation mechanism, coupled with solvent incorporated in the neutral film, and a structural rearrangement of polymer chains driven by the energy gain associated with increasing the delocalization of carriers. Finally ellipsometric results very clearly indicate that optical properties of polypyrrole are a sensitive function of the precise preparation protocol adopted. There is no such substance as standard polypyrrole, a fact that goes far to explaining the current confused state of much of the literature.

3.3. Polythiophene and Related Polymers

Potentiostatic thiophene electropolymerization was the subject of early efforts to employ three-parameter ellipsometry to investigate the detailed mechanism of film formation.(37,38) Interestingly a consistent observation was that no surface film formed at any potential until immediately after the minimum in the current-time behavior. Thus the charge passed during the initial current spike evidently generated oligomeric species in solution; this had not been obvious in earlier *in situ* electronic spectroscopy. This phenomenon was unequivocally established using a combination of ellipsometry and rotating disk studies, for poly(3-methylthiophene).(39)

The evolution of film properties with time, and the final optical properties of the films, were found to depend on the exact potential employed. Looking at data presented in Refs. 37 and 38, it is clear that all films grew in a manner essentially similar to that described for polypyrrole and polyaniline, namely nucleation followed by an overlap of growth sites, densification of the film, then outward growth of the (dense) film from the electrode. These stages can be discerned clearly for low-growth potentials, but at higher potentials (>2.0 V) the first two stages occur too rapidly to be resolved.

Some film-cycling experiments were conducted on the thicker polythiophene films grown at higher potentials,(37) and these were notable for the observation that film thickness decreased markedly on oxidation. This behavior is remarkably consistent with that discussed earlier for polypyrrole, and the reasons for it are doubtlessly similar.

Other workers have studied polythiophene electropolymerization with two-parameter ellipsometry.(40,41) The importance of experimental conditions in the growth of these films is emphasized by the fact that in their earlier work with 0.01-M monomer solutions, these authors obtained electroinactive films with optical properties very different from those of conventional conducting polymers and invariant with applied potential.[40] However on increasing the monomer concentration to 0.4 M, polymers with optical properties not dissimilar to those of Hamnett and Hillman were grown. In this work an increase in film thickness on oxidation was claimed, but it should be emphasized that absolute film thicknesses could not be unequivocally determined.[41]

An important means of altering physical properties of these heterocyclic conducting polymers is three-substitution. Arwin *et al.* used spectroscopic ellipsometry to investigate the dielectric function of spin coatings of poly(3-hexylthiophene) on gold electrodes as a function of potential and to see what effect light-induced polymer degradation had on this.[42] They found that films made by this method were rather more porous than films of other polymers grown electrochemically. Interestingly films did not degrade when exposed to light in a vacuum—only when exposed to both light and oxygen, which bodes well for potential applications in electrochromic devices.

Another polythiophene derivative with unusual properties is poly(benzo[c]thiophene). Theoretical predictions[43] suggested that this material would have a low band gap (*ca.* 1.0 eV). This was subsequently confirmed experimentally, using *in situ* electronic spectroscopy.[44] On oxidizing to the conducting form, the polymer bleaches from a dark blue color (due to the band-to-band transition) to almost colorless. A transparent, electronically conducting, organic material is obviously a fascinating prospect. However the growth of films of this polymer was beset by difficulties, partly due to the relative ease of formation of poly(1,3-dihydrobenzo[c]thiophene), an insulator, during electrochemical experiments. The early work[44] suggested that films could be produced only by using a tetraalkylammonium (or tetraarylphosphonium) bromide and relying on an electrocatalytic mechanism. Later it was shown that satisfactory films could be produced in conventional electrochemical experiments (i.e., by monomer oxidation in R_4NBF_4/CH_3CN electrolyte) with careful control of conditions.[45] Three-parameter ellipsometry was used to investigate differences in the mechanism of film formation by these methods. This showed that films grown by the latter route followed the now familiar mechanism of nucleation, consolidation, and steady outward growth, although at longer time spans, film thickness levels off, even though a steady anodic current flows. Films grown by electrocatalysis involving Br_2 were very different. Here results were consistent with a mechanism involving the formation of quite long oligomers in solution. These precipitated rapidly onto the electrode giving a thick (*ca.* 900 Å) but very diffuse film, which then densified, but only very slowly. Most of the charge passed in these experiments was evidently generating species in solution that were not subsequently incorporated into the growing film.

Although much of the earlier work on the infrared spectroscopy of conducting polymer films was carried out on polythiophene; *in situ* work has been rather sparser.[46–50] The main conclusion from this latter work was identification of carrier type from an examination of the types of electronic and IRAV band seen on oxidation of the polymer. Interestingly the intensities of IRAV and electronic bands do *not* generally track each other on oxidation, in spite of the fact that theoretical analysis of the intensity of the IRAV band shows that[51]

$$\xi_I \propto \frac{\rho e_o^2}{M_d} \tag{1}$$

where ρ is the charge density, M_d the dynamic mass of the carrier and $\Sigma_b(A_I)_b$ is the total intensity of the IRAV transitions. An analogous formula exists for electronic transitions:

$$\xi_E \propto \frac{\rho e_o^2}{m_e} \tag{2}$$

where m_e is the electron-band-effective mass. It follows that a plot of ξ_I versus ξ_E should be linear, with slope M_d/m_e. However this is generally not the case, since it appears that M_d in particular is itself a function of the carrier density. This is not entirely surprising: carrier–carrier interactions reduce carrier mobility and hence increase carrier-effective mass. In addition it has been suggested[48] that as chains become oxidized, they are forced by exchange interactions into more planar configurations, thereby enhancing the mobility of carriers and reducing the effective mass. We can model these two effects in first order by assuming a linear dependence of M_d on ρ: If we write

$$M_d = M_d^o(1 - [\kappa - \kappa']\rho) \tag{3}$$

then κ represents the effects of increasingly favorable chain configuration, or dekinking, and κ' the effects of carrier–carrier interaction. Combining Eq. (1–3) we obtain

$$\frac{1}{\xi_I} = \left(\frac{M_d^o}{m_e \xi_E}\right) - \left(\frac{M_d^o[\kappa - \kappa']}{e_o^2}\right) \tag{4}$$

It is clear that a plot of $1/\xi_i$ versus $1/\xi_E$ should be linear with slope M_d^o/m_e and an intercept of either sign or even zero. A negative intercept implies that oxidation-induced morphology changes were more significant than carrier–carrier interactions; a positive intercept implies the reverse. It is evident that the precise growth protocol has a critical effect on both κ and κ' as well as the magnitude of M_d^o and different films are likely to show considerable variability in both the sign of the intercept and the value of the slope.

It has indeed been found that Eq. (4) is applicable to thiophene-based polymers: Figs. 7.10(a) and (b) show *in situ* FTIR spectra taken during oxidation of a polythiophene film on platinum in acetonitrile solution. Electronic absorption and IRAVs are clearly evident, and Fig. 7.10(c) shows a plot of $1/\xi_I$ versus $1/\xi_E$. From Fig. 7.10(c), we note that the intercept is negative, indicating morphology changes dominate; the slope has a value of 1360, giving us, as expected, a large value of the effective carrier mass, consistent with its small-polaron nature. Similar plots can be obtained during the reduction of the film, and Fig. 7.10(d) shows such a plot for the rereduction of the film whose oxidation is shown in Fig. 7.10(c). The negative intercept has been reduced in magnitude, suggesting an increase in carrier interactions and an associated decrease in the importance of morphology changes. This probably arises because at least some of the morphology changes that take place during oxidation are irreversible—at least on the time scale of the experiment. Consistent with this is the fact that the slope has a value of 1029, again lower than that found for the film during the initial stages of oxidation.

4. CONCLUSION

The application of ellipsometry and *in situ* FTIR has revealed a wealth of detail about the electrochemical behavior of simple conducting polymers that would have been inaccessible by other techniques. There is however still much to be done to refine our understanding, some of which depends on advances in instrumentation and theory that are currently taking place. The advent of fast *spectroscopic* ellipsometry will certainly add to our confidence in models used to explain the formation of films on the electrode surface. Theoretical advances in understanding bonding in conducting polymers will also lead to better models of vibrational modes whose intensity is enhanced during the oxidation process. This understanding will give us in turn further insight into the nature of mobile carriers and their strong interactions with the polymer backbone. It is also becoming increasingly apparent that different anions and growth conditions give rise to remarkably different polymer morphologies. In particular films appear to show substantial difference in anisotropy and/or optical absorption properties that needs urgently to be explored with faster forms of ellipsometry.

REFERENCES

1. R. Murray, ed., in *Techniques of Chemistry*, vol. 22 (A. Weissberger, ed.) (Wiley, New York, 1992).
2. *Handbook of Conducting Polymers*, T. A. Skotheim, (ed.) (Marcel Dekker, New York, 1986).
3. R. Greef, in *Comprehensive Treatise of Electrochemistry* (R. E. White, J. O'M. Bockris, B. E. Conway, and E. Yeager, eds.) (Plenum Press, New York, 1984) p. 339.

4. P. A. Christensen and A. Hamnett in *Comprehensive Chemical Kinetics*, vol. 29 (R. G. Compton and A. Hamnett, eds.) (Elsevier, Amsterdam, 1989), chap. 1.
5. R. M. A. Azzam and N. M. Bashara, *Ellipsometry and Polarised Light* (North-Holland Publ., Amsterdam, 1987).
6. S. Gottesfeld, in *Electroanalytical Chemistry: A Series of Advances*, vol. 15 (A. J. Bard, ed.) (Marcel Dekker, New York, 1989), p. 143.
7. C. M. Carlin, L. J. Kepley, and A. J. Bard, *J. Electrochem. Soc.* **132**, 353 (1985).
8. B. Beden and C. Lamy in *Spectroelectrochemistry* (R. J. Gale, ed.) (Plenum Press, New York, 1988), chap. 5.
9. B. Beden, C. Lamy, A. Bewick, and K. Kunimatsu *J. Electroanal. Chem.* **121**, 343 (1981).
10. G. R. J. Robertson, J. L. Ord, D. J. De Smet, and M. A. Hopper, *J. Electrochem. Soc.* **136**, 3380 (1989).
11. A. Hamnett and S. J. Higgins, *Electrochimica Acta* **36**, 2123 (1991).
12. R. Greef, M. Kalaji, and L. M. Peter, *Faraday Discuss. Chem. Soc.* **88**, 277 (1989).
13. S. Gottesfeld, A. Redondo, and S. W. Feldberg, *J. Electrochem. Soc.* **134**, 271 (1987).
14. A. Redondo, E. A. Ticianelli, and S. Gottesfeld, *Mol. Cryst. Liq. Cryst.* **160**, 185 (1988).
15. A. Redondo, E. A. Ticianelli, and S. Gottesfeld, *Synth. Met.* **29**, E265 (1989).
16. I. Rubenstein, J. Rishpon, E. Sabatani, A. Redondo, and S. Gottesfeld, *J. Amer. Chem. Soc.* **112**, 6135 (1990).
17. C. Lee, J. Kwak, and A. J. Bard, *J. Electrochem. Soc.* **136**, 3720 (1989).
18. A. Hamnett, S. J. Higgins, P. R. Fisk, and W. J. Albery, *J. Electroanal. Chem.* **270**, 479 (1989).
19. S. J. Higgins and A. Hamnett, *Electrochimica Acta* **36**, 2123 (1991).
20. P. A. Christensen and A. Hamnett, *Electrochimica Acta* **36**, 1263 (1991).
21. Y-T. Kim, D. L. Allara, R. W. Collins, and K. Vedam, *Thin Solid Films* **193/194**, 350 (1990).
22. Y-T. Kim, R. W. Collins, K. Vedam, and D. L. Allara, *J. Electrochem Soc.* **138**, 3267 (1991).
23. (a) P. Pfluger, M. Krounbi, G. B. Street, and G. Weiser, *J. Chem. Phys.* **78**, 3212 (1983); (b) K. Yakushi, L. J. Lauchlan, T. C. Clarke, and G. B. Street, *J. Chem. Phys.* **79**, 4774 (1983).
24. J. C. Scott, P. Pfluger, M. T. Krounbi, and G. B. Street, *Phys. Rev.* B. **28**, 2140 (1983).
25. F. Genoud, M. Guglielmi, M. Nechtschein, E. Genies, and M. Salmon, *Phys. Rev. Lett.* **55**, 118 (1985).
26. M. Nechtschein, F. Devreux, F. Genoud, E. Vieil, J. M. Pernaud, and E. Genies, *Synth. Met.* **15**, 59 (1986).
27. F. Devreux, F. Genoud, M. Nechtschein, and B. Villeret, *Synth. Met.* **18**, 89 (1987).
28. W. J. Albery, *Farad. Disc. Royal Society Chem.* **88**, 247 (1989).
29. A. M. Waller and R. G. Compton, *J. Chem. Soc. Farad. Trans. I.* **85**, 977 (1989).
30. F. Moraes, H. E. Schaffer, M. Kobayashi, A. J. Heeger, and F. Wudl, *Phys. Rev. B.* **30**, 2948 (1984).
31. W. Hayes, F. L. Pratt, K. S. Wong, K. Kaneto, and K. Yoshino, *J. Physics C, Sol. State Physics* **18**, L555 (1985).
32. Y. H. Kim, S. Hotta, and A. J. Heeger, *Phys. Rev.* **B36**, 7486 (1987).
33. K. Kaneto, F. Uesugi, and K. Yoshino, *J. Phys. Soc. Jap.* **56**, 3703 (1987).
34. E. W. Tsai, G. W. Jang, and K. Rajeshwar, *J. Chem. Soc. Chem. Comm.* 1776, (1987).
35. J. Heinze, M. Storzbach, and J. Mortensen, *Ber. Bunsenges. Phys. Chem.* **91**, 960 (1987).
36. P. A. Christensen, A. Hamnett, A. R. Hillman, M. J. Swann and S. J. Higgins, *J. Chem. Soc. Far. Trans.* **89**, 921 (1993).
37. A. Hamnett and A. R. Hillman, *Ber. Bunsenges. Phys. Chem.* **91**, 329 (1987).
38. A. Hamnett and A. R. Hillman, *J. Electrochem. Soc.* **135**, 2517 (1988).
39. P. Lang, F. Chao, M. Costa, E. Lheritier, and F. Garnier, *Ber. Bunsenges. Phys. Chem.* **92**, 1528 (1988).
40. J. Zerbino, W. J. Plieth, C. Lahmann, and G. Kossmehl, *J. Electroanal. Chem.* **274**, 213 (1989).
41. J. Zerbino, W. J. Plieth, and G. Kossmehl, *J. Electroanal. Chem.* **260**, 361 (1989).

42. R. Jansson, H. Arwin, G. Gustafsson, and O. Inganaas, *Synthetic Metals* **28**, C371 (1989).
43. J. L. Brédas, R. Thémans, J. M. André, A. J. Heeger, and F. Wudl, *Synthetic Metals* **11**, 343 (1985).
44. M. Kobayashi, C. Colaneri, M. Boysel, F. Wudl, and A. J. Heeger, *J. Chem. Phys.* **82**, 5717 (1985).
45. P. A. Christensen, J. C. H. Kerr, S. J. Higgins, and A. Hamnett, *Faraday Disc. Chem. Soc.* **88**, 261 (1989).
46. H. Neugebauer, G. Nauer, A. Neckel, G. Tourillon, F. Garnier, and P. Lang, *J. Phys. Chem.* **88**, 652 (1984).
47. S. I. Yaniger and D. W. Vidrine, *Appl. Spectr.* **40**, 174 (1986).
48. P. A. Christensen, A. Hamnett, A. R. Hillman, M. J. Swann, and S. J. Higgins, *J. Chem. Soc. Farad. Trans.* **88**, 595 (1992).
49. P. A. Christensen, A. Hamnett, A. R. Hillman, M. J. Swann, and S. J. Higgins, *J. Chem. Soc. Farad. Trans.* **89**, 921 (1993).
50. P. A. Christensen, A. Hamnett, and D. C. Read, *Electrochimica Acta* **39**, 187 (1994).
51. Z. Vardeny, J. Tanaka, H. Fujimoto, and M. Tanaka, *Solid State Commun.* **50**, 937 (1984).

8

Osmium and Ruthenium Poly(pyridyl) Redox Polymers as Electrode Coatings

David M. Kelly and Johannes G. Vos

1. INTRODUCTION

The properties of modified electrodes are intricately linked to properties of the modifying layer applied. Of prime importance is of course the redox potential of the electroactive component in the layer. The mediating properties of the modified electrodes and their application as sensors or in other devices, are closely linked to the redox potential of the coating. The second factor to be considered is the structure of the layer and the manner in which this structure is affected by the contacting electrolyte. This last factor strongly influences electrochemical processes which are controlled by factors such as ion and substrate penetration into the layer. The working of the modified electrode is therefore an interplay between the redox properties of the mediator and the structural properties of the medium in which it is encapsulated. In preparing modified electrodes, as much attention should therefore be paid to the physical nature of the coating as to the choice of the redox active species. Layers that contain the same redox couple, but where the matrix is different, can show quite different electrochemical properties. For example, electrodes coated with polymers prepared by electrochemical polymerization can show quite different properties from those prepared from preformed polymers by solvent evaporation.

David M. Kelly and Johannes G. Vos • School of Chemical Sciences, Dublin City University, Dublin 9, Ireland.

Electroactive Polymer Electrochemistry, Part 2: Methods and Applications, edited by Michael E. G. Lyons. Plenum Press, New York, 1996.

A modified electrode is therefore an ideal example of a supramolecular assembly; the two components of this structure are the redox active center and the polymer layer. Both maintain their intrinsic properties, but overall properties of the modifying layer are more than the simple addition of these properties.

For the last number of years, we have been involved in studying electrochemical properties of a series of polymers of the type $[M(bipy)_2(POL)_nCl]Cl$ and $[M(bipy)_2(POL)_n]Cl_2$, where M = Os or Ru, bipy = 2,2′-bipyridyl, POL is poly-4-vinylpyridine (PVP), poly-N-vinylimidazole (PVI), etc.; the subscript n gives the polymer monomer units per electroactive center and is therefore a measure for the redox loading of the polymer. For example, in the material $[Ru(bipy)_2(PVP)_5Cl]Cl$, $n = 5$, but of these five monomer units only one is coordinated to the metal center; the other four are free. For bissubstituted complexes of the type $[M(bipy)_2(POL)_n]Cl_2$, two monomer units are coordinated, and $n - 2$ are free. In Fig. 8.1 the structure of two such polymers is given. These materials are prepared by reaction of the appropriate osmium or ruthenium polypyridyl complex with the preformed polymer, such as PVP or PVI. The resulting metallopolymer is soluble, so coatings can therefore be prepared by applying a known amount of a solution of known concentration with subsequent evaporation of the solvent. With this approach a systematic variation of polymer properties can be carried out, and the effect of a change in loading (n) and of the coordination mode, e.g., (POL)Cl versus $(POL)_2$ can be investigated.

Since preformed polymers often swell in contact with solvents, the electrochemical properties are expected to depend on the interplay between the polymer matrix and the electrolyte. We have investigated this interplay by using a number of different techniques. Investigations of this kind are intrinsically difficult to

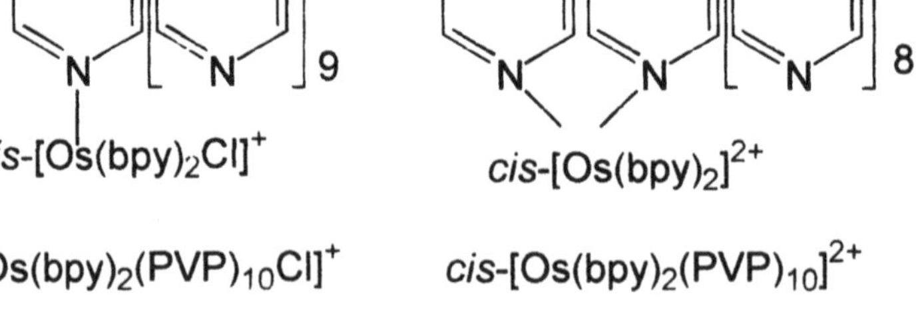

cis-$[Os(bpy)_2(PVP)_{10}Cl]^+$ cis-$[Os(bpy)_2(PVP)_{10}]^{2+}$

FIGURE 8.1. Coordination sphere of the metal center for (a) mono substituted and (b) bis substituted poly-4-vinylpyridine metallopolymers.

interpret because it is not possible to change just a single parameter in a particular system. For example, if the electrolyte concentration is changed, simple electrochemical considerations show that the electrochemical response is affected; however, a change in electrolyte can also change the structure of the polymer, which may itself affect ion availability within the film. Similarly, a change in the loading of the polymer changes the distance between electroactive centers but also affects the polymer structure. All trends observed when such variations are made should therefore be interpreted with care.

In this chapter the synthetic aspects of the earlier mentioned $[M(bipy)_2(PVP)_nCl]^+$ polymers (where M = Os,Ru) are discussed. The main part of the chapter is devoted to the effect of electrolyte and polymer loading on the electrochemistry observed at electrodes modified with these materials. Interaction between the polymer layer and the electrolyte is investigated using electrochemical techniques such as cyclic voltammetry, potential step methods, and the electrochemical quartz crystal microbalance. Attention is also paid to mediation reactions using such modified electrodes. Finally, the implications of these observations for analytical applications of these materials are discussed.

2. SYNTHESIS AND CHARACTERIZATION

The attachment of osmium and ruthenium polypyridyl complexes to the electrode surface using polymer matrices can be achieved by electrostatically incorporating charged complexes into cationic/anionic polymer films.[1–11] Alternatively the reductive polymerization of vinyl-containing metal complex monomers can lead to the formation of films of varying thickness,[12–27] but extensive crosslinking is a feature of materials formed in this way. An important feature of all these approaches is that it enables redox and excited-state behavior of the monomeric analogs to be carried over to polymeric materials.[28] Preparing metallopolymers through covalent attachment to preformed polymers offers many advantages; these include synthetic control, reproducibility, and a high-level of structural control. In this overview we concentrate on metallopolymers whose redox centers are covalently bound to the polymer backbone.

The synthesis of metallopolymers containing ruthenium and/or osmium compounds by covalent attachment of the metal centers to a polymer backbone is based on the different lability of the chloride ions in the complex cis-$[M(bipy)_2Cl_2]$.[5,29–35] Removal of the first chloride ion occurs readily by refluxing in a low boiling point alcohol, such as methanol or ethanol. Removal of the second chloride requires aqueous solvent mixtures and/or higher temperatures. Consequently for the synthesis of metallopolymers of the type $[M(bipy)_2(POL)_nCl]Cl$, heating at reflux in ethanol is sufficient, whereas for the bis-substituted materials, i.e., $[M(bipy)_2$

$(POL)_n]Cl_2$, ethanol/water mixtures are used. A general reaction scheme can be summarized as in Reactions 1–4:

$$[M(bipy)_2Cl_2] + S \rightarrow [M(bipy)_2(S)(Cl)]^+ + Cl^- \quad (1)$$

$$[M(bipy)_2(S)(Cl)]^+ + nPOL \rightarrow [M(bipy)_2(POL)_n(Cl)]^+ + S \quad (2)$$

$$[M(bipy)_2(POL)_n(Cl)]^+ + H_2O \rightarrow [M(bipy)_2(POL)_n(H_2O)]^{2+} + Cl^- \quad (3)$$

$$[M(bipy)_2(POL)_n(H_2O)]^{2+} \rightarrow [M(bipy)_2(POL)_n]^{2+} + H_2O \quad (4)$$

where S is solvent, M = Os or Ru.

When these metal complexes are reacted with a preformed polymer, crosslinking during synthesis is not observed, and soluble polymers are obtained. The fact that metallopolymers containing bis-substituted complexes are also soluble suggests that intrachain rather than interchain coordination is taking place. Since these materials are soluble, they can be characterized in solution and their application as electrode modifiers is facilitated. For the synthesis of bis-substituted materials, complexes cis-$[M(bipy)_2(H_2O)_2]^{2+}$[30,36,37] or their precursors $[M(bipy)_2CO_3]$[18,33] can be used as an alternative starting material.[29,38,39] Furthermore, in this approach the nature of the backbone of the metallopolymers can be changed easily, so that the effect of polymer properties on the electrochemical behavior of a particular redox-active group can be investigated. Ruthenium polymers of the type $[Ru(bipy)_2(PVP)_nL]^{x+}$, where L = N_3^-, NO_2^-, OH^-, CN^- ($x = 1$), L = CH_3CN ($x = 2$), and L = NO^+ ($x = 3$), have also been prepared using thermal methods. The aquopolymer can be generated photochemically from the (POL)Cl and $(POL)_2$ ruthenium polymers.[29,38–42]

This synthetic method offers considerable flexibility. For example, metallopolymers with different metal loadings can be prepared by varying the relative molar amounts of the reactants.[29,30,36,43] In this manner polymer-to-metal loadings ranging from 1:5 ($n = 5$, one metal complex per 5 monomer units) to 1:40 can be easily obtained.[43] The metal loadings reported are based on the ratio of the starting materials, assuming complete reaction.[43] Although more dilute loadings have been synthesized,[29,36] the presence in these materials of both bis- and monosubstituted centers[29] prevents precise evaluation of the metallopolymer composition. This, coupled to their necessarily smaller electrochemical response, makes them unsuitable for detailed electrochemical investigations. Because of the larger ligand field splitting in osmium complexes, they are much less reactive than their ruthenium analogs and longer reaction times are therefore needed.[43]

One of the advantages of soluble electroactive materials is that they can be investigated in great detail. Preformed polymers can be characterized by conventional methods, such as spectroscopy and elemental analysis; molecular weights and glass transition temperatures can also be determined using well-known techniques. Although there is no evidence that molecular weight has a strong influence on the electrochemical properties of the electroactive coating, it has been shown that higher molecular weight materials adhere much better to the electrode surface and enhance the lifetime of the electrode. Thermal analysis of some osmium and ruthenium polymers based on PVP has shown that the glass transition temperature of these metallopolymers is well above 100° C.[(43,44)] This strongly suggests that glass transitions do not have to be taken into account when analyzing the electrochemical behavior of these coatings.

The nature of coordination around the central atom is of prime importance for the redox potential of the electroactive center. During synthesis of the metallopolymers, the reaction is therefore continuously monitored. This is most conveniently carried out using absorption spectroscopy or cyclic voltammetry. Electronic

TABLE 8.1
Spectroscopic and Electrochemical Data for Selected Metallopolymers and Their Mononuclear Analogs

Polymer/Monomer	λ_{max} (nm)[a] Absorbance	λ_{max} (nm)[b] Emission (RT)	$E_{1/2}$[c] (V versus SCE)
$[Ru(bpy)_3]^{2+}$	290, 452	605[b]	1.24
$[Ru(bpy)_2(py)_2]^{2+}$	340, 459	606[b]	1.24
$[Ru(bpy)_2(PVP)_2]^{2+}$	342, 459	609[b]	1.24
$[Ru(bpy)_2(py)(H_2O)]^{2+}$	341, 473	—	0.82
$[Ru(bpy)_2(PVP)(H_2O)^{2+}$	343, 475	—	0.84
$[Ru(bpy)_2(py)Cl]^{+}$	352, 494	702	0.80
$[Ru(bpy)_2(PVP)Cl]^{+}$	353, 494	704	0.76
$[Ru(bpy)_2(mim)_2]^{2+}$	340, 492	640	0.97
$[Ru(bpy)_2(PVI)_2]^{2+}$	345, 485	636	0.86
$[Ru(bpy)_2(mim)(H_2O)]^{2+}$	338, 487	—	1.00
$[Ru(bpy)_2(PVI)(H_2O)]^{2+}$	340, 482	—	1.02
$[Ru(bpy)_2(mim)Cl]^{+}$	352, 512	—	0.66
$[Ru(bpy)_2(PVI)Cl]^{+}$	348, 502	635	0.57
$[Os(bpy)_2(py)_2]^{2+}$	440, 613	769	0.75
$[Os(bpy)_2(PVP)_2]^{2+}$	444, 616	764	0.75
$[Os(bpy)_2(py)Cl]^{+}$	500, 735	—	0.35
$[Os(bpy)_2(PVP)Cl]^{+}$	486, 730	—	0.35
$[Os(bpy)_2(mim)_2]^{2+}$	512, 686	754	0.48
$[Os(bpy)_2(PVI)_2]^{2+}$	496, 666	758	0.48
$[Os(bpy)_2(mim)Cl]^{+}$	365, 430	—	0.23
$[Os(bpy)_2(PVI)Cl]^{+}$	431, 524	—	0.23

[a]In methanol, [b]In acetonitrile, [c]0.1-M TEAP in acetonitrile. (From Refs. 29, 36, 38, 43, 45, 47, 63)
py = pyridine; mim = N-methylimidazole.

spectroscopy has proven useful in characterizing ruthenium polymers. For these materials the visible absorption spectrum is characteristic of a particular coordination mode. By comparing the spectrum with those of mononuclear-model compounds, the coordination sphere of the ruthenium center in polymeric materials can be verified[28–30,36,38,39,43–47] (see Table 8.1). Absorption spectra of osmium metallopolymers, while comparable to mononuclear models,[43] are typically more complicated than those of analogous ruthenium compounds. Figure 8.2 shows some typical absorption spectra. In general the energy of the absorption maxima is unaffected by metal loading. Approximate extinction coefficients, based on mononuclear compounds, have been evaluated,[28,43,45–47] and these are useful in verifying metal loading. Due to complications from different levels of hydration in the homopolymer and the subsequent metallated polymer, extinction coefficients can give only an approximation. Emission spectroscopy has also been used extensively to characterize these materials[28,29,38,39,43,44,47] (see Table 8.1). With this technique more detailed information can often be obtained, since the variation in emission wavelengths and intensity of different species that may be present is much greater than the variation observed in absorption spectra. With emission techniques small amounts of strongly emitting species can therefore be identified, for example the presence of $[Ru(bipy)_2(PVI)_2]^{2+}$ in a mostly $[Ru(bipy)_2(PVI)Cl]$-containing material.[29]

Spectroscopic investigations suggest that the photophysical properties of the osmium and ruthenium centers are carried over relatively unchanged from monomeric analogs to the metallopolymers. This is confirmed by the photochemical reactivity of the metallopolymers. Irradiation of the mononuclear model compounds results in a photochemically induced ligand exchange process via population of an antibonding triplet metal-centered, ^{3}MC, state.[48–50] For ruthenium polymers, the chloride ion or one of the pyridine (or imidazole) groups of the polymer is lost in aqueous acids and replaced by H_2O,[29,39,42] as in Reaction 5:

$$[Ru(bipy)_2(PVP)_nCl]^+ + H_2O \xrightarrow{h\nu} [Ru(bipy)_2(PVP)_n(H_2O)]^{2+} + Cl^- \qquad (5)$$

This can be observed conveniently for thin films of the metallopolymer coated on glassy carbon electrodes by a change in the oxidation potential of the Ru(II)/(III) couple due to the photochemically induced change in the coordination sphere. Figure 8.3 shows an example of the photolysis of an electrode modified with the mixed metal polymer $[Os(bipy)_2(PVP)_{10}Cl]Cl/[Ru(bipy)_2(PVP)_{10}Cl]Cl$. Photolysis of a glassy carbon electrode coated with this polymer results in a shift of the ruthenium redox from about 0.55 V to 0.68 V versus SCE while the osmium-based redox couple at 0.16 V is not affected.[43] Since the ^{3}MC state for the osmium poly(pyridyl) complexes is approximately 30% higher in energy than for ruthenium,[51,52] and the ^{3}MLCT state lower, the energy needed for population of the ^{3}MC is much greater. Osmium complexes are therefore normally photostable.

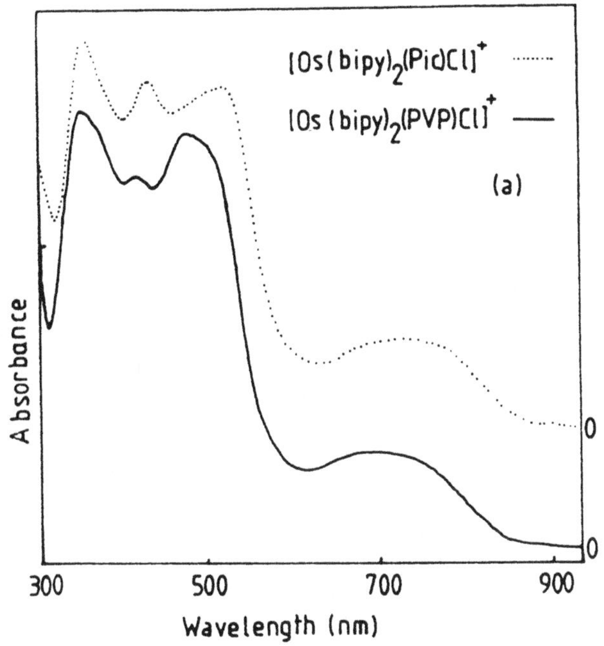

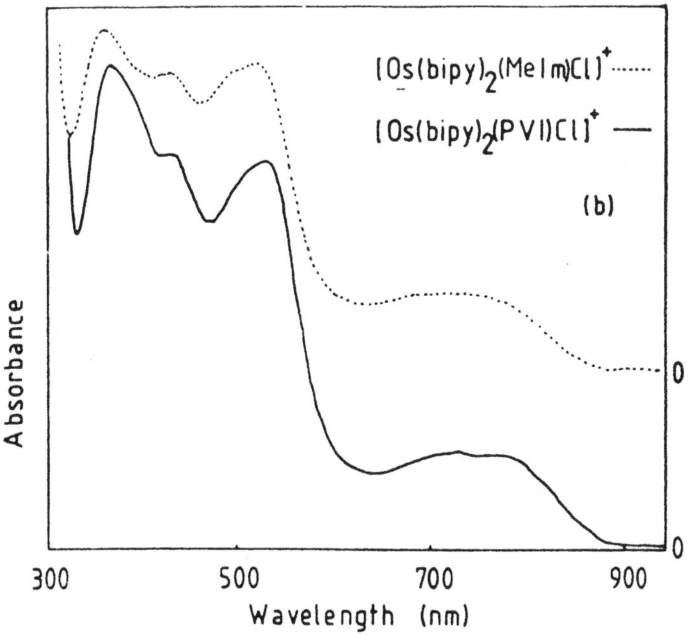

FIGURE 8.2. Ultraviolet/visible spectra of a selection of metallopolymers and model compounds. Spectra of the metallopolymers were recorded in methanol and model compounds in acetonitrile. Sample concentration is 10^{-4} M. Pic = 4-methylpyridine; MeIm = N-methylimidazole. (Reproduced with permission from Ref. 43.)

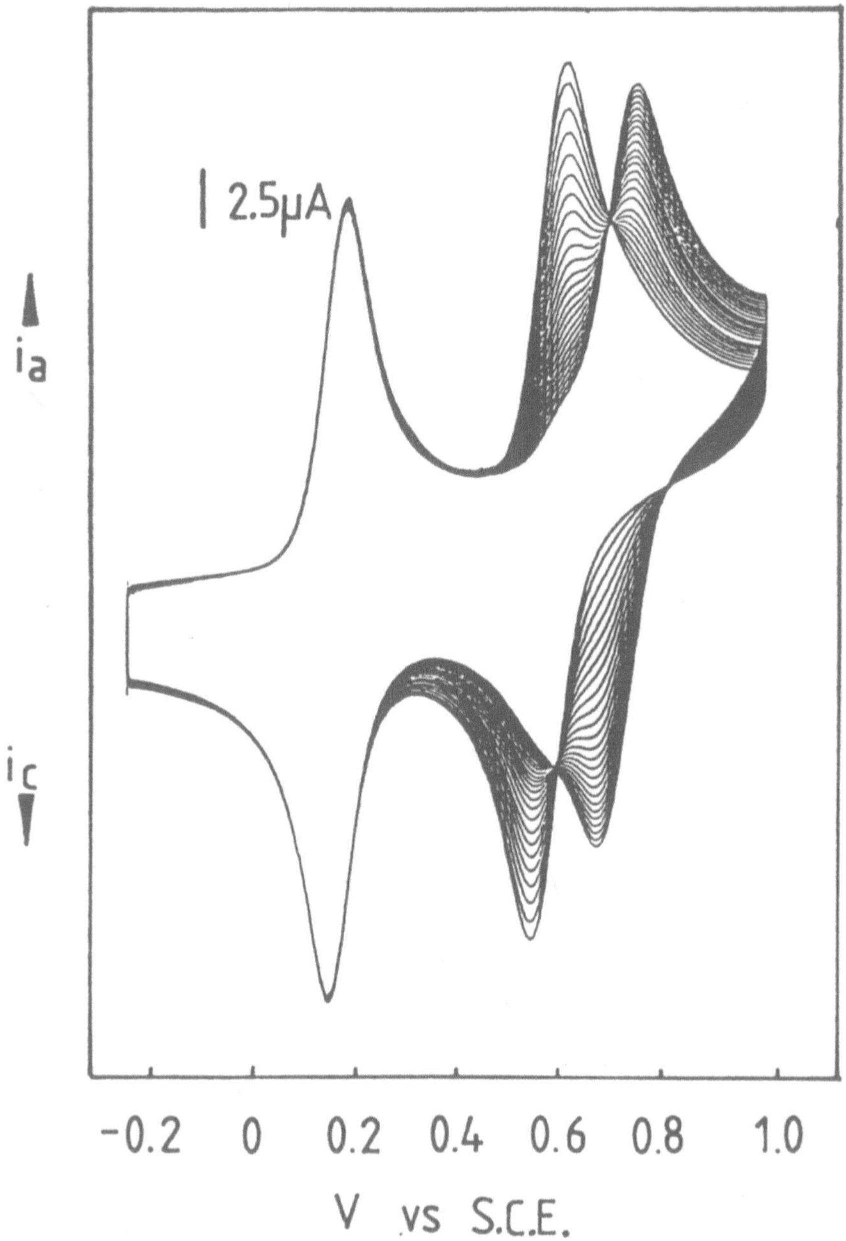

FIGURE 8.3. Photochemically induced ligand exchange reactions in a thin, electrode-bound film of the metallopolymer $[Os(bipy)_2(PVP)_{10}Cl]Cl/[Ru(bipy)_2(PVP)_{10}Cl]Cl$ using 1.0 M aqueous $HClO_4$ as an electrolyte; scan rate: 100 mV s^{-1}, surface coverage 1×10^{-8} mol cm^{-2}. The interval between scans is 30 s. (From Ref. 43.)

Upon photolysis of the ruthenium polymers in CH_3CN-based electrolytes, this solvent is the incoming ligand. In CH_3CN the counterion ClO_4^- can also bind under certain conditions.(41) For the analogous PVI polymer, similar reactions are observed. Interestingly, on photolysis of $[Ru(bipy)_2(PVI)_5Cl]Cl$ in sulphuric acid, an equilibrium is established between the aquo and a sulphato complex. The latter species converts fully reversible into the aquo species in the dark.(29)

The most important characterization for these metallopolymers is the investigation of their electrochemical behavior. The ruthenium- and osmium-containing polymers show a metal-based M(II)/(III) oxidation potential and bipy-based reduction potentials. Since the later are found at quite negative potentials, they have very limited use and are therefore not included in this discussion. The formal potential of the M(II)/(III) oxidation in these metallopolymers is very sensitive to the type of complex obtained.(28–30,38–46,53–62) The nature of the polymer bound complex can therefore be identified by comparing the redox potential with those of analogous polymeric and monomeric compounds (see Table 8.1). Typically oxidation of the Os(II)/(III) is easier by 0.3–0.5 V compared to the analogous Ru(II)/(III) oxidation.(31,35,63) The Os(III)/(IV) oxidation is also at a much lower potential, and may be observed.(43) For ruthenium complexes however, the separation in potential of the Ru(II)/(III) and Ru(III)/(IV) couples is greater, and the Ru(III)/(IV) oxidation is rarely observed.(64,65) These differences in redox behavior can be explained in terms of differences between a second-row transition metal, ruthenium, and a third-row transition metal, osmium, and differences in the nature of the metal–ligand bond. The nature of the coordination sphere surrounding the metal also determines the level of electron density at the metal center.(66–77) The presence of a chloride ligand, an electron donor in monosubstituted materials, increases electron density around the metal center and so increases the ease of metal oxidation compared to the bis-substituted polymers.(29,42,43) Being a better σ-donor than PVP, PVI also increases electron density in the metal *d*-orbitals. Metallopolymers using PVI backbones therefore have lower M(II)/(III) oxidation potentials than PVP counterparts(29,39,43) (see Table 8.1).

Figure 8.4 shows the importance of the coordination mode around the metal ion for the electrochemical properties of the layer. In Fig. 8.4 the mediation of the Fe(II)/(III) oxidation is studied by using a rotating disk electrode. Initially a thin film of $[Ru(bipy)_2(PVP)_5Cl]^+$ is used and with this coating the current potential curve I is obtained (see Fig. 8.4b). On photolysis of the coating and formation of the aquocomplex (according to Reaction 5); curve II is obtained. Rotating disk behavior very clearly shows that the redox potential of the modifying layer is of prime importance to the electrochemical properties of the modified electrode.

As already mentioned, higher oxidation states of ruthenium are rarely observed in these systems. However, the presence of the aquoligand in $[Ru(bipy)_2(PVP)_nH_2O]^{2+}$ leads to additional redox behavior, and the Ru(III)/(IV) couple is seen (Fig. 8.5).(29,38–42,53) The peak height of this couple depends on the cyclic

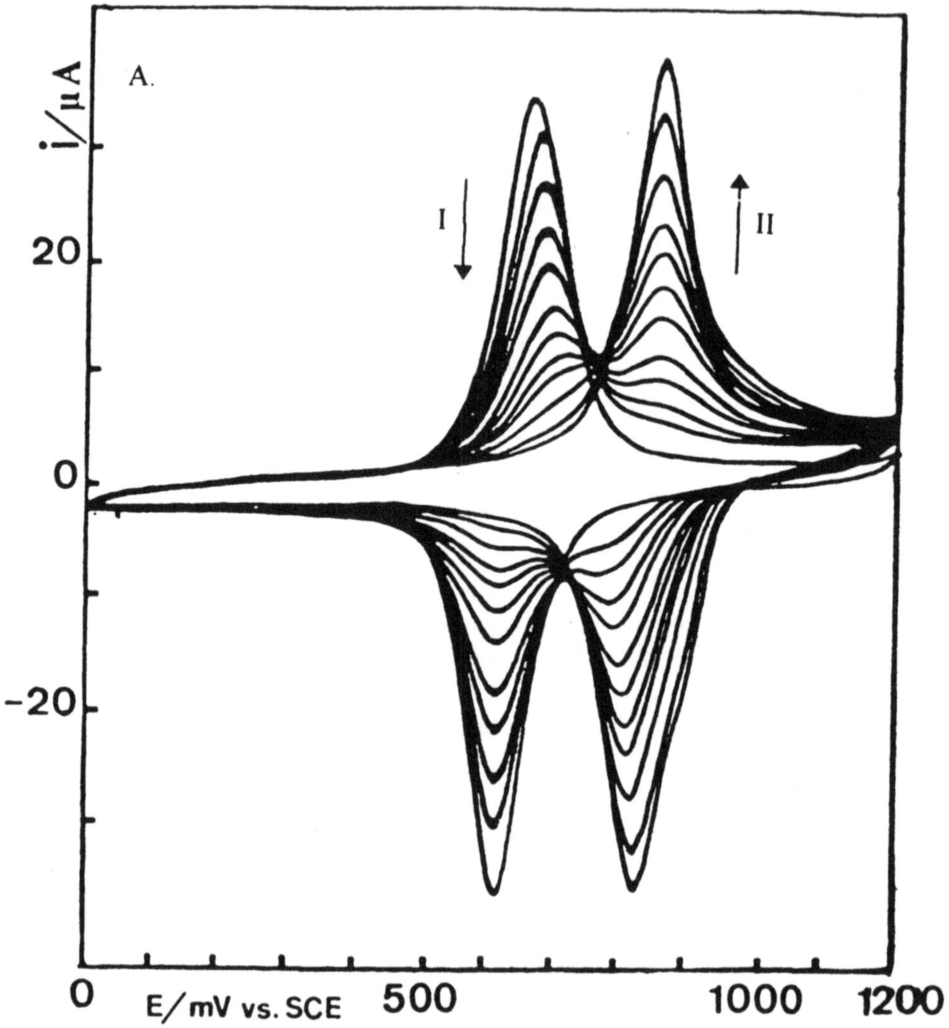

FIGURE 8.4. (a) Photochemically induced ligand exchange reaction in a thin film of $[Ru(bipy)_2(PVP)Cl]^+$; electrolyte: 1.0 mol dm^{-3} $HClO_4$, scan rate: 100 mV s^{-1}, surface coverage: 1×10^{-8} mol cm^{-2}. (b) Current voltage curves of a 5×10^{-3} M aqueous solution of Fe(II) (I) before photosubstitution and (II) after photosubstitution, rotation speed: 16.6 Hz, scan rate: 2 mV s^{-1}.

voltammetric scan rate, and it can be clearly discerned only from the Ru(II)/(III) couple at slow sweep rates.[38,42] In accordance with the mechanism proposed for mononuclear analogs,[78–81] Ru(III)/(IV) oxidation in the metallopolymer is explained by an EC electrochemical process. The Ru(III) centers disproportionate to give Ru(II) and Ru(IV), with Ru(II) centers then being reoxidized[38] (see Eq. 6 and 7).

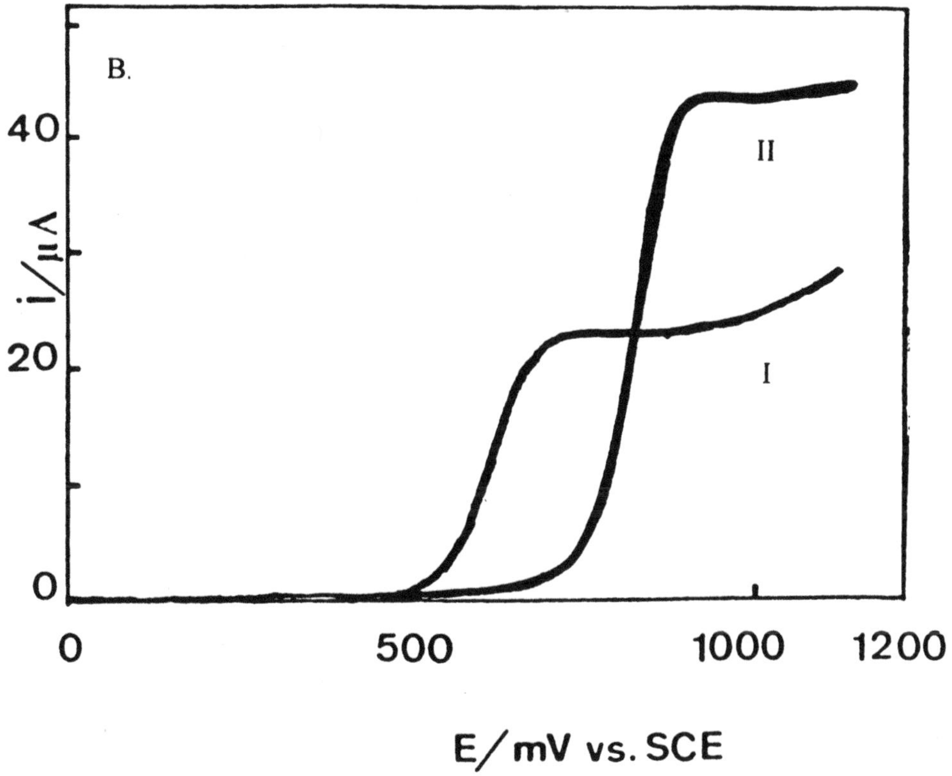

FIGURE 8.4. *(Continued)*

At pHs below the pK_a of the coordinated aquo ligand in the Ru(III) complex the redox behavior is

$$[Ru^{II}(bipy)_2(PVP)_n(H_2O)]^{2+} \rightarrow [Ru^{III}(bipy)_2(PVP)_n(H_2O)]^{3+} + e^- \qquad (6)$$

$$2[Ru^{III}(bipy)_2(PVP)_n(H_2O)]^{3+} \rightarrow [Ru^{II}(bipy)_2(PVP)_n(H_2O)]^{2+} + [Ru^{IV}(=0)(bipy)_2(PVP)_n]^{2+} + 2H^+ \qquad (7)$$

Because of coupled electron and proton exchange in these processes, reduction and oxidation potentials of Ru(II)/(III) and Ru(III)/(IV) couples are pH-dependent.[38,42,53]

Analysis of thin layers of the metallopolymers was carried out using normal surface analysis techniques and both scanning tunneling spectroscopy (STM)[82] and electron spectroscopy for chemical analysis (ESCA)[28] have been used to

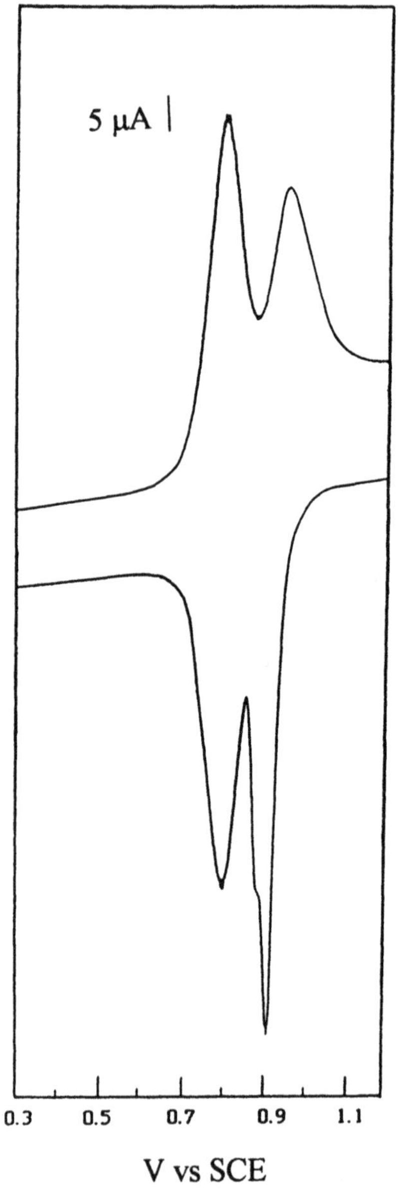

FIGURE 8.5. Slow sweep rate cyclic voltammogram of $[Ru(bipy)_2(PVP)(H_2O)]^{2+}$; electrolyte: 1.0 M $HClO_4$, scan rate: 5 mV s^{-1}, surface coverage: *ca.* 2×10^{-8} mol cm^{-2}.

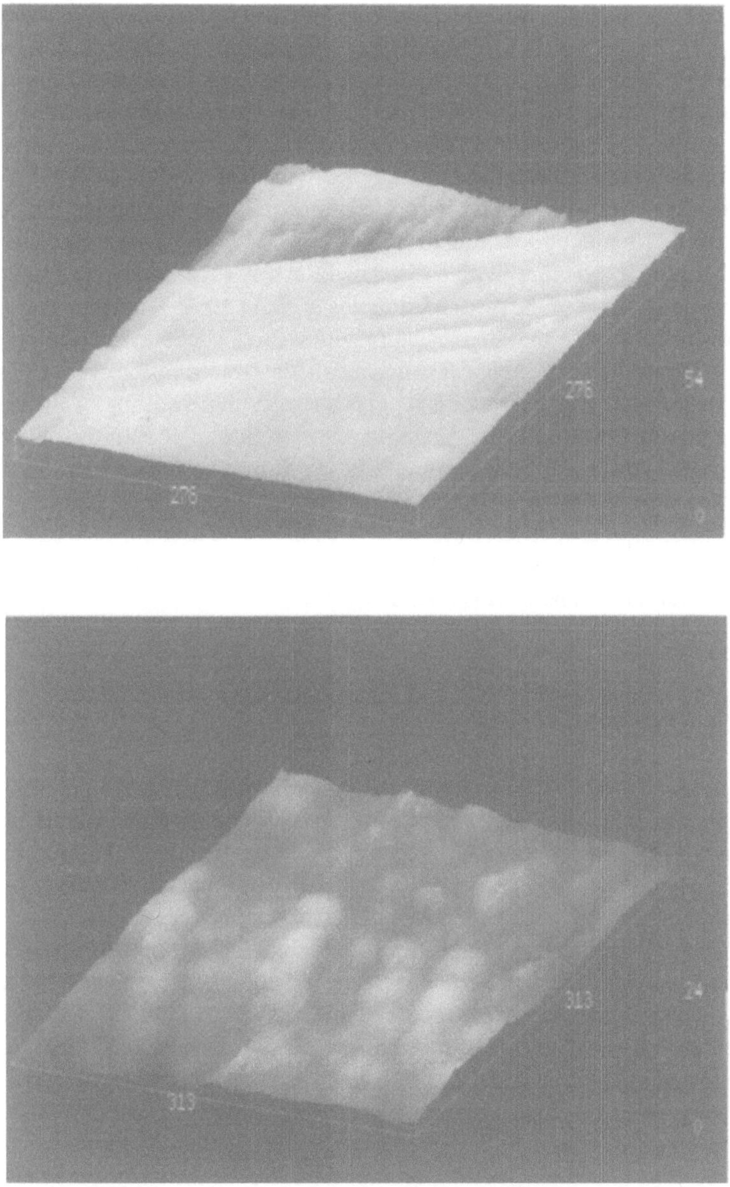

a

b

FIGURE 8.6. The STM analysis of the polymer $[Os(bipy)_2(PVP)_{10}Cl]Cl$ (a) before electrochemical cycling and (b) after electrochemical cycling in 0.4 M H_2SO_4. (From Ref. 82.)

investigate the surface structure. The STM analysis of the polymer $[Os(bipy)_2(PVP)_{10}Cl]Cl$ (see Fig. 8.6) showed that the surface of coatings does change considerably after taking electrochemical measurements in sulphuric acid. Figure 8.6(a) shows that before electrochemical cycling, there are straight fibrils on the surface of the coating about 20 Å from each other. There are also a series of steps with an edge-to-edge distance of 20 Å and with step heights of 8–9 Å. These distances agree with dimensions expected for a polymer containing osmium bis (bipy) units. After electrochemical cycling in 0.4-M H_2SO_4, the surface structure adopts a granular structure, with a distance of about 40 Å between granules. The side-by-side distance of the rows of granules is about 40 Å. Further magnification shows the presence of a fine step structure with smooth edges and step heights from 30–60 Å. Patterns observed appear to depend on the concentration of the electrolyte used. Changes in surface structure suggest strongly that swelling of the polymer occurs, and after electrochemical cycling, the structure of at least the outside part of the layer has changed. This shows that these coatings have clear memory effects. If one therefore wishes to investigate the effect of electrolyte concentration, etc., on electrochemical properties, fresh electrodes should be used for each experiment. More evidence of the importance of memory effects is given in the electrochemical quartz crystal microbalance (EQCM) discussion.

3. HOMOGENEOUS AND HETEROGENEOUS CHARGE TRANSPORT PROPERTIES

A feature of modified electrodes that has received particular attention is the mechanism and rate-of-charge propagation through the modifying layer, because the dynamics of this process have a critical effect on the practical applicability of modified electrodes. It is generally accepted that the mode of charge transport through redox polymers is electron-hopping between adjacent redox sites.[83–87] It has however become increasingly apparent, that for certain systems, the rate determining step for charge transport is not necessarily the aforementioned self-exchange mechanism. In many cases charge transport through modifying layers has been found to be limited by ion transport processes, such as permeation and the availability of counterions, and polymer chain movement required for orientation of the redox sites into close proximity for electron hopping to occur.[83,84]

In a series of papers,[57–62] we examined the effects of redox site loading, electrolyte type and concentration, and temperature on the homogeneous charge transport process and the heterogeneous electron transfer reaction for electrodes modified with thin films of the type $[Os(bipy)_2(PVP)_nCl]Cl$ and $[Os(bipy)_2(PVI)_{10}](Cl)_2$. This section presents results of the investigation of the $M^{II/III}$ oxidation in PVP-based metallopolymers.[58–62] The profound effect of the aforementioned variables on rate-limiting processes of homogeneous charge transport is highlighted. What has emerged from these studies is the clear dependence the charge transport properties of these modifying materials on physicochemical properties of the polymer phase and on the interplay between the polymer phase and the contacting electrolyte. In this chapter we concentrate on the interplay between PVP backbone and electrolyte in perchloric and sulphuric acid electrolytes, although other electrolytes have been examined. In sulphuric acid it is anticipated that protonation of the polymer backbone (pK_a of PVP is *ca.* 3.3)[88] is likely to result in extensive film swelling.[58,59] Perchloric acid provides an ideal comparison because of the insolubility of the perchlorate salts of the metallopolymer. Due to the reported crosslinking properties of the perchlorate anion via the formation of ion pair associations,[89] a rigid nonswollen matrix is expected.[90–92]

To determine charge transport parameters of the modified electrodes under investigation, both short time scale potential-step methods, such as chronoamperometry, chronocoulometry, and sampled current voltammetry, and longer time scale cyclic voltammetry techniques have been applied. This enables one to investigate the influence of the time domain used in the measurement on charge transport parameters observed. This is of particular interest, since it was recently suggested[59–61,93] that the time scale needed to reach equilibrium after a redox process is not necessarily the same for all species involved in the electrochemical reaction. Of further interest is the heterogeneous charge transport process between the polymer coating and the underlying electrode. This has been investigated using sampled current voltammetry.

3.1. Theoretical Framework

3.1.1. Homogeneous Charge Transport

Cyclic voltammetry (CV) has found widespread application in investigating and characterizing modified electrode processes.[94] Characterizing modifying layers under conditions of thin layer behavior has received particular attention.[95] In the absence of diffusional limitations and under conditions of complete oxidation/reduction of electroactive centers, thin layer/surface-type behavior prevails. The ideal model for voltammetric behavior under such conditions was considered, and the following features are characteristic[96,97]:

$$i_{(p)} = \frac{n^2 \mathrm{F}^2 A \Gamma v}{4 \mathrm{R} T} \tag{8}$$

$$E_{pa} = E_{pc} \tag{9}$$

$$FWHH = \frac{90.6}{n} \mathrm{mV \ at\ 25° \ C} \tag{10}$$

where $i_{(p)}$ is the peak current; n the number of electrons passed; F the Faraday constant; Γ the total amount of electroactive material on the electrode surface; A the electrode v the scan rate; R the gas constant; T the temperature; E_{pa} and E_{pc} the anodic and cathodic peak potentials, respectively; and *FWHH* the full peak width at half-height. The compounds under investigation typically exhibit surface behavior at scan rates less than 5 mV/sec.[57,59–61] Figure 8.7 highlights this behavior, where the anodic and cathodic branches converge to almost zero peak-to-peak separation and the ratio of anodic to cathodic peak current is 1. Under these conditions of exhaustive oxidation/reduction of the modifying layer, the total amount of electroactive material on the surface, Γ, can be determined from the expression:

$$\Gamma = \frac{Q}{n\text{F}A} \tag{11}$$

where Q is the total amount of charge passed, which may be evaluated by integration of the area under the oxidation or reduction peak.[44,58,59,62,94] Surface coverages estimated in this way ranged from 5×10^{-10} to 5×10^{-8} mol/cm^2.

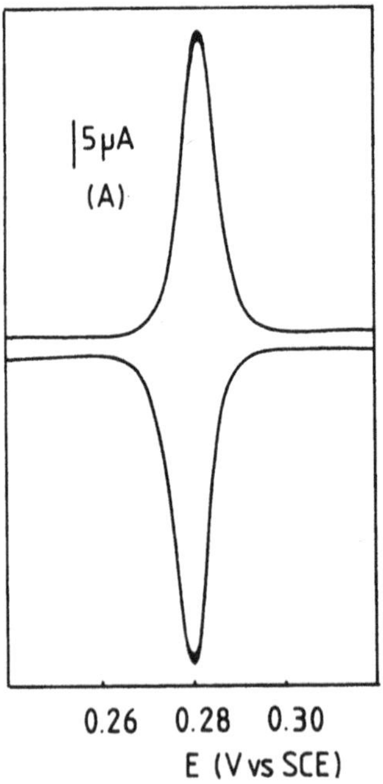

FIGURE 8.7. Cyclic voltammetry behavior of $[Os(bipy)_2(PVP)_{25}Cl]Cl$ in 1.0 M NaCl; sweep rate: 5 mV/s, surface coverage: 1.8×10^{-8} mol cm^{-2}. (From. Ref. 60.)

At increased scan rates, a transition is observed from the thin layer regime, where the peak current varies linearly with scan rate, to semi-infinite diffusion control, where a $v^{1/2}$ dependence is observed. Under semi-infinite diffusion conditions, voltammograms are characteristic of those obtained for solution species. The general voltammogram characteristics are

$$E_{pa} - E_{pc} = \frac{0.0592}{n} \text{V at } 25° \text{ C} \tag{12}$$

$$\frac{i_{pa}}{i_{pc}} = 1 \tag{13}$$

where i_{pa} and i_{pc} are the anodic and cathodic peak currents, respectively.

The peak current under these conditions for a reversible electrode reaction is given by the Randles–Sevčik equation[98]:

$$i_p = \frac{[0.4463 n^{3/2} \mathrm{F}^{3/2} A (D_{CT} v)^{1/2} C]}{(\mathrm{R}T)^{1/2}} \tag{14}$$

where D_{CT} is the apparent charge transport diffusion coefficient, A the area of electrode and C the concentration of electroactive species; all other parameters were defined previously. For the work reported here, the evaluation of charge transport rates using cyclic voltammetry was carried out at high sweep rates, typically 50–500 mV/s under semi-infinite conditions.[58–61]

An alternative approach combining both surface and semi-infinite diffusional behavior was proposed by Aoki.[99] In this theoretical model peak current is given by:

$$i_p = 0.446 n\mathrm{F}A[D_{CT}C/L]W^{1/2}\tanh Y \tag{15}$$

where

$$W = n\mathrm{F}L^2 v / D_{CT}RT \tag{16}$$

$$Y = 0.56W^{1/2} + 0.05W \tag{17}$$

Therefore, from a knowledge of surface coverage and peak current, diffusion coefficients can be obtained over a large range of sweep rates. For solution-type behavior, i.e., when W is large, this equation reduces to the semi-infinite diffusion Randles–Sevčik equation.

Values determined for diffusion coefficients/charge transport rates using both cyclic voltammetry techniques were essentially the same (see Fig. 8.8), and these are denoted as $D_{CT}(CV)$. These charge transport rates typically ranged from 10^{-11}–10^{-10} $\mathrm{cm^2/s}$[58–61] (see Tables 8.2 and 8.3).

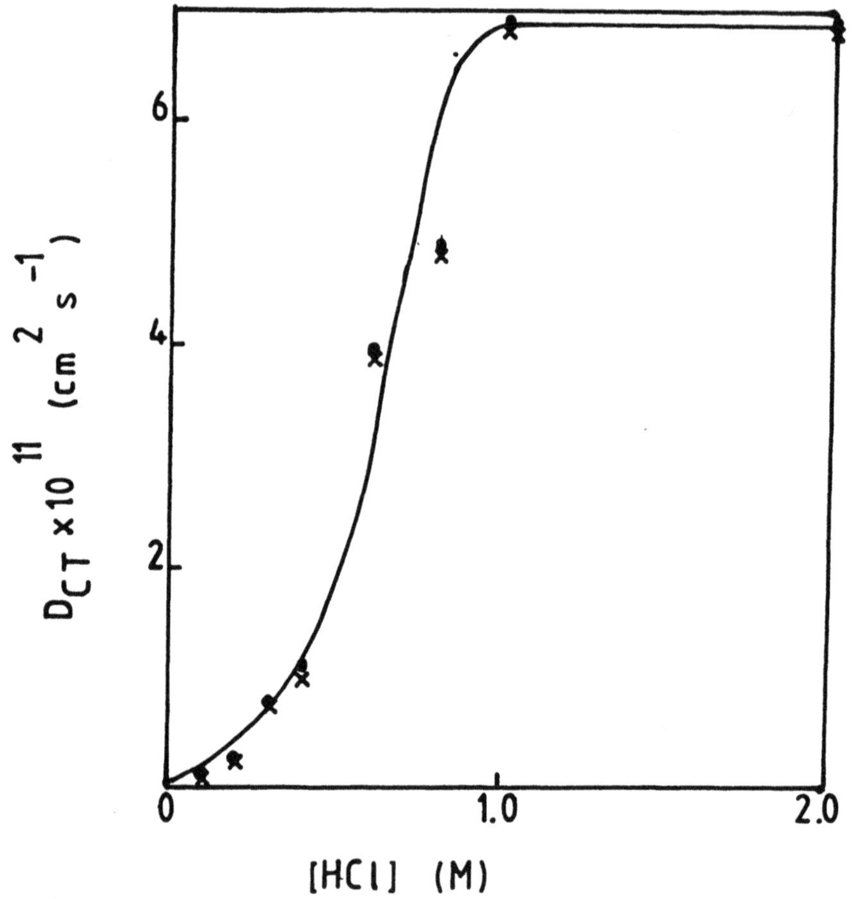

FIGURE 8.8. The effect of HCl concentration on the rate of charge transport measured as D_{CT} through films of $[Os(bipy)_2(PVP)_{10}Cl]Cl$, as measured by cyclic voltammetry using the Randles–Sevčik (**o**) and Aoki (**x**) equations. (From Ref. 58.)

Potential-step chronoamperometry (CA), chronocoulometry (CC), and sampled current voltammetry were used to probe the nature of the charge transport process close to the electrode surface. By manipulating the time scale of the current decay response at sufficiently short times, semi-infinite diffusion is observed. The resulting current response conforms to the Cottrell equation[100]:

$$i(t) = \frac{nFAD_{CT}^{1/2}C}{\pi^{1/2}t^{1/2}} \tag{18}$$

Linear Cottrell plots with zero intercepts were obtained for CA experiments for times up to 20 msec; the slopes of which were used to evaluate the rate of

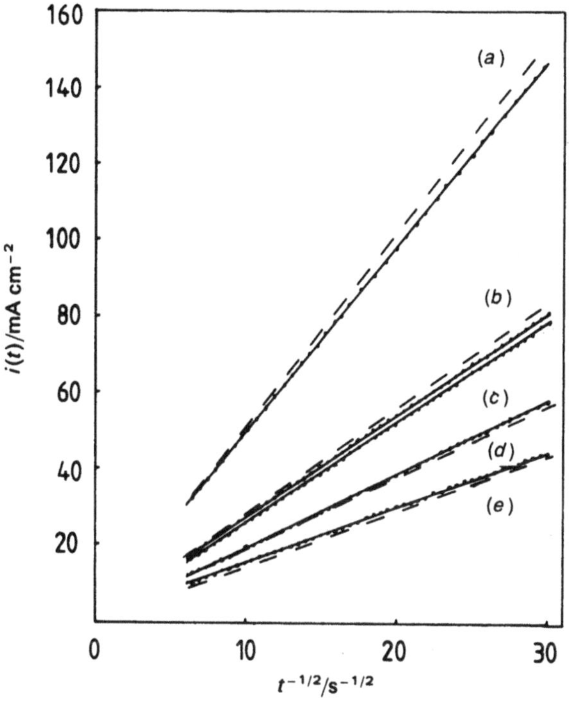

FIGURE 8.9. The effect of redox site loading on the chronoamperometric response for $[Os(bipy)_2(PVP)_nCl]Cl$ films on glassy carbon electrodes presented as Cottrell plots. Oxidation—and reduction - - - current response; 1.0 M HCl as supporting electrolyte; (a) $n = 10$, (b) $n = 15$, (c) $n = 5$, (d) $n = 20$, (e) $n = 25$. (From Ref. 61.)

homogeneous charge transport, which is denoted as $D_{CT}(PS)$. Figure 8.9 highlights typical Cottrell behavior observed for these polymers, $[Os(bipy)_2(PVP)_nCl]Cl$, where $25 \geq n \geq 5$. Charge transport rates determined by short time-scale techniques ranged from 10^{-9}–10^{-8} cm^2/s[58–61] (see Tables 8.2 and 8.3).

Accurately determining charge transport rates requires a diffusional response and the absence of migrational effects. Perturbation of the diffusion current by other components is indicated by a nonzero intercept of the short-time portion of the Cottrell plot[101,102] or by a peak in the $it^{1/2}$ versus $t^{1/2}$ plot.[103] Such features were not apparent in our analysis. The effect of redox site loading on the rate of homogeneous charge transport can also reveal migration. Andrieux and Saveant[101] demonstrated that in the presence of migration, the dependence of D_{CT} on redox site loading goes beyond simple proportionality, eventually reaching a cubic dependence when the mobility of counterions is much lower than the mobility of electrons. Such a relationship does not arise in the analysis presented here.[61]

TABLE 8.2
Effects of Redox Site Loading and Sulphuric Acid Concentration on the Charge Transport Properties of $[Os(bipy)_2(PVP)_nCl]Cl$ Films

Metal Loading	Concentration M	D_{ct}(PS) × 10^{10} cm^2/s	E_a(PS) kJ/mol	$\Delta S^{\neq}$(PS) $Jmol^{-1}K^{-1}$	D_{ct}(CV) × 10^{11} cm^2/s	E_a(CV) kJ/mol	$\Delta S^{\neq}$(CV) $Jmol^{-1}K^{-1}$
$n = 5$	0.1	13	23	–82	32	20	–103
	1.0	12	8	–133	108	25	23
$n = 10$	0.1	30	24	–83	25	11	–147
	1.0	92	6	–134	167	116	221
$n = 15$	0.1	89	15	–111	17	21	–123
	1.0	620	39	–14	78	90	121
$n = 20$	0.1	63	14	–122	21	22	–123
	1.0	301	35	–38	75	80	82
$n = 25$	0.1	38	15	–126	210	65	36
	1.0	85	35	–53	116	75	65

(From Refs. 58, 59).
All parameters are as defined in the text.

TABLE 8.3
Effects of Redox Site Loading and Perchloric Acid Concentration on the Charge Transport Properties of $[Os(bipy)_2(PVP)_nCl]Cl$ Films

Metal Loading	Concentration M	D_{ct}(PS) × 10^{10} cm^2/s	E_a(PS) kJ/mol	$\Delta S^{\neq}$(PS) $Jmol^{-1}K^{-1}$	D_{ct}(CV) × 10^{11} cm^2/s	E_a(CV) kJ/mol	$\Delta S^{\neq}$(CV) $Jmol^{-1}K^{-1}$
$n = 5$	0.1	9	16	–108	26	49	–9
	1.0	14	10	–123	9	42	–39
$n = 10$	0.1	37	15	–113	5		
T > 285 K						224	548
T < 285 K						35	–106
	1.0	67			0.5	92	92
T > 285 K			134	293			
T < 285 K			51	–38			
$n = 15$	0.1	10	41	–42	11	163	333
	1.0	25	44	–23	5	151	295
$n = 20$	0.1	10	44	–36	11	160	320
	1.0	21	39	–46	4	154	291
$n = 25$	0.1	8	43	–47	12	158	304
	1.0	16	35	–65	4	157	298

(From Refs. 58, 61).
All parameters are as defined in the text.

Furthermore the reduction/oxidation current ratio is also expected to be nonunity in the presence of migration.[104] In all cases the i_{pa}/i_{pc} ratio is unity, and identical current responses were obtained for fully oxidized and reduced films.

Evaluating activation parameters associated with the homogeneous charge transport of process can provide valuable information about rate-limiting processes of the charge transport mechanism.[55,83,84,94,105,106] Activation energies were evaluated by measuring the charge transport rate over the temperature range 278–308 K in conjunction with the following equation[57–61]:

$$D_{CT} = D^0_{CT} \exp\left(\frac{-E_a}{RT}\right) \tag{19}$$

where D^0_{CT} is a preexponential factor, E_a is activation energy, and T is the absolute temperature. The magnitude of activation energies can aid diagnosis of the rate-determining step of the charge transport process,[94,106] i.e., electron hopping (about 5–10 kJ/mol), counterion motion (10–50 kJ/mol), and polymer chain motion (60–120 kJ/mol). The activation enthalpy can be evaluated from the following expression:

$$\Delta H^{\neq} = E_a - RT \tag{20}$$

The associated activation entropy can be determined from the Eyring equation[107]:

$$D^0_{CT} = e\delta^2\left(\frac{K_b T}{h}\right)\exp\left(\Delta S^{\neq}/R\right) \tag{21}$$

where e is the base of the natural log, K_b the Boltzman constant, h is Planck's constant, and δ is the equilibrium separation of vicinal redox sites. A separation of 2.5 Å between adjacent monomers was estimated for the poly(4-vinylpyridine) backbone using crystal structure data.[108,109] For the work presented here, a rigid-rod structure of the metallopolymer backbone is assumed; e.g., osmium centers are assumed to be separated by 25 Å when $n = 10$. The accuracy of these activation parameters is somewhat limited, given that intersite separation, concentration of osmium sites, and the charge transport rate are likely to be temperature- and solvent-dependent.[60] Despite this limitation, the trends observed should remain valid, and rate-limiting cases of ion diffusion and polymer chain movement can be distinguished.[110] In the particular case of entropy terms, variation in intersite separation arising from polymer swelling, does not change the sign of the entropy term.[59–61] Estimated entropies can therefore be assigned to disordering polymer chain motions in the case of positive entropies[83,55] and to ion motion or electron hopping in the case of negative entropies.[84]

3.1.2. Heterogeneous Charge Transport

Sampled current voltammetry (SCV) involves applying a potential-step wave form of increasing amplitude, covering a potential window where initially no redox reaction occurs, to one where the current response is diffusion-controlled. The current decay from the exciting pulses is then recorded instantaneously at several different times. In the work under discussion, plots of $i_{\lim}$ versus $t^{-1/2}$ were linear in accordance with the Cottrell equation.(59,111) Charge transport rates evaluated using this technique were essentially the same, within experimental error, as those evaluated using chronamperometry.

SCV was used to study heterogeneous processes between the electrode surface and modifying layers. Figure 8.10 shows a typical sampled current voltammogram for Os(II)/(III) oxidation. These sigmoidal-shaped waves are similar to those observed at unmodified electrodes for solution-phase reactants.(112) The increase in anodic currents with increasing sampling times and at the same time, the shift in half-wave potentials in voltammograms for the oxidation process to more positive potentials with decreasing sampling time is accepted as being indicative of Butler–Volmer kinetics.(113)

An expression derived by Matsuda *et al.*(114) for the current–potential relationship for SCV for the simple electrode process: red = ox + ne,$^-$ allows us to evaluate the standard heterogeneous rate constant k^0 and the transfer coefficient α:

$$E = E^* \pm \frac{RT}{\alpha nF} \ln \left(X \left\{ \frac{1.75 + X^2[1 + \exp(\pm \xi)]^2}{1 - X[1 + \exp(\pm \xi)]} \right\}^{1/2} \right) \tag{22}$$

$$E^* = E^0 \pm \frac{RT}{\alpha nF} \ln \left[\frac{4k^0(\tau)^{1/2}}{(3)^{1/2}(D_{CT})^{1/2}} \right] \tag{23}$$

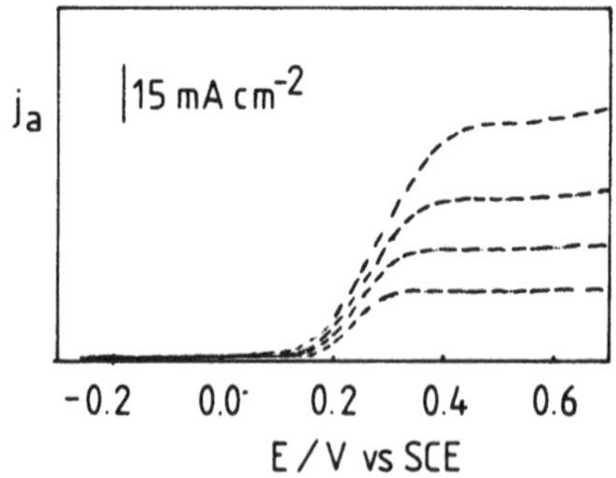

FIGURE 8.10. Typical sampled current voltammogram for the Os(II/III) oxidation within $[Os(bipy)_2(PVP)_{10}Cl]Cl$ films; electrolyte: 0.1 M H_2SO_4; sampling times (*top to bottom*). 1, 2, 4, and 10 ms; surface coverage: 2.1×10^{-8} mol cm^{-2}. (From Ref. 59.)

where R is the gas constant, E is the electrode potential, E^0 the reversible half-wave potential, τ is sampling time, ξ is a dimensionless parameter expressed as $[(nF/RT)(E-E^0)]$, X is the ratio of the current at the potential E to the anodic limiting current {Cottrell current}; all other symbols are defined previously. A plot of the right-hand side of the equation against E is linear, with a slope of $(\alpha nF/RT)$ and intercept E^*. Thus k^0 and α can be evaluated by observing that $D_{CT}(PS)^a = D_{CT}(PS)^c$ (57,59,60,62) and using the value of the formal potential obtained from cyclic voltammetry.

Temperature effects on k^0 were examined to evaluate enthalpic and entropic parameters for the heterogeneous electron transfer reaction. The effect of temperature on the standard heterogeneous rate constant is considered to obey the following relationship(115):

$$\log(k^0) = \log\left\{Z_{el}\,\mathrm{kp}\,\exp\left[\frac{(\Delta S\neq{}^0)_{\mathrm{real}}}{\mathrm{R}}\right]\right\} - \frac{(\Delta H\neq{}^0)_{\mathrm{real}}}{2.303RT} \tag{24}$$

where Z_{el} is the heterogeneous collision frequency (assumed to have negligible temperature dependence), k and p are constants (assumed equal to unity); $(\Delta S\neq^0)_{\mathrm{real}}$ and $(\Delta H\neq^0)_{\mathrm{real}}$ represent the entropic and enthalpic barriers at the standard potential after correcting for entropic and enthalpic driving forces, $\Delta H^0_{rc}(=T\Delta S^0_{rc})$ and ΔS^0_{rc}, respectively. Figure 8.11 shows the temperature dependence of k^0, whose slope and intercept can be used to evaluate $(\Delta S\neq^0)_{\mathrm{real}}$ and $(\Delta H\neq^0)_{\mathrm{real}}$, respectively (assuming

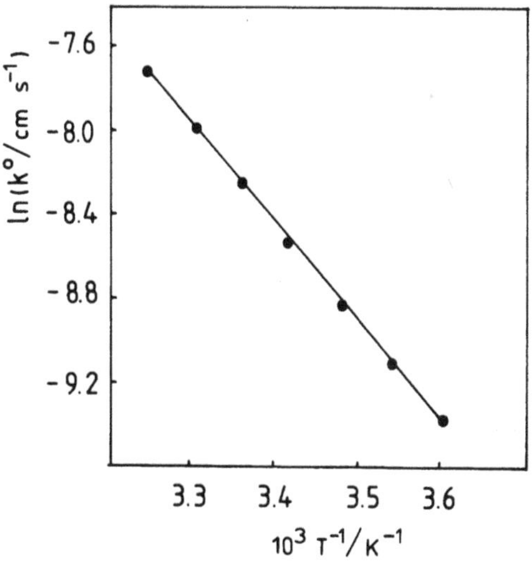

FIGURE 8.11. Temperature dependence of k^0 for an $[Os(bipy)_2(PVP)_{10}Cl]Cl$-modified electrode in 0.1 M H_2SO_4. (From Ref. 59.)

$Z_{el} \cong 10^{-4}$cm/s). The reaction entropy ΔS^0_{rc} can be calculated from the temperature dependence of the formal potential of the redox couple[116,117]:

$$\Delta S^0_{rc} \approx F(dE^0/dT) \tag{25}$$

This reaction entropy difference between reduced and oxidized forms of the redox couple can also be evaluated using the Born electrostatic model[118]:

$$(\Delta S^0_{rc})_{\text{Born}} = 47.88(z^2_{\text{ox}}/a_{\text{ox}} - z^2_{\text{red}}/a_{\text{red}}) \tag{26}$$

where z_{ox} and z_{red} are the net charges and a_{ox} and a_{red} are the equivalent radii in angstroms of the oxidized and reduced species, respectively.

The temperature dependence of k^0 can also be used to calculate $(\Delta H^0)_{\text{ideal}}$ and $(\Delta S^0)_{\text{ideal}}$, which give the enthalpic and entropic barriers to electron transfer at a particular potential where they are evaluated from:

$$(\Delta H^0)_{\text{ideal}} = (\Delta H^0)_{\text{real}} + \alpha T \Delta S^0_{rc} \tag{27}$$

$$(\Delta S^0)_{\text{ideal}} = (\Delta S^0)_{\text{real}} + \alpha \Delta S^0_{rc} \tag{28}$$

3.2. Experimental Results and Discussion

3.2.1. Homogeneous Charge Transport

The influence of sulphuric acid concentration on charge transport rates for a series of osmium loadings, $25 \geq n \geq 5$, is shown in Table 8.2. As noted previously, the rate of charge transport can be limited by one of three processes[94]:

The intrinsic barrier to electron self-exchange

The rate of ion diffusion required to maintain electroneutrality

The rate of polymer chain motion required to decrease the intersite separation allowing electron self-exchange to occur

Electron transport rates have been evaluated independent of ion or polymer motion limitations using steady-state techniques. These rates range from 10^{-8}–10^{-6} cm^2/s.[119,120] These values are considerably greater than charge transport rates presented here, suggesting that electron self-exchange is not the rate-determining step of charge propagation through these modifying layers.[59,61] Differentiating between polymer movement and electron hopping or ion movement as rate-limiting steps for charge transport is possible on the basis of the magnitude of the activation energy (large for polymer movement) and the sign of the entropy term (positive for polymer movement). It is, however, more difficult to decide between ion movement or electron hopping as rate-determining steps, since both activation energy and entropy changes can be similar. On the basis of the magnitude of D_{CT} and its strong variation with electrolyte concentration, it seems likely that ion movement rather than electron hopping is rate-limiting. In principle it is however possible that a variation in D_{CT} with the electrolyte is related to changes in polymer structure leading to a systematic variation in the redox site distance. In that case the possibility of electron hopping would have to be reconsidered. Such a change in structure with electrolyte concentration is not supported by perchloric acid EQCM data.

D_{CT}(PS) increases approximately linearly with increased osmium loading for $25 \geq n \geq 15$, then decreases for the higher $n = 10$ and $n = 5$ loadings. With increasing sulphuric acid concentration, D_{CT}(PS) increases more significantly with increased redox site loading for $20 \geq n \geq 10$. However, D_{CT}(PS) is insensitive to increasing electrolyte concentration for both $n = 25$ and $n = 5$ loadings. Activation energies determined by potential-step experiments, E_a(PS), are all less than 40 kJ/mol, and these are coupled to negative entropy terms independent of redox site loading and sulphuric acid concentration (see Table 8.2). These activation energies[(121)] and negative entropy[(84)] terms, suggest ion diffusion limitations. The associated negative entropy terms are accepted as indicating ion diffusion limitations, which require ordering the polymer backbone due to film expansion and reduced chain mobility.[(59)] The increase in D_{CT}(PS) with increasing electrolyte concentration likely reflects the greater availability of charge-compensating counterions because of the more swollen nature of the film.[(58,59)] If electron hopping were rate-determining, one would expect a decrease in D_{CT} with an increase of electrolyte concentration, due to increased redox site separation.

In 0.1-M H_2SO_4 both D_{CT}(CV) and E_a(CV) are insensitive to redox site loading for $20 \geq n \geq 5$. Low activation energies and associated negative entropy terms suggest that ion diffusion limits D_{CT}(CV) for these loadings. The D_{CT}(CV) is however significantly greater for the $n = 25$ loading by almost a factor of 10 in most cases. E_a(CV) for this metallopolymer is 70 kJ/mol, and it is associated with a positive entropy term, both of which suggest that polymer chain motion is the rate-limiting step.[(55,105)] These limitations most likely reflect the more swollen nature of this lower loaded metallopolymer.[(58,59)] The positive entropy term is consistent with polymer chain motions being required to reduce the intersite separation for the electron self-exchange process.[(59)] D_{CT}(CV) increases with increasing electrolyte concentration for $20 \geq n \geq 5$. In contrast the effect of increasing electrolyte concentration for the $n = 25$ loading is to decrease D_{CT}(CV). This decrease likely reflects increased swelling of the layer at higher sulphuric acid concentrations, which results in an increased intersite separation.[(59)]

For $20 \geq n \geq 10$ in a 1.0-M electrolyte, D_{CT}(CV) increases with increasing osmium loading. E_a(CV) is large and coupled to positive entropy terms. D_{CT}(CV) decreases for $n = 5$, with an associated decrease in E_a(CV) to 25 kJ/mol, which is coupled to a negative entropy term. This suggests that for low loadings, films are extensively swollen and polymer motion is required for electron self-exchange to occur. With increased metal loading, the modifying films become less swollen, thus reducing the requirement for polymer chain motion and allowing D_{CT}(CV) to increase. Further increasing in osmium loading to $n = 5$ reduces film swelling, resulting in an ion diffusion limitation.[59]

The influence of perchloric acid concentration on the charge transport rate for the same series of osmium metallopolymers is shown in Table 8.3. Both D_{CT}(PS) and D_{CT}(CV) are relatively insensitive to redox site loading for $25 \geq n \geq 15$. D_{CT}(PS) increases for the $n = 10$ polymer before decreasing for the $n = 5$ polymer. The effect of increasing concentration is quite different from that reported for sulphuric acid electrolyte. While D_{CT}(PS) increases as expected with increasing electrolyte concentration, the opposite effect is noted for D_{CT}(CV), which decreases with increasing electrolyte concentration, independent of metal loading (see Table 8.3). The relatively smaller charge transport rates obtained in perchloric acid, compared with those obtained in sulphuric acid, likely reflect compaction of films in perchloric acid electrolyte, resulting in hindered counterion diffusion. This interpretation is consistent with the reported crosslinking properties of the perchlorate ion.[89–92] Furthermore, a recent report dealing with electrodes modified with quaternized PVP films containing [Re(CO)$_3$(phen)], phen = 1,10-phenanthroline, suggests that nearly dehydrated, compact films are formed in perchlorate electrolytes.[122] Electrochemical quartz crystal microbalance studies of the $n = 10$ loaded polymer clearly indicate that in comparison to other electrolytes, in perchlorate electrolytes very little solvent accompanies the incoming counterion during oxidation of metal centers (*vide* infra). The difference in behavior observed for D_{CT}(PS) and D_{CT}(CV) as a result of a change in electrolyte concentration shows the importance of the measurement time scale. In this case different processes are clearly observed to be rate-limiting.

For $25 \geq n \geq 15$ E_a(PS) is approximately 41 kJ/mol in both 0.1- and 1.0-M perchloric acid. The activation energy is reduced to 15.9 and 9.9 kJ/mol in a 0.1-M and 1.0-M electrolyte, respectively, for the $n = 5$ loading. These values, coupled with negative entropy terms, reflect ion diffusion limitation. The fact that, when compared to results obtained in H_2SO_4, both D_{CT}(PS) and D_{CT}(CV) change relatively little with increasing electrolyte concentration reflects the compact and nonswollen nature of the film.[61] The $n = 10$ polymer exhibits a similar value of 14.5 kJ/mol in 0.1-M electrolyte to that obtained for the $n = 5$ loading. In 1.0-M perchloric acid an unusual dual-sloped Arrhenius behavior is observed (see Fig 8.12). Similar behavior was previously reported for ruthenium containing copolymers.[55] Two rate determining processes are characterized: at temperatures below 285K, E_a(PS) is similar to that observed for lower loadings with a value of 51 kJ/mol, and it is again indicative of ion diffusion limitations. At temperatures above 285K, E_a(PS) is decidedly larger, with a value of 134 kJ/mol, which suggests that polymer chain motions are rate-limiting. This duality may reflect ion diffusion limitations

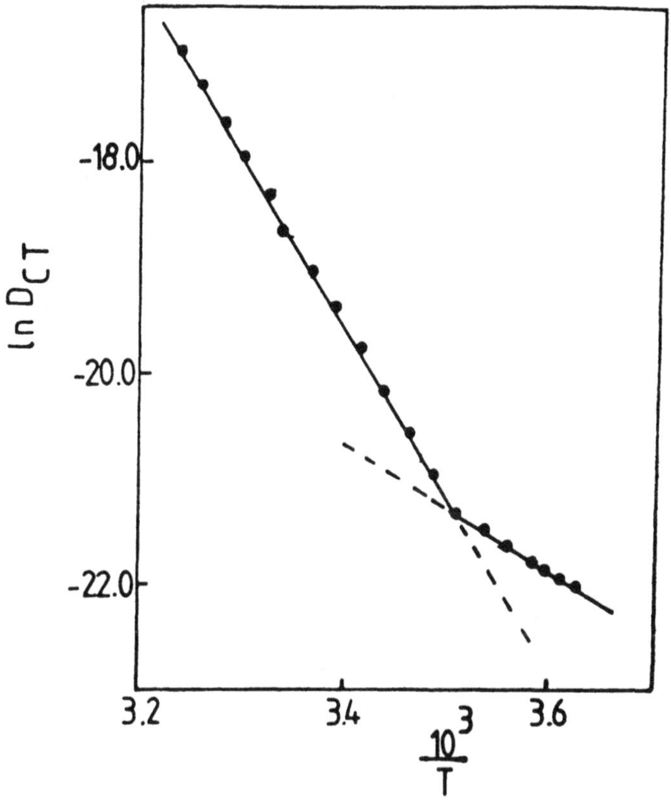

FIGURE 8.12. Temperature dependence of D_{CT} as measured from chronoamperometry in 1.0 M $HClO_4$. (From Ref. 58.)

imposed by the compacted film at low temperatures and expansion of the film at higher temperatures, requiring polymer motion to restore the intersite separation for electron self-exchange to occur.[61]

The E_a(CV) values are considerably larger than those observed using potential-step techniques (see Table 8.3). For $25 \geq n \geq 15$ E_a(CV) ranges from 151–163 kJ/mol, and it is associated with large positive entropies for both 0.1- and 1.0-M perchloric acid electrolyte concentrations. These indicate segmental chain motion where chain motion is restricted due to considerable film compaction.[58] For $n = 5$, little variation of the activation energy with electrolyte concentration is observed, and E_a(CV) is coupled to a negative entropy term, suggesting ion motion limitations. These trends possibly reflect the more compact nature of the polymer film with decreasing metal loading, i.e., more unbound pyridine units available for cross-linking. In 0.1-M perchloric acid a dual–Arrhenius slope behavior, similar to that observed for E_a(PS) in a 1.0-M electrolyte, is noted. The small activation value coupled with a negative entropy term suggest that at low temperatures ($T < 285$K) counterion diffusion limits the charge transport process. At high temperatures ($T >$ 285K), the large activation energy and associated positive entropy suggest that polymer chain motion controls the charge transport process. It would appear that the polymer matrix has become so swollen that polymer chain motion is needed to reduce the intersite separation.[58]

The rate of charge transport through these $[Os(bipy)_2(PVP)_nCl]Cl$ films clearly depends on the experiment time scale, with potential-step techniques giving consistently larger charge transport values than cyclic voltammetry. The extent of film oxidation over which the Cottrell equation is valid is closely related to the relative difference in D_{CT} between the two techniques; e.g., for $n = 5$ in 1.0-M H_2SO_4 where the Cottrell equation is valid for 80% of the determined film thickness, the observed difference between D_{CT}(PS) and D_{CT}(CV) is negligible.[59] However, when the difference in D_{CT} is more significant, at, e.g., the lower loadings in sulphuric acid and all loadings in perchloric acid, the Cottrell equation is valid for less of the total film thickness.[59,61] This variation in D_{CT} may reflect the extent of film oxidation and the associated motion of charge-compensating counterions. The longer time scale cyclic voltammetry experiments oxidize a greater portion of the modifying layer, thus requiring counterion diffusion to occur over a greater portion of the layer than potential-step techniques.[59,60,123] Furthermore, it was reported that for polyvinylferrocene, electroneutrality is satisfied mainly by counterion motion at short times, while only at longer times do neutral species including salt and solvent move to satisfy activity relationships[93] that may suppress D_{CT}(CV) relative to D_{CT}(PS). This variation between D_{CT}(CV) and D_{CT}(PS) depends on electrolyte type and concentration and the redox site loading.

A feature of these materials is that the Gibb's energy term remains constant at 50–60 kJ/mol regardless of the metal loading, electrolyte type, or concentration, and the rate-determining step, for a given polymer and metal center.[44,57–61] This indicates that variations observed in activation energies, following changes in the aforementioned variables, are a direct consequence of entropy effects imposed by the polymer matrix, and not the result of variations in the fundamental electron transfer process, which remains essentially unaffected.[59]

3.2.2. Heterogeneous Kinetics

The dependence of k^0, the heterogeneous electron rate constant, and α, the anodic transfer coefficient, on both the concentration of sulphuric acid and redox site loading is shown in Table 8.4. For $n \geq 20$ k^0 increases with increasing sulphuric acid concentration. The $n = 25$ loading however exhibits a decrease in k^0 with increasing electrolyte concentration.[59] For $25 \geq n \geq 15$, k^0 increases with increased redox site loading before decreasing for the $n = 10$ and $n = 5$ metal loadings. In

TABLE 8.4
Effects of Redox Site Loading and Sulphuric Acid Concentration on the Heterogeneous Kinetics of $[Os(bipy)_2(PVP)_nCl]Cl$ Films

Metal Loading	Concentration M	$k^0/10^{-4}$ cm/s	α	$(\Delta H\neq^0)_{real}$ kJ/mol	$(\Delta S\neq^0)_{real}$ $Jmol^{-1}K^{-1}$	ΔS_{rc}^0 $Jmol^{-1}K^{-1}$
$n = 5$	0.1	3.0	0.31	40	−9	20
	1.0	4.1	0.48	22	−67	18
$n = 10$	0.1	14.1	0.32	25	−55	19
	1.0	20.9	0.46	34	−27	20
$n = 15$	0.1	12.9	0.35	20	−71	21
	1.0	32.5	0.49	38	−13	19
$n = 20$	0.1	9.1	0.35	18	−80	20
	1.0	19.1	0.48	39	−8	20
$n = 25$	0.1	6.3	0.42	16	−82	19
	1.0	5.1	0.38	41	−1	20

(From Ref. 59.)

0.1-M sulphuric acid, the enthalpy $[(\Delta H\neq^0)_{real}]$ and entropy $[(\Delta S\neq^0)_{real}]$ of activation for heterogeneous electron transfer become more positive and less negative, respectively, with increasing redox site loading. In 1.0-M electrolyte this behavior is reversed, with both terms decreasing with increased metal loading.[59]

Table 8.5 shows the dependence of k^0 and α on the concentration of perchloric acid and the redox-site loading. By increasing the perchloric acid concentration, k^0 increases for all loadings. The effect of increased metal loading for $25 \geq n \geq 10$ is to increase k^0 followed by a decrease in k^0 for the $n = 5$ loaded metallopolymer. For $n \geq 10$, α increases with increased concentration but remains relatively constant for $25 \geq n \geq 15$ loadings unlike in sulphuric acid. The effect of increasing the redox-site loading on the enthalpy and entropy terms is similar in both electrolyte concentrations. Both $(\Delta H\neq^0)_{real}$ and $(\Delta S\neq^0)_{real}$ increase in a linear manner with increasing redox-site loading. For $n = 15$ a dual-slope behavior is observed, which is, possibly related to the behavior observed for homogeneous charge transport.[111]

Results obtained highlight the dependency of the heterogeneous kinetics of electrodes modified with thin layers of $[Os(bipy)_2(PVP)_nCl]Cl$ on the electrolyte and redox-site loading. For all electrolytes and redox-site loadings, k^0 is typically an order of magnitude smaller than values reported for electroactive species in solution.[124] Such behavior was previously reported[125,126] and explained in terms of a partial blocking of the electrode surface by the modifying layer.[127] In this approach k^0 is considered to be the product of $(1 - \theta)$ and k^0 for heterogeneous electron transfer at the active sites, where θ is the fraction of sites that are available for electron transfer. Since the polymer backbone is, for the osmium containing

TABLE 8.5
Effects of Redox Site Loading and Perchloric Acid Concentration on the Heterogeneous Kinetics of $[Os(bipy)_2(PVP)_nCl]Cl$ Films

Metal Loading	Concentration M	$k^0/10^{-4}$ cm/s	α	$(\Delta H\neq^0)_{real}$ kJmol^{-1}	$(\Delta S\neq^0)_{real}$ Jmol^{-1}K^{-1}	ΔS_{rc}^0 Jmol^{-1} K^{-1}
$n = 5$	0.1	2.2	0.30	69	96	17
	1.0	3.9	0.32	53	30	19
$n = 10$	0.1	3.1	0.14	52	25	21
	1.0	5.0	0.24	41	−12	21
				T < 285 K		
$n = 15$	0.1	2.6	0.14	10	−113	17
				T > 285 K		
				50	20	21
	1.0	3.8	0.18	36	−26	17
$n = 20$	0.1	2.5	0.14	43	1	19
	1.0	3.5	0.18	34	−32	20
$n = 25$	0.1	2.5	0.12	38	−15	19
	1.0	3.2	0.16	33	−35	19

(From Ref. 62.)

polymer films presented here, nonconducting, heterogeneous electron transfer can occur only between the electrode material and the osmium center. As the redox site concentration is increased, intersite separation decreases, so the number of sites at which heterogeneous electron transfer can take place should increase, thus increasing k^0. However, k^0 increases only with redox-site loading up to $n = 10$ or $n = 15$, depending on the electrolyte concentration, which suggests that this model cannot fully explain the present results.[59,61]

In both electrolytes, for a 0.1-M electrolyte concentration (except for $n = 15$ in perchloric acid), $(\Delta H\neq^0)_{real}$ and $(\Delta S\neq^0)_{real}$ increase with increasing redox-site loading, the enthalpy term becomes more positive, and the entropy term becomes less negative (see Tables 8.4 and 8.5). In 1.0-M perchloric acid a trend similar to that observed in a 0.1-M electrolyte is noted. However, in 1.0-M sulphuric acid this trend is reversed, with the enthalpy and entropy decreasing with increased redox-site loading. The fact that homogeneous (for potential-step experiments) and heterogeneous entropy terms are both negative suggests that both processes are related and most likely associated with the availability of charge-compensating counterions within the layer.[62] This interpretation is supported by the observed linear correlation between D_{CT}(PS) and k^0. This suggests that both the homogeneous charge transport process, as measured by potential-step experiments, and the heterogeneous electron transfer reaction are affected similarly by changes in

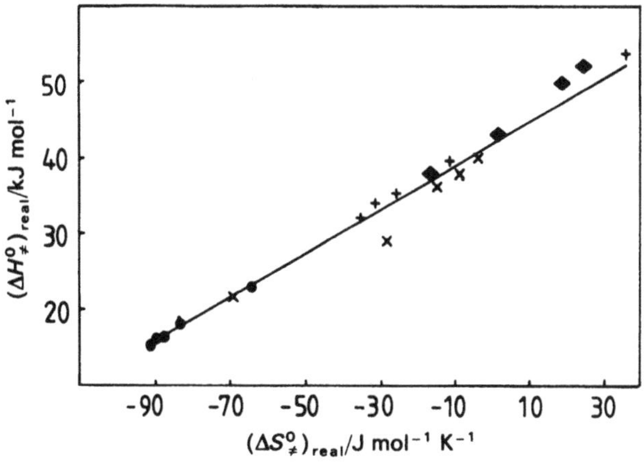

FIGURE 8.13. Correlation of $(\Delta H^0\neq)_{real}$ and $(\Delta S^0\neq)_{real}$ for the heterogeneous electron transfer reaction of Os(II/III) within $[Os(bipy)_2(PVP)_nCl]Cl$-modified electrodes; •: 0.1-M HCl, **x**: 1.0-M HCl, ♦: 0.1-M $HClO_4$, and +: 1.0-M $HClO_4$. (From Ref. 62.)

redox-site loading and electrolyte concentration.[59,62,128] In perchloric acid the similarity of the trends observed in 0.1-M and 1.0-M electrolyte concentrations suggests that the double-layer formation is similar in both cases.[62] The observed linear relationship between $(\Delta H\neq^0)_{real}$ and $(\Delta S\neq^0)_{real}$ is significant because, despite varying trends in 1.0-M electrolytes, values obtained for all electrolytes lie on a single line,[59,62,111] suggesting that a single process is responsible for the observed behavior, which is highlighted in Fig. 8.13 for perchloric and hydrochloric acid electrolytes.

The effect of temperature on E^0 can be use to calculate the reaction entropy ΔS^0_{rc} (see Fig. 8.14). In both electrolytes examined, a positive reaction entropy of similar magnitude is obtained for all loadings.[59,62] This term is considered sensitive to changes in the solvent polarization required to allow electron transfer.[116,117] Previously reported positive entropy terms were attributed to more extensive solvent ordering in the oxidized state.[128] In the case presented here, we postulate that increased polarization of water molecules adjacent to the aromatic rings of the osmium complex occurs in the higher oxidation state.[59] The Born model yields a value of 27 $JMol^{-1}K^{-1}$ for the reaction entropy, where $a_{ox} = a_{red} = 7.5$ Å[62] in the case of perchloric acid electrolytes, and 16 $JMol^{-1}K^{-1}$[59] in the case of sulphuric acid electrolytes. The relatively constant nature of (ΔS^0_{rc}) with increased redox-site loading and electrolyte concentration suggests that the local chemical microenvironment of osmium centers remains largely unchanged as redox-site loading changes.[62] Furthermore the effect of increasing the concentration of electrolyte does not appear to effect the solvation of the metal center significantly, thus

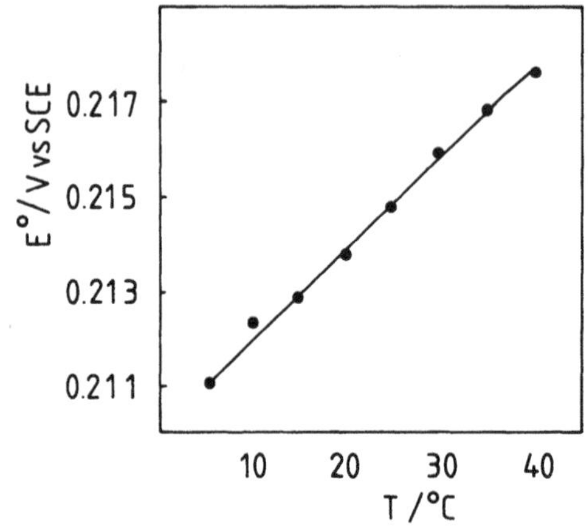

FIGURE 8.14. Temperature dependence of the formal potential of Os(II/III) oxidation within $[Os(bipy)_2(PVP)_{20}Cl]Cl$ in 0.1 M H_2SO_4; surface coverage: 1.8×10^{-8} mol cm^{-2}. (From Ref. 59.)

emphasizing the importance of considering not only redox properties of the metal center, but also physicochemical properties of the polymer backbone.

A transfer coefficient of 0.5 ± 0.05 was obtained for a model monomeric compound ($[Os(bipy)_2(pic)Cl]PF_6$) at a bare glassy carbon electrode. Values similar to this were obtained only in sulphuric acid at high concentrations.[59] However, it is important to note that where low transfer coefficients are observed, e.g., perchloric acid, the sum of α_b and α_c is still close to unity.[52] This suggest that Butler–Volmer kinetics still operate in these conditions. Given that the transfer coefficient can yield information about the position of the reaction site with respect to the outer Helmholtz plane, the similarity of the transfer coefficient in sulphuric acid electrolyte for the solution and polymer bound species suggests that the position of the reacting transition state with respect to the electrode is similar in both cases. In general, the point of electron exchange is considered to be the outer Helmholtz plane; therefore heterogeneous electron exchange occurs over several angstroms, and the double layer is found at the electrode/film interface.[59,62]

The correlation between homogeneous charge transport and heterogeneous electron transfer suggests that processes such as ion diffusion affect both processes similarly. The consistency of the determined reaction entropy term with increased redox-site loading and concentration suggests that the chemical microenvironment of the redox centers remains unaltered. This, in conjunction with the observed dependence of $D_{CT}(PS)$ and k^0 on the aforementioned variables, highlights again the profound effect of structural properties of the polymer phase on the charge transport behavior of these films.

4. ELECTROCHEMICAL QUARTZ CRYSTAL MICROBALANCE

In this section the equilibrium (or thermodynamic) mass changes within films of $[Os(bipy)_2(PVP)_{10}Cl]^+$ as a function of $HClO_4$ concentration and polymer redox state are considered.[129,130] In previous sections the importance of the structural properties of the polymer phase on the rate of charge transport through these electrode-modifying layers was highlighted. The determination of ion and solvent content in these modifying layers is fundamental to successfully characterizing these systems. Experimentally, *in situ* measurement of mobile species transfer into and out of electroactive polymers demands a technique capable of quantitatively determining small changes in the polymer film population of species in the presence of vast excesses of the same species in the bathing solution. One simple experimental technique meeting this requirement is the *in situ* electrochemical quartz crystal microbalance (EQCM) technique.[131–141]

A number of redox polymers were investigated using EQCM. Ion and solvent population changes within films of poly(vinyl ferrocene) (PVF) during redox processes were studied,[93,131,142–148] as were the ion exchange of ferri (ferro) cyanide[149] and open circuit bimolecular reactions[150] in PVF. Mass changes accompanying redox processes, obtained at slow cyclic voltammetric scan rates for PVF films in perchlorate electrolyte, indicated that mass transfer was due to the interfacial movement of a single perchlorate counteranion and 4–5 solvent molecules.[142,144,146] Coion participation in PVF redox processes in perchlorate was not in evidence.[146] The importance of the counteranion type in PVF was addressed, and an almost linear relationship between the observed mass change and anion molecular weight was observed[146] (slight deviations were attributed to variations in the polymer structure affecting solvent movement). However, in a more recent study,[142] more pronounced changes in PVF morphology with changes in the counteranion type were observed. This is graphically illustrated by the behavior of PVF in PF_6^- and Cl^- electrolytes.[151] In the PF_6^- electrolyte no solvent movement was observed, while in the Cl^- electrolyte extensive delamination of the polymer from the electrode during oxidation occurred due to increased solvation of the polymer layer.

Counteranion effects on the $[Os(bipy)_2(PVP)_{10}Cl]^+$ redox polymer were studied.[152,153] In para-toluene sulphonate (pTS^-), NO_3^-, SO_4^-, and Cl^- electrolytes, significant solvent transfer was observed that was thought to be related to the increased solvent content of the polymer layers in the initial state when converted into these salt forms. The large movement of solvent in pTS^- electrolytes was confirmed from isotopic substitution of H_2O with D_2O.[153] In perchlorate no solvent transfer occurs, with the layers behaving permselectively.[153]

The importance of polymer swelling and solvation was clearly demonstrated in the break-in and memory effects observed in many electroactive polymer systems. Break-in effects in PVF,[142,154] tetracyanoquinodimethane

(TCNQ),[155,156] nitrated polystyrene,[157] poly (3-alkylthiophene),[158] and polyaniline[159] were observed. In a recent paper,[160] we highlighted the existence and long duration of memory effects following the transfer of a $[Os(bipy)_2(PVP)_{10}Cl]^+$-modified electrode between two electrolyte solutions. On transferring such films from perchloric to toluenesulphonic acid solutions, film structural changes occur very slowly when compared to the time scale in conventional electrochemical experiments.[160] These results demonstrate that extended redox cycling is required to eradicate the mass transfer characteristics associated with previously employed electrolytes (see Fig. 8.15). This work highlights that equilibrium (including structural changes) between the polymer film and contacting electrolyte may be surprisingly slow and that cyclic voltammetry is insensitive to these structural changes, whereas the EQCM provides a means of monitoring their progress.[160]

Polymer swelling during the electrochemical deposition of PVF in CH_2Cl_2 was investigated.[161] For PVF transferred to aqueous electrolytes, where rigid layers exist,[161] the solvation of the polymer in the reduced and oxidized layer was

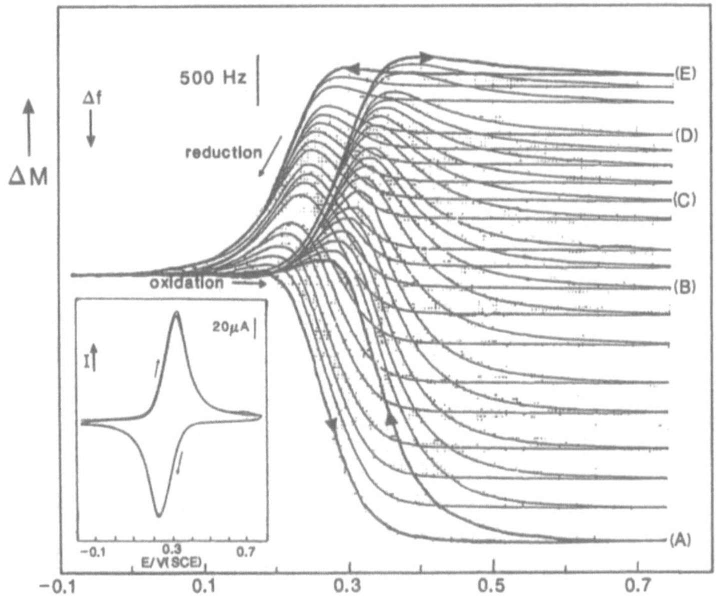

FIGURE 8.15. The EQCM data for an $[Os(bipy)_2(PVP)_{10}Cl]^+$ film in a 0.1 M p-TSA solution. The film was previously exposed to a 1 M $HClO_4$ solution. Main figure shows selected mass-(frequency-) potential curves after (a) 0, (b) 45, (c) 120, (d) 180, and (e) 240 minutes of potential cycling. Inset shows simultaneously acquired I–E curves; scan rate: 50 mV s^{-1}; Os surface coverage: 2×10^{-8} mol cm^{-2}. (From Ref. 160.)

studied by the combined use of EQCM and ellipsometry measurements.[144] In nitrated polystyrene films, increased solvation of the polymer with increasing redox cycling was observed.[162] The effect of electrolyte concentration[155,156] and temperature[156] on polymer swelling in TCNQ films was studied at the EQCM. In TCNQ films cation motion maintains electroneutrality within the polymer during redox cycling, and mass changes observed were found to be dominated by the concomitant movement of water of hydration.[155,156]

The resident ion and solvent population within the polymer layer was found to be of great importance in determining mass changes observed at the EQCM.[163] For thermodynamic mass changes in polythionine films, mass changes observed were rationalized in terms of an ion aggregation model, in which ions resident within the polymer layer provided a source of counteranion.[163] The transfer of large quantities of solvent in response to activity gradients was also evoked. This model was an extension of the approach outlined by Bruckenstein and Hillman, which describes the general features of mass and charge transfer processes.[132] Finally, the kinetics of mobile species transfer was addressed for nitrated polystyrene,[157] and was systematically studied for PVF[93,131,147,148] and polythionine.[147,148,164]

The following sections highlight more pertinent aspects of the EQCM technique, its application to analysis, and results obtained for the $[Os(bipy)_2(PVP)_{10}Cl]Cl$ polymer in perchloric acid electrolyte.[129,130] For a more extensive discussion of this technique, the reader is referred to Refs. 135–138.

4.1. *Theoretical Framework*

Changes in the attached mass alter the resonant frequency of a quartz crystal oscillator.[136] One measures the change in resonant frequency (Δf) of the crystal oscillator from its base value, resulting from a change in the mass of the electrode (substrate plus attached film and included electrolyte.[139] When the attached mass is rigidly coupled to the crystal, the frequency change (Δf/Hz) caused by a change in attached mass (Δm/g/cm^2) is given by the Sauerbrey equation[140]:

$$\Delta f = -\left(\frac{2}{\rho v}\right)\Delta m f_0^2 \tag{29}$$

where ρ is the density of the quartz crystal and v is the wave velocity within it. Substituting for the constants $\rho = 2.65$ gcm^{-3} and v = 3.34×10^5 cmHz,[133] the variation in frequency of an oscillating crystal as a result of a small mass change is given as:

$$\Delta f = -2.26 \times 10^6 f_0^2 \Delta m \tag{30}$$

When an oscillating crystal is immersed in a liquid, an oscillating boundary layer is set up due to viscous coupling of the liquid medium. It was shown that upon immersion the change in frequency of the crystal Δf can be expressed as:

$$\Delta f = f_0^{3/2}\left(\frac{n_L \rho_L}{\pi \rho_q \mu_q}\right)^{1/2} \quad (31)$$

where n_L is the liquid viscosity, ρ_L is the liquid density, and μ_q is the shear modulus of quartz. For thin rigid films attached to the immersed side of a crystal, the overall frequency change is simply the sum of the frequency changes due to the attached film [Eq. (30)] and the presence of the liquid [Eq. (31)].

The EQCM cannot differentiate between the interfacial movement of charged and neutral species. However, for a known electrochemical reaction, the simultaneous capture of charge data defines the extent of the reaction and provides information on electroneutrality requirements for the movement of mobile ions. Thus, the combination of charge and mass transfer data may yield additional insight into redox processes occurring within the polymer layer. In this section the equilibrium mass of the layer is investigated as a function of perchloric acid electrolyte concentration and the oxidation state of redox centers. Mass changes observed are discussed in terms of polymer morphology, as dictated by the contacting electrolyte. Further insight into mass transfer processes accompanying redox switching is obtained by using a diagnostic scheme developed for interpreting EQCM data.[93,131,138]

This scheme requires the mass–charge relationship[93,131,138,141] to be defined and of eliminating enables a selected mobile ion from mass changes observed. The net total mass change per unit area of the electrode, occurring at the polymer/electrolyte interface, is given by:

$$\Delta M = \sum m_m \Delta \Gamma_m \quad (32)$$

where $\Delta\Gamma_m$ is the change in the molar population of species m within the polymer film and m_m is the molar mass. The change in mass of the polymer film is due to two components, the movement of ionic species and the movement of net neutrals; therefore:

$$\Delta M = \sum m_i \Delta \Gamma_i + \sum m_n \Delta \Gamma_n \quad (33)$$

where subscripts i and n denote ionic and neutrals species, respectively. The total mass change is calculated from frequency measurements using the Sauerbrey equation. The net charge passed at the polymer/electrolyte interface is

$$Q = \int i dt = -\sum z_i F \Delta \Gamma_i \quad (34)$$

where z_iF is the charge carried per mole of the ionic species i. In this analysis the charge is defined as positive for film oxidation. Combining Eqs. (33) and (34), the total mass change per mole of redox site can be obtained

$$\frac{\Delta MF}{Q} = \frac{(\sum m_n \Delta\Gamma_n + \sum m_i \Delta\Gamma_i)}{(\sum z_i \Delta\Gamma_i)} \tag{35}$$

This describes the net mass change per mole of redox sites converted, and the units are g/mol. For a film in a quasi-equilibrium state, Eq. (35) is referred to in future discussions as the normalized mass change.

To evaluate the contribution of a particular ion j to the observed mass change, an expression Φ_j is used. This is a linear combination of the mass change observed and the mass change associated with the movement of the ion j if it alone maintains electroneutrality. Φ_j is therefore the weighted difference of mass and charge responses in units of g/cm^2 and is defined by:

$$\Phi_j = \Delta M + Q\left(\frac{m_j}{z_jF}\right) = \sum m_n \Delta\Gamma_n + \sum\left[m_i - \left(\frac{m_j z_i}{z_j}\right)\right]\Delta\Gamma_i \tag{36}$$

where Φ_j represents the additional mass transfer (either positive or negative), of all mobile species to that anticipated for the motion of the chosen ion j.

4.2. EQCM Analysis of $[Os(bipy)_2(PVP)_{10}Cl]Cl$ in $HClO_4$ Electrolytes

4.2.1. $HClO_4$ Concentration Effects on Os^{II} Layer Mass and Rigidity

A successful analysis of mass changes in polymer-modified electrodes using the EQCM strongly depends on the rigidity of the layer, since the Sauerbrey equation is valid only under this condition. It is therefore of prime importance to establish this rigidity, and this has been investigated using admittance measurements.(129,130) Figure 8.16 highlights the open-circuit change in the admittance-frequency response for an $[Os(bipy)_2(PVP)_{10}Cl]Cl$-coated crystal as the contacting solution concentration is increased from H_2O to 1-M $HClO_4$. No voltammetry was performed in the course of acquiring the data.

Figure 8.16 shows two important features. Firstly, the significant decrease in resonant frequency of the crystal as the $HClO_4$ concentration is increased represents an increase in the layer mass resulting from protonation of uncoordinated pyridine moieties(88) of the polymer backbone and the influx of perchlorate counterions required to maintain electroneutrality. Secondly, excluding 0.01-M $HClO_4$, the resonant shape remains constant (if not becoming sharper) as the electrolyte

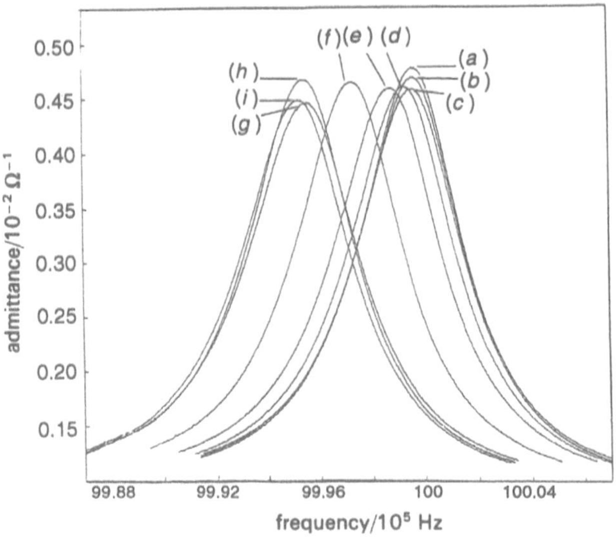

FIGURE 8.16. Admittance spectra for an $[Os(bipy)_2(PVP)_{10}Cl]^+$-coated Au EQCM on exposure to increasing concentrations of an $HClO_4$ electrolyte in the absence of electrochemical cycling; gravimetric surface coverage: 2×10^{-8} mol cm^{-2}. (a) H_2O, (b) 1×10^{-5}, (c) 1×10^{-4}, (d) 4×10^{-4}, (e) 1×10^{-3}, (f) 4×10^{-3}, (g) 1×10^{-2}, (h) 0.1, (i) 1.0 M $HClO_4$. (From Ref. 129.)

concentration is increased from 10^{-4} to 0.1 M $HClO_4$, suggesting that the layer remains rigid. The maintenance of a rigid polymer structure, despite an increase in the ionic content of the polymer phase, most likely represents compaction of the polymer layer due to crosslinking of polymer segments by the perchlorate anion.[89–92] Furthermore, it was shown that the disruption of hydrogen bonding between water molecules by ClO_4^-[165] can result in dehydrated and deswollen polymer structures.[105,122] At concentrations below 0.01 M, polymer expansion resulting in a structural change is noted. This break-in effect is highlighted by the fact that admittance spectra obtained with decreasing electrolyte concentration (following voltammetric cycling), do not correspond to those obtained on initial exposure (see Table 8.6). Similar break-in effects were previously observed,[142,154–156,158,159,166,167] and these may be attributed to difficulty in attaining equilibrium levels of solvation in the inner polymer layer. This can be facilitated only by extensive counteranion and solvent transfer, which occurs during electrochemical cycling.[129] The fact that this transition occurs in 0.01-M $HClO_4$ and polymer expansion occurs in this $HClO_4$ concentration, even without redox switching, highlights the unique behavior of the polymer film in this electrolyte concentration.

So although there are changes in the morphology of this metallopolymer with changes in $HClO_4$ concentrations, these changes are small and allow the application

TABLE 8.6
Admittance Spectra at Open Circuit for a $[Os(bpy)_2(PVP)_{10}Cl]^+$ Coated Crystal in $HClO_4$

	Admittance[a,b]/$10^{-2}\Omega^{-1}$		FWHM[a]/kHz		Δf[a,c]/Hz	
$[HClO_4]$/M	A[d]	B	A	B	A	B
H_2O	0.478	0.402	6.096	7.681	0	–518
1×10^{-5}	0.468	0.402	6.096	7.681	–52	–518
1×10^{-4}	0.456	0.406	6.304	7.681	–104	–1036
4×10^{-4}	0.460	0.412	6.304	7.526	–466	–1192
1×10^{-3}	0.460	0.420	6.304	7.372	–933	–1554
4×10^{-3}	0.464	0.414	6.356	7.681	–2435	–3057
0.01	0.444	0.412	6.825	7.836	–4249	–4404
0.10	0.468	0.470	6.304	6.444	–4508	–4663
1.00	0.448	0.442	6.721	6.908	–4663	–4560

Uncoated electrode (air): 10.010881 MHz; coated electrode (air): 10.003180 MHz. Coated electrode (H_2O): 9.999150 MHz (zero value, Column A)
[a]Data interpolated from points of 20-Hz resolution. [b]Admittance maximum (at resonant frequency). [c]Δf is the shift in center frequency with respect to the value on initial exposure to H_2O (corrected for viscous load of electrolyte).
[d]Column A: Data for first exposure to electrolyte without electrochemical cycling; Column B: Open circuit data after electrochemical cycling at each concentration. Gravimetric surface coverage 20×10^{-8} mol cm^{-2}.
(From Ref. 129.)

of the Sauerbrey equation. The maintenance of rigid polymer structures for polymer films that are only partially protonated suggests that crosslinking occurs at all concentrations and indicates that perchlorate must exist in clusters within the polymer matrix when present in small quantities.[(129)] This is in agreement with the observed behavior of perchlorate in quaternary ammonium anion exchangers.[(168)]

Figure 8.17 highlights the mass data obtained at open circuit as the electrolyte concentration is increased from pure H_2O to 4-M $HClO_4$. To ensure the expulsion of the original chloride counterion from the polymer layer and the conversion of the polymer into the perchlorate salt form, the film was continually scanned voltammetrically for several hours at 1 mV/s, at pH 3.5.[(129)] The electrode was then rinsed, stored in Milli-Q water, and the experiment continued at this pH. All mass changes are in addition to the initially imbibed solvent in pure H_2O, and are corrected for electrolyte viscosity changes.[(129,130)] Imbibition of *ca.* 44(±5) molecules of H_2O per polymer unit (based on the gravimetric evaluation of polymer surface coverage) was observed on immersing coated electrodes in deoxygenated demineralized water and noting the frequency difference due to the solvent.[(129)]

If the insertion of a proton with a solvated perchlorate anion is assumed, then by using the solution hydration number of 2.6,[(134)] an upper limit of *ca.* 7 anions per polymer unit is obtained.[(129)] This suggests incomplete protonation of the polymer backbone, which is considered unlikely at pHs much more acidic than the

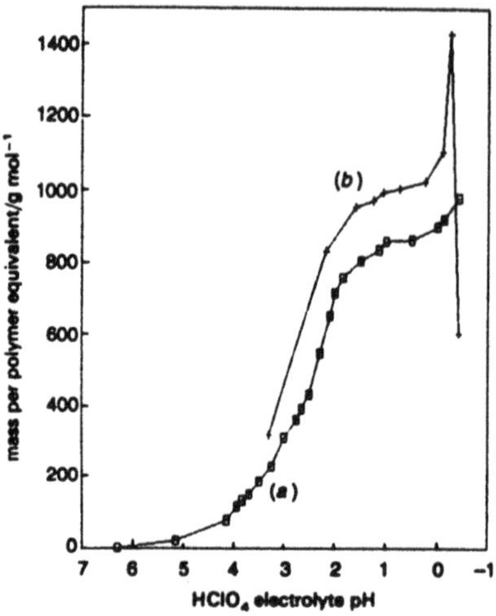

FIGURE 8.17. Change in resident $[Os(bipy)_2(PVP)_{10}Cl]^+$ layer mass as a function of the $HClO_4$ concentration and Os oxidation state; gravimetric surface coverage: 8.7×10^{-8} mol cm^{-2}. (a) Os^{II}, (b) Os^{III}. (From Ref. 129.)

pK_a of the PVP homopolymer ($pK_a = 3.3$).[88] For the incorporation of H^+ with an unsolvated anion, an upper limit of *ca.* 9 perchlorate anions per polymer unit is obtained. This corresponds to the electroneutrality requirement for the 9 protonated, uncoordinated pyridine moieties of each polymer unit.

4.2.2. Effect of $HClO_4$ Concentration on Redox-Induced Mass Changes

Figure 8.18 shows a typical cyclic voltammetric response for the $Os^{II/III}$ couple at 5 mV/s in 0.1-M $HClO_4$. Figure 8.16 also includes the mass change within the polymer layer associated with maintaining electroneutrality during the charge transport process.[129] While true surface behavior is not observed in $HClO_4$ electrolytes at these cyclic voltammetric time scales, there is no significant difference in the normalized mass changes obtained at 1 and 5 mV/s for most electrolytes. These time scales are thus sufficient for all mobile species to establish quasi-equilibrium. The mass potential plot shows the establishment of a quasi-equilibrium mass after oxidation of the layer and the reestablishment of the original baseline mass following rereduction.

For a permselective film, the variation of E'_{app} with anion activity in the external solution is given by the Nernstian expression[169]:

$$E'_{app} = E^0 - \left(\frac{RT}{nF}\right) \ln\left[\frac{(a_{A-s})}{(a_{A-p})}\right] \qquad (37)$$

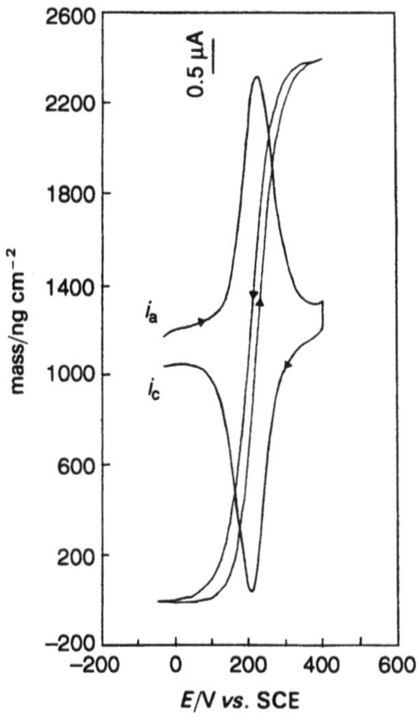

FIGURE 8.18. Cyclic voltammogram and mass potential plot for an $[Os(bipy)_2(PVP)_{10}Cl]^+$-coated electrode at 5 mVs^{-1} in 0.1 M $HClO_4$; coulometric surface coverage: 1×10^{-8} mol cm^{-2}. (From Ref. 129.)

where E^0 is the formal potential of the redox couple, a_{A-s} and a_{A-p} are the activities of the counteranion in the external solution and the polymer phase, respectively. Provided that a_{A-p} remains essentially constant as the electrolyte concentration is varied and there is no significant change in polymer swelling, Eq. 33 has a predicted slope of −59 mV/decade of anion activity. The change in formal potential of the $Os^{II/III}$ couple, (E'_{app}), with the logarithm of the bathing solution ClO_4^- activity is shown in Fig. 8.19. E'_{app} is taken as the average of the anodic and cathodic peak potentials of the cyclic voltammogram at 1 mV/s. Activities at concentrations less than 0.03 M were calculated using the Debye–Huckel equation,[170] while at higher $HClO_4$ concentrations, activities were obtained from Ref. 171. The slope of the plot in Fig. 8.19 for the concentration range 0.001–1.0 M $HClO_4$, is −61±3 mV/decade. The near-Nernstian slope indicates permselective behavior of the osmium metallopolymer in this concentration range, with little coion participation in the charge transport process. Similar behavior was observed in near-neutral pH perchlorate electrolyte for this polymer[153] and for PVF.[142,146]

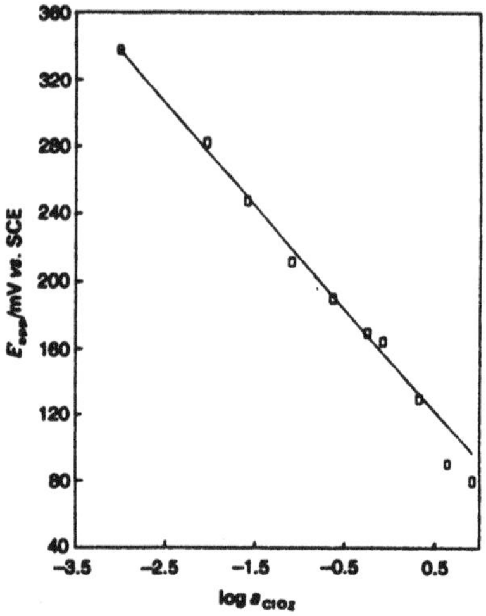

FIGURE 8.19. Variation of E'_{app} with the solution ClO_4^- anion activity. (From Ref. 129.)

4.2.3. Redox Switching in the Concentration Range 0.1–1.0 M $HClO_4$

Taking into account mechanical forces involved in anion insertion[90–92] and the movement of a solvent in response to activity gradients,[132] one might expect significant changes in layer morphology to occur during redox processes in this polymer,[129] which would prevent using the Sauerbrey equation. To determine whether the layer is rigid, the admittance frequency response of the crystal was measured during redox switching. Figure 8.20 shows the change in the admittance spectrum during polymer oxidation in 1.0-M $HClO_4$. Within the 0.1–1.0 M concentration range, the monotonic decrease/increase in resonant frequency as redox switching occurs indicates anion and solvent insertion/expulsion, while the constant resonant shape indicates that there is little change in the viscoelasticity of the polymer layer and no significant morphology changes occur.[129,130] Thus, in this concentration range, rigidity of the polymer layer is also maintained during redox switching, and the frequency mass relationship is not compromised.

For nonrigid polymer layers, viscoelasticity often becomes more pronounced as surface coverage is increased.[139] The evaluation of mass from the dampened oscillation, although not strictly related to frequency changes measured due to violation of the Sauerbrey equation, is less than the true mass change occurring.[139] For the $[Os(bipy)_2(PVP)_{10}Cl]^+$ polymer, a plot of the variation of mass change with surface coverage yields a linear plot with a zero intercept in 0.1- and 1.0-M electrolyte concentrations.[129] A significant change in polymer structure in different

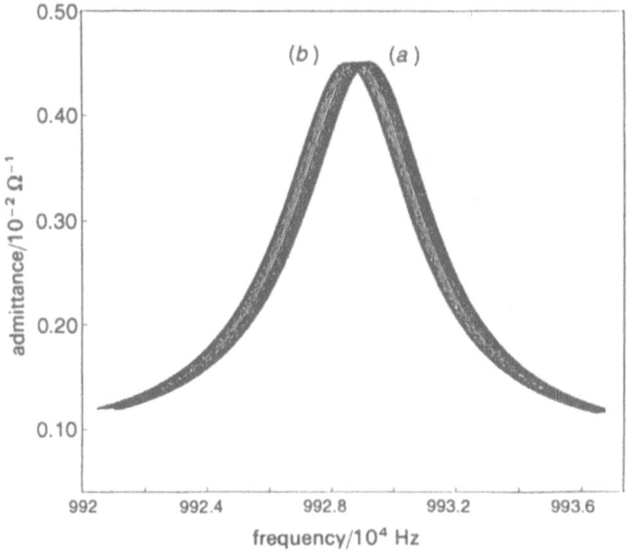

FIGURE 8.20. Admittance spectra during redox switching in 1.0 M $HClO_4$ at 5 mV s^{-1}; coulometric surface coverage: 2×10^{-8} mol cm^{-2}. (a) Os^{II}, (b) Os^{III}. (From Ref. 129.)

oxidation states results in a change in the coupled viscous liquid load[172–174] and a nonzero intercept. The linear relationship between the frequency of oscillation and the film thickness confirms that the polymer behaves as a rigid, thin layer for the range of film thicknesses employed, and illustrate mass changes can be accurately evaluated from frequency measurements.

In the concentration range of 0.1–1.0 M $HClO_4$, there is little variation in the normalized mass change of *ca.* 144 g/mol associated with the oxidation of each osmium center. Potentiometric data in Fig. 8.19 indicate the existence of permselective films within this concentration range. This reflects the layer properties in the reduced Os^{II} state (Fig. 8.16), where in this concentration range, the layer mass also remains invariant and represents a permselective film in which there is full protonation of the PVP backbone. Because of this permselectivity, the normalized mass change of *ca.* 144 g/mol can be equated to the movement of a single perchlorate counteranion and an average of 2.6 solvent molecules for every osmium site oxidized. Although this level of solvent transfer is close to the hydration number of ClO_4^- of 2.6,[175] it was shown that solution values for anion hydration are not necessarily applicable to ions within polymer films. In fact for mass transfer within this polymer in $NaClO_4$, no solvent transfer with the counteranion was observed.[152,153]

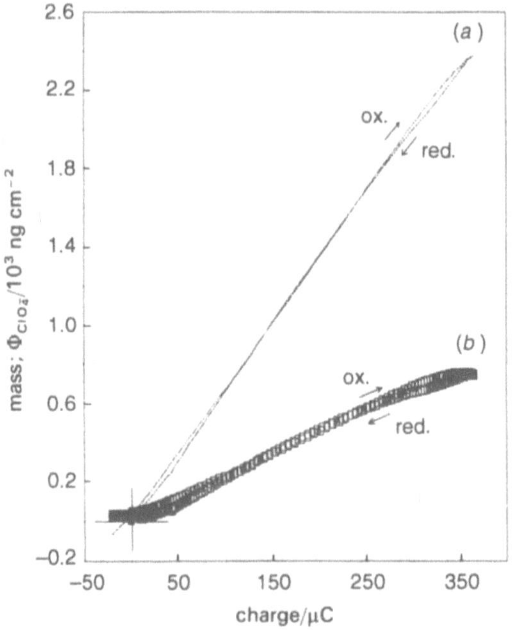

FIGURE 8.21. (a) Mass and (b) $\Phi_{ClO_4^-}$ charge plots in 0.1 M $HClO_4$ at 1 mV s^{-1}; coulometric surface coverage: 2×10^{-8} mol cm^{-2}. (From Ref. 129.)

Figure 8.21 shows the $\Phi_{ClO_4^-}$ [see Eq. (36)] and mass charge plots for a 1 mV/s scan rate in 0.1-M $HClO_4$. Analysis in a 1.0-M electrolyte is similar, although nonclosure of the $\Phi_{ClO_4^-}$ plot is observed. This nonclosure is a feature of the behavior of this metallopolymer in $HClO_4$, and it was previously observed in a study of polybithiophene.[138] Nonclosure of the plot may arise from trace oxygen in solution, or instrumental drift in the integrator offset, which can become significant during slow scan rates used in this work. However it is thought that the drift in charge is most probably due to incomplete charge recovery from the compact polymer structures that exist in $HClO_4$ electrolytes. Based on these plots, the mechanism of coion expulsion from deprotonation of the polymer backbone can be discounted. At all times during oxidation, $\Phi_{ClO_4^-}$ is positive, indicating the net movement of mobile species into the layer. The monotonic change in $\Phi_{ClO_4^-}$ represents the concomitant movement of all mobile species with the anion. At all but the lowest and highest electrolyte concentrations, the half-total mass change $E_{1/2(m)}$ and the half-total charge change $E_{1/2(q)}$ coincide for both anodic and cathodic processes,[129,130] indicating concomitant mass and charge transfer.[147,152]

This, in conjunction with the $\Phi_{ClO_4^-}$ data, indicates that mass changes are due to a single mass transfer mechanism, and are not the result of several different processes.

4.2.4. Redox Switching at Concentrations below 0.1-M $HClO_4$

The apparent pK_a of protonated pyridine groups in the immobilized layer was estimated to be 2.50 (±0.10).[129] Significantly, it is at the electrolyte concentration corresponding to this pH (*ca.* 0.01-M $HClO_4$), that morphology changes within this polymer are found. The pK_a^{app} obtained for immobilized metallopolymer is more acidic than the pK_a of the homopolymer in solution (3.3). At pHs greater than pH 2.50, the contacting electrolyte is 10^{-5}–10^{-3} M, while the concentration of fixed osmium sites is *ca.* 0.7 M.[58,59] Consequently there is a significant Donnan potential at the membrane–solution interface[176] that excludes the electrolyte. This Donnan potential results in difficulty in protonating the polymer backbone due to the electroneutrality requirement for counteranion insertion with the finite void volume of the rigid, compact polymer film.[129] Subsequent swelling of the polymer layer in this electrolyte concentration resulting from reduced levels of protonation is highlighted in Fig. 8.22, which shows that this change in morphology is exacerbated during oxidation of the polymer layer.

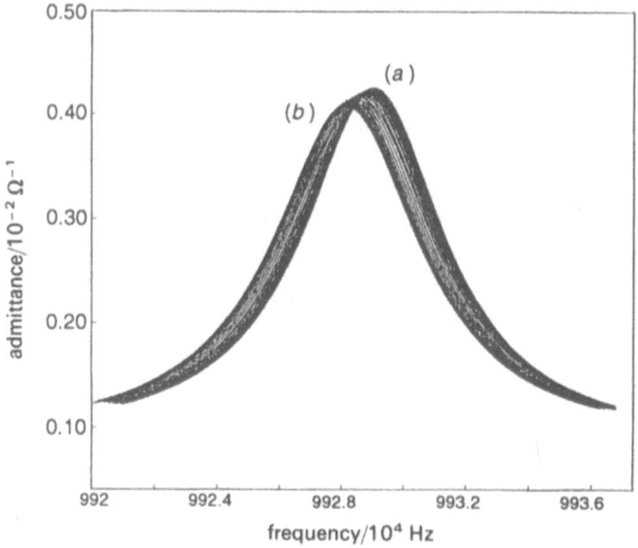

FIGURE 8.22. Admittance spectra during redox switching in 0.01 M $HClO_4$ at 5 mV s^{-1}; coulometric surface coverage: 2×10^{-8} mol cm^{-2}. (a) Os^{II}, (b) Os^{III}. (From Ref. 129.)

This change in film structure has a profound influence on normalized mass changes occurring during redox processes, which increase from *ca.* 144 g/mol for the concentration range 0.1–1.0 M to 186 g/mol for 0.01 M $HClO_4$. The interrelationship between polymer morphology and the extent of solvent transfer is evident in Fig. 8.22. The normalized mass change of 186 g/mol represents the greater level of solvent movement facilitated by change in polymer structure, and it is in turn the cause of further swelling.(130)

4.2.5. Redox Switching in the Concentration Range of 1.0–4.0-M $HClO_4$

As the concentration is increased from 1.0 to 4.0 M $HClO_4$, there is a breakdown in the Nernstian relationship between ClO_4^- activity and E'_{app} (see Fig. 8.19). This breakdown is reflected in the increased normalized mass change observed at concentrations from 1.0–3.0-M electrolyte, i.e., in 2-M $HClO_4$ Δm is equal to 186 and in 3-M $HClO_4$, Δm is equal to 479 g/mol.(129) Assuming that solvent transfer due to activity considerations is the same as for lower concentrations (i.e., 2–3 solvent molecules per redox site conversion), this increase in normalized mass is most likely due to neutral perchloric acid transfer. Therefore, in 3.0-M $HClO_4$, the transfer of undissociated acid is about 2.8 moles of $(H_3O^+)(ClO_4^-)$ per mole of osmium sites.

Figure 8.23 highlights the mass–charge and $\Phi_{ClO_4^-}$–charge plots for the metallopolymer in 4-M $HClO_4$ at 5 mV/s. The mass transport process has changed, and a net mass loss is observed for the overall oxidation. The initial 30–40% of oxidation shows a mass influx into the polymer layer equivalent to a mass change of *ca.* 594 g/mol. This yields a value of 3.8 for $(H_3O^+)(ClO_4^-)$ transfer. However, during continued oxidation, there is a surprising mass loss from the polymer layer. The fixed site concentration of osmium centers is *ca.* 0.7 M,(130) and it is anticipated that on complete protonation of the PVP backbone, the concentration of positively charged sites increases tenfold to *ca.* 7.0 M. Film compaction in $HClO_4$ increases this concentration even further.(168) The concentration of bathing electrolyte is therefore a significant fraction of the polymer phase fixed-site concentration at this $HClO_4$ electrolyte concentration. In a model developed for ion exchange membranes by Gregor,(177) the partial exclusion of free electrolyte from the membrane, in contact with bathing solution concentrations approaching the fixed-site concentration, results in the movement of solvent from the low free electrolyte concentration within the membrane to high concentrations in the contacting solution. This counter-Donnan osmotic flux not only results in deswelling, but also serves to exclude electrolyte further and thus cause further flux of material from the layer.(177)

The initial mass increase in 4.0 M $HClO_4$ most likely reflects the presence of a threshold level of charge density within the polymer that can be tolerated before the osmotic flux of solvent becomes thermodynamically favored, which leads to a

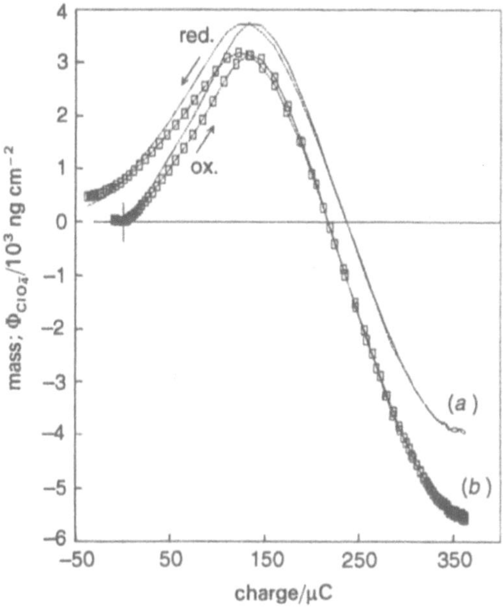

FIGURE 8.23. (a) Mass and (b) $\Phi_{ClO_4^-}$–charge plots in 4.0 M $HClO_4$ at 5 mV s^{-1}; coulometric surface coverage: 1.5×10^{-8} mol cm^{-2}. (From Ref. 129.)

mass loss. This osmotic flux not only reduces the polymer phase solvent, but may also include the movement of neutral electrolyte from the polymer layer and result in a partial reattainment of permselectivity.[129]

5. MEDIATED ELECTROCATALYSIS AT REDOX POLYMERS

The range of applications envisaged for modified electrodes is considerable; their unique features are already exploited in such diverse areas as electrochemical sensors,[178] heterogeneous electrocatalysis,[179,180] solar energy conversion,[181,182] electrochromics,[183,184] energy storage,[185,186] and information storage.[19,187] Of these, no area of modified electrode research is more prolific than electroanalysis. The use of polymeric-modifying materials in electroanalysis offers considerable advantages compared to other chemically modified electrodes. Polymer materials offer synthetic flexibility and inherent stability as well as ease of application to electrode substrates. Polymers also offer the possibility of up to 10^5 molecular layers of electroactive species[178] at high surface concentration (up to 10-M) at the electrode surface compared to conventional monolayer-derivatized surfaces.[188] Polymer-modified electrodes can therefore react with substrates in a three-dimensional zone rather than a two-dimensional region at monolayer-modified electrodes. This provides the possibility of heterogeneous catalysis at the polymer-modified electrode competing with the solution phase homogenous catalysis and the resulting advantages associated with this development.[189]

It is often necessary to apply potentials in excess of that required by thermodynamics to drive a particular electrochemical reaction when the kinetics of a redox reaction are slow at a bare electrode. This overpotential can be lowered by accelerating the rate of electrochemical reaction by modifying the electrode surface. For surface-immobilized redox sites, electrocatalysis is accomplished by shuttling electrons by repeated cycling between the catalytic and precatalytic state between the electrode and the electroactive substrate; this process is known as mediated electrocatalysis.[190]

The theory of mediated electrocatalysis has been studied extensively by many authors, and established theories allow mechanistic elucidation and kinetic parameters to be evaluated. An extensive discussion of these appears in Ref. 188. The mediation process can be described by the following reactions:

$$A \rightarrow B + e^- \tag{38}$$

$$B + Y \rightarrow A + Z \tag{39}$$

In this example, the mediating process involves the oxidation of the surface-bound redox species A, but for a mediation process based on the reduction of the surface-bound redox species, the same theory applies. Depending on such factors as film morphology, film thickness, the partition coefficient, diffusion rate of the electroactive species through the film, and electron transfer rates, four types of limiting cases can be identified

Electron and substrate diffusion (D_E and D_Y, respectively) are so rapid that the rate-controlling step is the rate k of the catalytic reaction.

The catalytic reaction is so rapid that the rate is controlled by the two diffusion processes, i.e., diffusion of electrons and substrate.

When diffusion of electrons is faster than diffusion of the substrate, a pure kinetic situation may arise by mutual compensation of the latter process and the catalytic reaction.

In the opposite case, when diffusion of the substrate is faster than diffusion of electrons, a pure kinetic situation may again arise, resulting in the mutual compensation of electron diffusion and catalytic reaction.

The fundamental processes involved in the mediating process are identified as charge introduction at the modifying/electrode interface, charge introduction at the layer/electrolyte interface, and reaction of the target analyte with the modifying layer. Coupled to these reactions, one may observe substrate diffusion into the film as dictated by the partition coefficient, K. If the substrate is capable of penetrating the film, then the diffusion rate of the substrate, D_Y, within the layer will in all but a few cases be considerably less than the solution value D_s.

The two models most frequently used to investigate the mediating properties of modified electrodes are those developed by Albery *et al.*[191–194] and Saveant *et al.*[195–199] The two differ in that Albery and coworkers introduced the concept of reaction layers to describe two different regions within the polymer film: a region where permeating substrate Y is converted into product Z and a region where B is reconverted into A. These layers are usually located close to the film–electrolyte and film–electrode interfaces, respectively. In Saveant and coworkers' approach the mediation process is described in terms of characteristic currents. Other models have also been developed.

For both models the rotating disk electrode provides necessary experimental data. This technique allows control of substrate diffusion. The effect of the modifying layer on the limiting current can be obtained by analyzing results obtained using the Koutecky–Levich formalism. The rate-limiting step of the mediating process can be identified from the nature of the Koutecky–Levich plots and from the effect that changes in film thickness, mediator concentration, and rotation rate have on these plots. In this section we simply use the framework developed in Albery and Mount (Chapter 4) to analyze mediation properties of these electrodes and to determine rate-limiting processes and/or the location of the mediating process within the polymer layer.

First of all, we must realize that the most important property of the electroactive polymer, as far as mediation is concerned, is its redox potential. To mediate a reduction of a solution species, the redox potential of the electroactive layer must be less positive than that of the analyte; for the mediated oxidation process, the reverse is the case. This means that the osmium polymers under consideration here which have a redox potential of about 250 mV, are thermodynamically able to mediate the reduction of Fe(III) to Fe(II),[200,201] but not the reverse process (see Fig. 8.24), since the formal potential of the Fe(III/II) couple is 450 mV. The difference in the two redox potentials can be considered the driving force for the mediating process. On the basis of these considerations, it is clear that the mediated reduction of Fe(III) [as in Eq. (38)] is irreversible.

$$\text{Fe(III)} + \text{Os(II)} \rightarrow \text{Fe(II)} + \text{Os(III)} \tag{40}$$

Characteristics of this process are discussed here as a function of another important parameter, the electrolyte. It has been shown in Section 8.3.2. that charge

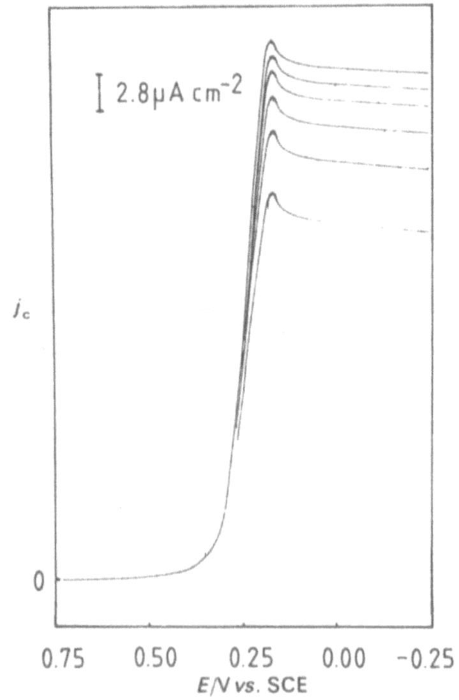

FIGURE 8.24. Rotating disk voltammograms showing reduction of a 0.2 mM $[Fe(H_2O)_6]^{3+}$ solution in 0.1 M H_2SO_4 at an electrode modified with $[Os(bipy)_2(PVP)_{10}Cl]Cl$. Rotation rates (*bottom to top*): 500, 1000, 1500, 2000, 2500, and 3000 rpm; surface coverage: 5×10^{-9} mol cm^{-2}. (From Ref. 201.)

transport behavior of these redox polymers strongly depends on the nature and concentration of the electrolyte. We therefore compare the mediating behavior of the polymer $[Os(bipy)_2(PVP)_{10}Cl]Cl$ in both sulphuric acid and perchloric acid electrolytes.

A typical set of rotating disk voltammograms for the mediated reduction of $[Fe(H_2O)_6]^{3+}$ in 0.1 M H_2SO_4 is shown in Fig. 8.24. These plots show that the mediating process occurs in the potential region of the Os(III/II) redox couple. Koutechy–Levich plots obtained from such curves are shown in Fig. 8.25 as a function of the layer surface coverage. These Koutechy–Levich plots are linear, and the catalytic reaction increases with increasing film thickness. Furthermore slopes obtained are the same as the one obtained at the bare electrode. Since the intercepts of these plots are equal to k'_{ME}, the electrochemical rate constant clearly depends on the layer thickness L. In fact for surface coverages up to 1×10^{-7} mol/cm^2, this

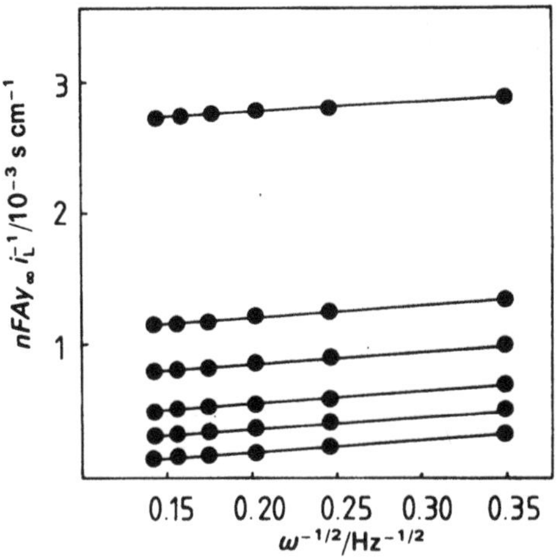

FIGURE 8.25. Typical Koutechy–Levich plots for the reduction of 0.2 mM $[Fe(H_2O)_6]^{3+}$ in 0.1 M H_2SO_4 at an $[Os(bipy)_2(PVP)_{10}Cl]Cl$-modified electrode; surface coverages in mol cm^{-2} (*top to bottom*): 7.0×10^{-10}, 1.8×10^{-9}, 2.7×10^{-9}, 5.0×10^{-9}, 1.1×10^{-8}, and bare platinum. (From Ref. 201.)

relationship is linear. For thicker layers however k'_{ME} is independent of layer thickness. Controlled potential coulometry experiments, which yield the concentration of the mediator Os(II) as a function of the potential, and an analysis of the rising portion of the rotating disk voltammograms show that k'_{ME} is also linearly dependent on b_0, the mediator concentration. Since the limiting current depends on the rotation rate of the electrode and the Koutechy–Levich plots are linear, the St_e, LSt_e, and *LEK* cases can be eliminated as models to describe the mediation process. Given furthermore that the slope of these plots is the same as that obtained for bare electrodes, the $LRZt_et_y$ case can be eliminated. If then the dependence of the k'_{ME} on the surface coverage is taken into account, we conclude that for surface coverages between 1×10^{-8} and 7×10^{-10} mol/cm^2, the kinetic zone for the mediation reaction is of the *LK* type. The reaction occurs therefore throughout the polymer layer, leading to a three-dimensional catalysis. For surface coverages of more than 1×10^{-7} mol/cm^2, an *LSk* mechanism is operating. So in this case mediation is taking place in a relatively thick layer at the electrolyte/film interface, and it is limited by substrate diffusion from the bulk solution to the electrode surface.

In 1.0 M $HClO_4$ quite a different situation is observed: Koutecky–Levich plots in this electrolyte for the same mediation reaction are shown in Fig. 8.26 as a function of the layer thickness. As in 0.1-M sulphuric acid, limiting currents depend on the rotation rate. However, although Koutecky–Levich plots are linear and have

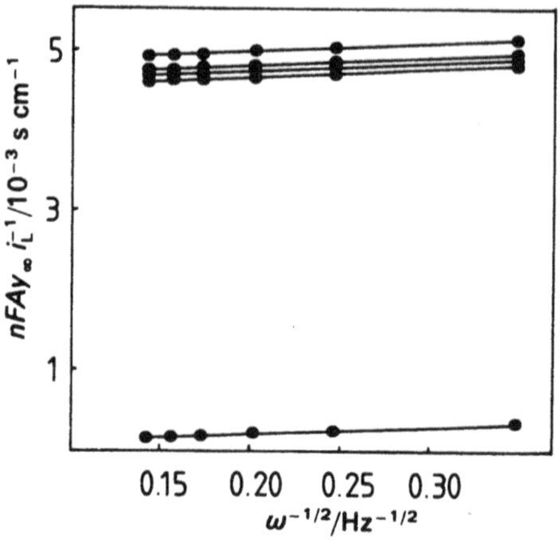

FIGURE 8.26. Typical Koutechy–Levich plots for the reduction of 0.2 mmol dm^{-3} $[Fe(H_2O)_6]^{3+}$ in 1.0 mol dm^{-3} $HClO_4$ at an $[Os(bipy)_2(PVP)_{10}Cl]Cl$-modified electrode; surface coverages in mol cm^{-2} (*top to bottom*): 7.0×10^{-10}, 1.8×10^{-9}, 2.7×10^{-9}, 5.0×10^{-9}, 1.1×10^{-8}, and bare platinum. (From Ref. 201.)

a slope similar to one obtained for a bare electrode, the intercept of these plots does not depend significantly on layer thickness. This, and the fact that the reaction order with respect to the concentration of the mediator b_0 is unity suggest that in this case, the kinetic zone is of the *Sk* type. In this case the mediation reaction takes place within a thin layer at the film–electrolyte interface. Most likely under these conditions, the substrate cannot penetrate the film. These observations have obvious implications for the analytical application of such electrodes. The choice of an electrolyte is clearly not an obvious one, and measurement conditions must be considered carefully.

The mediation behavior of this redox polymer for the reduction of nitrite is quite different.[202] This mediation process was studied in acidic sulphate electrolytes, and it most likely occurs through reduction of the intermediate NO^+ by Os(II) sites. This reduction process is not observed at unmodified electrodes. Interestingly the limiting current obtained using coated rotating disk electrodes is independent of rotation speed. This leads to Koutechy–Levich plots with zero slope (see Fig. 8.27). These plots show however that the intercept depends on film thickness. This behavior corresponds to the kinetic situation *Lk* according to the Albery notation and R, using Saveant's model. Since the slope of the Koutechy–Levich plots is zero, the diffusion of substrate in the solution does not seem to exert control over the limiting current. This is consistent with a slow rate for the cross-exchange reaction. This rate can be determined from the analytical expression for k'_{ME}, which can be derived from the *Lk* kinetic case:

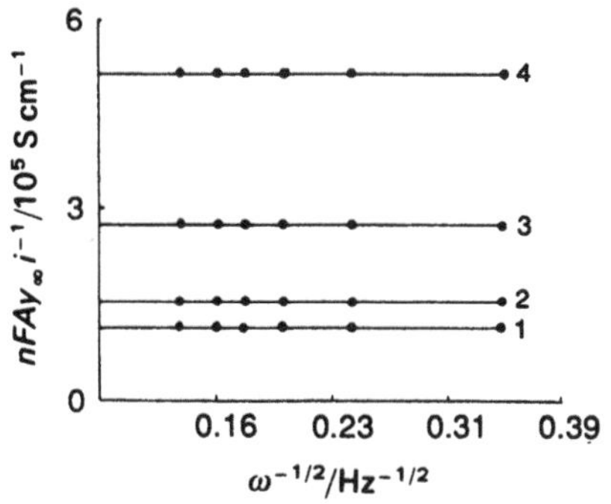

FIGURE 8.27. Koutechy–Levich plots for the mediated reduction of 2 mM nitrite. Curves 1–4 represent surface coverages of 1.8×10^{-8}, 6.2×10^{-9}, 4.0×10^{-9}, and 1.5×10^{-9} mol cm^{-2}, respectively. (From Ref. 202.)

$$k'_{ME} = kKb_0L \quad (41)$$

where the symbols have their usual meanings. From oxidation experiments with the analogous ruthenium polymer, which is not able to mediate the oxidation of nitrite, it has been estimated that the partition coefficient is about 1. By using the appropriate values for b_0 and L, k was estimated to be $(9.4 \pm 2) \times 10^{-1}$ dm^{3}mol^{-1}s^{-1}. This rate constant is indeed small when compared to the value obtained for the previously described reduction of Fe(III) under the same kinetic regime (2.8×10^3 dm^{3}mol^{-1}s^{-1}). The mediation reaction therefore takes place throughout the film, and it is controlled by the rate of the cross-reaction.

As mentioned earlier the mediation properties of the coatings are strongly related to the redox potential of the electroactive couple. For this reason osmium polymers are not able to mediate the Fe(II) oxidation. However, for the corresponding ruthenium polymer, which has a redox potential of about 0.7 V versus SCE, the mediation of the Fe(II) oxidation reaction is possible.[203] In 1.0-M HCl for polymer surface coverages in the range of 10^{-10}–6.7×10^{-10} mol cm^{-2}, the kinetic control of this reaction is of the R type, i.e., catalytic reaction is the sole rate-determining step, since the rate of charge propagation and substrate diffusion are sufficiently rapid so as not to interfere kinetically.[203] For thicker films there is a shift in kinetic control from the R type to the SR type. For coverages of 10^{-8} mol cm^{-2}, kinetic control is entirely of the SR type; i.e., catalytic reaction takes place in a reaction layer adjacent to the coating–solution boundary.[203] Based on this work the simultaneous detection of both Fe(II) and Fe(III) in a single sample aliquot, using a dual-electrode assembly modified with ruthenium and osmium polymers was reported.[200]

6. CONCLUSION

The examples described in this chapter clearly show the potential of modified electrodes based on redox-active osmium-containing polymers. The redox potentials of these materials can be manipulated by varying the nature of the polymer-bound redox couple, which allows us to tailor polymers to particular application, especially in the sensors area. Furthermore changes can also be made in the polymer composition both with respect to polymer loading and the nature of the polymer backbone. This will allow control of such parameters as substrate diffusion and charge transport through the layer. This flexibility allows the systematic investigation of electrochemical properties of electrodes modified with such materials.

Charge transport and mediation parameters show a strong dependence on the nature and concentration of the electrolyte. This points to a coupling between electrochemical properties of the layer and physicochemical properties of the polymer phase, making these structures good examples of supramolecular chemistry. It appears from these studies that for such metallopolymers, requirements for rapid electron transport rates may be mutually exclusive, i.e., a high concentration of redox sites and facile ion diffusion to maintain electroneutrality. The higher loaded ($n = 5$ and $n = 10$) metallopolymers exhibit ion diffusion limitations, and the fastest charge transport rates are obtained for intermediate loadings at high electrolyte concentrations. The EQCM data further emphasize the profound relationship between the contacting electrolyte and the morphology of the polymer phase. Much more work is needed to understand fully the nature of the electrochemical processes in detail. In particular the observed time dependence of the charge transport process needs more detailed investigation. Finally, since electrochemical properties are strongly coupled to the interaction of electrolyte and polymer, statements about the charge transport and mediation properties of such modified electrodes are valid only under the experimental conditions used.

ACKNOWLEDGMENT. The financial support of the Irish Science and Technology Agency (Forbait) is gratefully acknowledged.

Note Added in Proof. During the preparation of this chapter a series of papers dealing with the metallopolymers under discussion has appeared. Crosslinking of the polymer $[Os(bipy)_2(PVP)_{10}Cl]Cl$ with *p*-dibromobenzene, 1,5-dibromopentane and 1,10-dibromodecane leads to improved stability of the modifying layer. The effect of crosslinking on the charge transport parameters and activation energies for these systems is reported.[204–207] It has been shown that such crosslinking techniques can be used in an analytical device to differentiate between normally interfering analyses such as nitrite and Fe(III).[208]

The effect of a change in the nature of the polymer backbone has also been reported.[209] The introduction of styrene moieties yields stable modifying layers but the charge transport and the partition of substrate from solution into the polymer are negatively affected. It has also been shown that the addition of organic solvents such as acetonitrile and methanol can significantly increase the charge transport properties of $[Os(bipy)_2(PVP)_{10}Cl]Cl$.[210]

Further studies have also appeared in the mass transfer properties of the modifying layers.[211–213] The mass transfer behavior of the polymer $[Os(bipy)_2(PVP)_{10}Cl]Cl$ in para-toluene sulphonic acid has been studied in detail.[212] In this study the effect of the non-rigidity of the layer on the mass transport data is discussed in some detail. A detailed study of the pH dependence of the mass transfer of a glassy carbon electrode modified with the $[Ru(bipy)_2(PVP)_{10}(H_2O)]^{2+}$ is reported.[213] This study shows that the EQCM can be used very effectively for the study of redox reactions which show complicated (in this case pH dependent) features. (See also Section 8.2).

Finally, an important application of these materials is in the area of biosensors. Heller and co-workers have carried out detailed elegant investigations on the application of osmium based redox polymers in glucose sensors.[214–217] (For a detailed discussion of biosensors see Chapter 11.) Recently they have studied the effect of crosslinking of the polymer $[Os(bipy)_2(PVP)_6Cl]Cl$ with the diepoxide poly(ethylene glycol) diglycidyl ether.[218] In this study the charge transport properties of quaternized and crosslinked hydrogels similar to those used in the preparation of their biosensors was reported. The results obtained suggest that charge transport increases with protonation and quaternization. For unquaternized systems charge transport also is shown to increase with crosslinking. An increase of the charge transport properties with the degree of crosslinking was also observed by in other studies.[204] Finally, it has recently been reported that for biosensors based on immobilized osmium polypyridyl mediators the direct reaction between the polymer bound osmium mediator and ascorbic acid needs to be considered.[219]

REFERENCES

1. N. Oyama, T. Ohsaka, T. Ushirogouchi, S. Sanpel, and S. Nakamura, *Bull. Chem. Soc. Jpn.* **61**, 3103 (1988).
2. H. S. White, J. Leddy, and A. J. Bard, *J. Am. Chem. Soc.* **104**, 4811 (1982).
3. Y-M. Tsou and F. C. Anson, *J. Electrochem. Soc.* **131**, 595 (1984).
4. E. T. T. Jones and L. R. Faulkner, *J. Electroanal. Chem. Soc.* **222**, 201 (1987).
5. M. Majda and L. R. Faulkner, *J. Electroanal. Chem.* **137**, 149 (1982).
6. M. Majda and L. R. Faulkner, *J. Electroanal. Chem.* **169**, 77 (1984).
7. M. Majda and L. R. Faulkner, *J. Electroanal. Chem.* **169**, 97 (1987).
8. X. Chen, P. He, and L. R. Faulkner, *J. Electroanal. Chem.* **222**, 223 (1987).
9. M. Sharp, B. Lindholm, and E. L. Lind, *J. Electroanal. Chem.* **274**, 35 (1989).
10. F. C. Anson, Y. M. Tsou, and J. M. Saveant, *J. Electroanal. Chem.* **178**, 113 (1984).
11. J. Leddy and A. J. Bard, *J. Electroanal. Chem.* **189**, 203 (1985).

12. H. D. Abruna, P. Denisevich, M. Umana, T. J. Meyer, and R. W. Murray, *J. Am. Chem. Soc.* **103**, 1 (1981).
13. P. Denisevich, K. W. Willman, and R. W. Murray, *J. Am. Chem. Soc.* **103**, 4727 (1981).
14. T. Ikeda, C. R. Leidner, and R. W. Murray, *J. Am. Chem. Soc.* **103**, 7422 (1981).
15. T. Ikeda, R. H. Schmehl, P. Denisevich, K. W. Willman, and R. W. Murray, *J. Am. Chem. Soc.* **104**, 2683 (1982).
16. J. S. Facci, R. H. Schmehl, and R. W. Murray, *J. Am. Chem. Soc.* **104**, 4959 (1982).
17. P. Denisevich, H. D. Abruna, C. R. Leidner, T. J. Meyer, and R. W. Murray, *Inorg. Chem.* **21**, 2153 (1982).
18. J. M. Calvert, R. H. Schmehl, B. P. Sullivan, J. S. Facci, T. J. Meyer, and R. W. Murray, *Inorg. Chem.* **22**, 2151 (1983).
19. P. G. Pickup and R. W. Murray, *J. Am. Chem. Soc.* **105**, 4510 (1983).
20. P. G. Pickup and R.W. Murray, *J. Electrochem. Soc.* **131**, 833 (1984).
21. J. M. Calvert, D. L. Peebles, and R. J. Nowak, *Inorg. Chem.* **24**, 3111 (1985).
22. C. D. Ellis, W. R. Murphy, and T. J. Meyer, *J. Am. Chem. Soc.* **103**, 7480 (1981).
23. P. G. Pickup, W. Kutner, C. R. Leidner, and R. W. Murray, *J. Am. Chem. Soc.* **106**, 1991 (1984).
24. C. R. Leidner and R. W. Murray, *J. Am. Chem. Soc.* **106**, 1606 (1984).
25. A. G. Ewing, B. P. Feldman, and R. W. Murray, *J. Electroanal. Chem. Soc.* **172**, 145 (1984).
26. J. C. Jernigan, C. E. D. Chidsey, and R. W. Murray, *J. Am. Chem. Soc.* **107**, 2824 (1985).
27. J. Leddy, A. J. Bard, J. T. Maloy, and J. M. Saveant, *J. Electroanal. Chem. Soc.* **187**, 205 (1985).
28. J. M. Calvert and T. J. Meyer, *Inorg. Chem.* **20**, 27 (1981).
29. S. M. Geraty and J. G. Vos, *J. Chem. Soc. Dalton Trans.* 3073 (1987).
30. J. M. Clear, J. M. Kelly, C. M. O'Connell, and J. G. Vos, *J. Chem. Res,(M)* 3039 (1981).
31. E. M. Kober, J. L. Marshall, W. J. Dressick, B. P. Sullivan, J. V. Casper, and T. J. Meyer, *Inorg. Chem.* **24**, 2755 (1985).
32. B. P. Sullivan, D. J. Salmon, and T. J. Meyer, *Inorg. Chem.* **17**, 3334 (1978).
33. E. C. Johnson, B. P. Sullivan, D. J. Salmon, S. A. Adeyemi, and T. J. Meyer, *Inorg. Chem.* **17**, 2211 (1978).
34. D. A. Buckingham, F. P. Dwyer, H. A. Goodwin, and A. M. Sargeson, *Aust. J. Chem.* **17**, 325 (1964).
35. E. M. Kober, J. V. Casper, R. S. Lumpkin, and T. J. Meyer, *J. Phys. Chem.* **90**, 3722 (1986).
36. J. M. Clear, J. M. Kelly, D. C. Pepper, and J. G. Vos, *Inorg. Chim. Acta* **33**, L139 (1979).
37. B. Durham, S. R. Wilson, D. J. Hodgson, and T. J. Meyer, *J. Am. Chem. Soc.* **102**, 600 (1980).
38. J. M. Calvert and T. J. Meyer, *Inorg. Chem.* **21**, 3978 (1982).
39. S. M. Geraty and J. G. Vos, *J. Electroanal. Chem.* **176**, 389 (1984).
40. O. Haas and J. G. Vos, *J. Electroanal. Chem.* **113**, 139 (1980).
41. O. Haas, M. Kriens, and J. G. Vos, *J. Am. Chem. Soc.* **103**, 1318 (1981).
42. O. Haas, H. R. Zumbrunnen, and J. G. Vos, *Electrochim. Acta* **30**, 1551 (1985).
43. R. J. Forster and J. G. Vos, *Macromolecules* **23**, 4372 (1990).
44. D. Leech, R. J. Forster, M. R. Smyth, and J. G. Vos, *J. Mater. Chem.* **1**, 629 (1991).
45. J. M. Calvert, J. V. Casper, R. A. Binstead, T. D. Westmoreland, and T. J. Meyer, *J. Am. Chem. Soc.* **104**, 6620 (1982).
46. J. M. Clear, J. M. Kelly, and J. G. Vos, *Makromol. Chem.* **184**, 613 (1983).
47. T. Shimidzu, K. Izaki, Y. Akai, and T. Iyoda, *Polym. J.* **13**, 889 (1981).
48. J. van Houten and R. J. Watts, *J. Am. Chem. Soc.* **98**, 4853 (1976).
49. J. van Houten and R. J. Watts, *Inorg. Chem.* **17**, 3381 (1978).
50. B. Durham, J. V. Casper, J. K. Nagle, and T. J. Meyer, *J. Am. Chem. Soc.* **104**, 4803 (1982).
51. A. B. P. Lever, *Inorganic Electronic Spectroscopy* (Elsevier, New York, 1968).
52. C. K. Balhausen, *Absorption Spectra and Chemical Bonding in Complexes* (Pergamon, New York, 1972), p. 114.

53. G. J. Samuels and T. J. Meyer, *J. Am. Chem. Soc.* **103**, 307 (1981).
54. O. Haas, N. Muller, and H. Gerischer, *Electrochim. Acta* **27**, 991 (1982).
55. M. E. G. Lyons, H. G. Fay, J. G. Vos, and A. J. Kelly, *J. Electroanal. Chem.* **250**, 207 (1988).
56. R. J. Forster and J. G. Vos, *Langmuir* **10**, 4330 (1994).
57. R. J. Forster and J. G. Vos, *J. Inorganic and Organometallic Polymers* **1**, 67 (1991).
58. R. J. Forster, A. J. Kelly, J. G. Vos, and M. E. G. Lyons, *J. Electroanal. Chem.* **270**, 365 (1989).
59. R. J. Forster and J. G. Vos, *J. Electroanal. Chem.* **314**, 135 (1991).
60. R. J. Forster and J. G. Vos, *Electrochim. Acta* **37**, 159 (1992).
61. R. J. Forster, J. G. Vos, and M. E. G. Lyons, *J. Chem. Soc. Faraday Trans.* **87**, 3761 (1991).
62. R. J. Forster, J. G. Vos, and M. E. G. Lyons, *J. Chem. Soc. Faraday Trans.* **87**, 3769 (1991).
63. E. M. Kober, J. V. Casper, B. P. Sullivan, and T. J. Meyer, *Inorg. Chem.* **27**, 4587 (1988).
64. B. P. Sullivan, D. J. Salmon, T. J. Meyer, and J. Peedin, *Inorg. Chem.* **18**, 3369 (1979).
65. J. G. Gaudiello, P. G. Bradley, K. A. Norton, W. H. Woodruff, and A. J. Bard, *Inorg. Chem.* **23**, 3 (1984).
66. A. Juris, V. Balzani, F. Barigelletti, S. Campagna, P. Belser, and A. von Zelewsky, *Coord. Chem. Rev.* **84**, 85 (1988).
67. S. D. Ernst and W. Kaim, *Inorg. Chem.* **28**, 1520 (1989).
68. F. Barigelletti, A. Juris, V. Balzani, P. Belser, and A. von Zelewsky, *Inorg. Chem.* **26**, 4115 (1987).
69. A. Juris, S. Campagna, V. Balzani, G. Gremaud, and A. von Zelewsky, *Inorg. Chem.* **27**, 3652 (1988).
70. A. Juris, P. Belser, F. Baribelletti, A. von Zelewsky, and V. Balzani, *Inorg. Chem.* **25**, 256 (1986).
71. F. Barigelletti, P. Belser, A. von Zelewsky, A. Juris, and V. Balzani, *J. Phys. Chem.* **89**, 3680 (1985).
72. F. Barigelletti, A. Juris, V. Balzani, P. Belser, and A. von Zelewsky, *Inorg. Chem.* **26**, 3335 (1983).
73. Y. Kawanishi, N. Kitamura, and S. Tazuka, *Inorg. Chem.* **28**, 2968 (1989).
74. H. Taube, *Pure Appl. Chem.* **51**, 901 (1979).
75. A. C. Sarupa and R. F. Fenske, *Inorg. Chem.* **14**, 247 (1975).
76. P. M. Treichel, H. J. Mueh, and B. E. Bursten, *Isr. J. Chem.* **15**, 253 (1977).
77. P. M. Treichel, H. J. Mueh, and B. E. Bursten, *J. Organomet. Chem.* **110**, C49 (1976).
78. B. A. Moyer and T. J. Meyer, *J. Am. Chem. Soc.* **100**, 3601 (1978).
79. B. A. Moyer and T. J. Meyer, *J. Am. Chem. Soc.* ***101***, 1326 (1979).
80. B. A. Moyer and T. J. Meyer, *Inorg. Chem.* **20**, 436 (1981).
81. B. Durham, J. L. Walsh, C. L. Carter, and T. J. Meyer, *Inorg. Chem.* **19**, 860 (1980).
82. N. M. D. Brown, H. X. You, R. J. Forster, and J. G. Vos, *J. Mater. Chem.* **1**, 517 (1990).
83. C. P. Andrieux, in *Electrochemistry, Sensors and Analysis*, vol. 25 (M. R. Smyth and J. G. Vos, eds.) Analytical Symposia Series (Elsevier, Amsterdam 1986), 235.
84. P. Daum, J. R. Lenhard, D. R. Rolison, and R. W. Murray, *J. Am. Chem. Soc.* **102**, 4649 (1980).
85. A. H. Shroeder, F. B. Kaufman, V. Patel, and E. M. Engler, *J. Electroanal. Chem.* **113**, 193 (1980).
86. T. P. Henning, H. S. White, and A. J. Bard, *J. Am. Chem. Soc.* **103**, 3937 (1981).
87. R. J. Nowak, F. A. Schultz, M. Umana, R. Lam, and R. W. Murray, *Anal. Chem.* **52**, 315 (1980).
88. P. Ferruti and R. Barbucci, *Adv. Polym. Sci.* **58**, 55 (1984).
89. A. Eisenberg, *Macromolecules* **3**, 147 (1970).
90. E. F. Bowden, M. F. Dautartas, and J. F. Evans, *J. Electroanal. Chem.* **219**, 49 (1987).
91. M. F. Dautartas, E. F. Bowden, and J. F. Evans, *J. Electroanal. Chem.* **219**, 71 (1987).
92. E. F. Bowden, M. F. Dautartas, and J. F. Evans, *J. Electroanal. Chem.* **219**, 91 (1987).
93. A. R. Hillman, D. C. Loveday, and S. Bruckenstein, *J. Electroanal. Chem.* **300**, 67 (1991).
94. A. R. Hillman, in *Electrochemical Science and Technology of Polymers* vol. 1 (R. G. Linford, ed.) (Elsevier, Amsterdam, 1987), p. 103.
95. H. Daifuku, K. Aoki, K. Tokuda, and H. Matsuda, *J. Electroanal. Chem.* **183**, 1 (1985).
96. A. J. Bard and L. R. Faulkner, *Electrochemical Methods, Fundamentals and Applications* (Wiley, New York, 1980).

97. D. M. Mohliner, *J. Electroanal. Chem.* **146**, 417 (1983).
98. A. Sevčik, *Collect. Czech. Chem. Commun.* **44**, 327 (1948).
99. K. Aoki, K. Tokuda, and H. Matsuda, *J. Electroanal. Chem.* **146**, 417 (1983).
100. F. G. Cottrell, *Z. Phys. Chem.* **42**, 385 (1902).
101. C. P. Andrieux and J. M. Saveant, *J. Phys. Chem.* **92**, 6761 (1988).
102. W. T. Yap and R. A. Durst, *J. Electroanal. Chem.* **216**, 11 (1987).
103. W. T. Yap, R. A. Durst, E. A. Blubaugh, and D. D. Blubaugh, *J. Electroanal. Chem.* **144**, 69 (1983).
104. J. M. Saveant, *J. Phys. Chem.* **92**, 4526 (1988).
105. S. M. Oh and L. R. Faulkner, *J. Electroanal. Chem.* **269**, 77 (1989).
106. R. W. Murray, in *Electroanalytical Chemistry*, vol. 13 (A. J. Bard, ed.) (Dekker, New York, 1983), p. 191.
107. R. C. Bowers and R. W. Murray, *Anal. Chem.* **38**, 461 (1966).
108. D. Ghesquiere, B. Ban, and C. Chachaty, *Macromolecules* **10**, 743 (1977).
109. D. P. Rillema, D. S. Jones, and H. A. Levy, *J. Chem. Soc., Chem. Commun.* 849 (1979).
110. R. Lange and K. Doblhofer, *J. Electroanal. Chem.* **237**, 13 (1987).
111. R. J. Forster, Ph.D. diss. Dublin City University, 1990.
112. T. Ohsaka, N. Oyama, S. Yamaguchi, and H. Matsuda, *Bull. Chem. Soc. Jpn.* **54**, 2475 (1981).
113. K. J. Vetter, *Electrochemical Kinetics* (Academic Press, New York and London, 1987), p. 107.
114. H. Matsuda, *Bull. Chem. Soc. Jpn.* **53**, 3439 (1980).
115. M. J. Weaver, *J. Phys. Chem.* **83**, 2645 (1976).
116. E. L. Lee, R. J. Cave, K. L. Guyer, P. D. Tyma, and M. J. Weaver, *J. Am. Chem. Soc.* **101**, 1131 (1979).
117. J. T. Hupp and M. J. Weaver, *J. Electroanal. Chem.* **131**, 619 (1984).
118. E. L. Lee and M. J. Weaver, *Inorg. Chem.* **19**, 1077 (1980).
119. M. E. G. Lyons, H. G. Fay, T. McCabe, J. Corrish, J. G. Vos, and A. J. Kelly, *J. Chem. Soc. Faraday Trans.* **86**, 2905 (1990).
120. C. E. D. Chidsey and R. W. Murray, *Science* **231**, 25 (1986).
121. P. Daum and R. W. Murray, *J. Phys. Chem.* **85**, 389 (1981).
122. S. M. Oh and L. R. Faulkner, *J. Am. Chem. Soc.* **111**, 5613 (1989).
123. D. Leech, Ph.D. diss. Dublin City University, 1991.
124. S. Yamaguchi, H. Matsuda, T. Ohsaka, and N. Oyama, *Bull. Chem. Soc. Jpn.* **56**, 2952 (1983).
125. T. Ohsaka, T. Okajima, and N. Oyama, *J. Electroanal. Chem.* **215**, 191 (1986).
126. T. Ohsaka, S. Kunimura, and N. Oyama, *Electrochim. Acta* **33**, 639 (1983).
127. C. Amatore, J. M. Saveant, and D. Tessier, *J. Electroanal. Chem.* **146**, 37 (1983).
128. T. Ohsaka, H. Yamamoto, and N. Oyama, *J. Phys. Chem.* **91**, 3775 (1987).
129. A. P. Clarke, J. G. Vos, A. Glidle, and A. R. Hillman, *J. Chem. Soc. Faraday Trans.* **89**, 1695 (1993).
130. A. P. Clarke, Ph.D. diss. Dublin City University, 1992.
131. A. R. Hillman, D. C. Loveday, M. J. Swann, S. Bruckenstein, and C. P. Wilde, *J. Chem. Soc. Faraday Trans.* **87**, 2047 (1991).
132. S. Bruckenstein and A. R. Hillman, *J. Phys. Chem.* **92**, 4837 (1988).
133. S. Bruckenstein and M. Shay, *Electrochim. Acta* **30**, 1295 (1985).
134. D. Orata and D. A. Buttry, *J. Am. Chem. Soc.* **109**, 3574 (1987).
135. M. R. Deakin and D. A. Buttry, *Anal. Chem.* **61**, 1147A (1989).
136. C. Lu and A. W. Czanderna, eds., *Applications of Piezoelectric Crystal Microbalances*, vol. 7 (Elsevier, New York, 1984).
137. D. A. Buttry, in *Electroanalytical Chemistry*, vol. 17 (A. J. Bard, ed.) (Marcel Dekker, New York, 1991), p. 1.
138. A. R. Hillman, M. J. Swann, and S. Bruckenstein, *J. Phys. Chem.* **95**, 3271 (1991).
139. A. Glidle, A. R. Hillman, and S. Bruckenstein, *J. Electroanal. Chem.* **318**, 411 (1991).
140. G. Sauerbrey, *Z. Phys.* **155**, 206 (1959).

141. G. S. Ostrom and D. A. Buttry, *J. Electroanal. Chem.* **256**, 411 (1988).
142. G. Inzelt and J. Bacskai, *Electrochim. Acta* **37**, 647 (1992).
143. W. D. Hinsburg and K. K. Kanazawa, *Rev. Sci. Instrum.* **60**, 489 (1989).
144. A. R. Hillman, R. M. Eales, D. C. Loveday, M. J. Swann, A. Hamnet, S. J. Higgins, S. Bruckenstein, and C. P. Wilde, *Faraday Discuss. Chem. Soc.* **88**, 151 (1989).
145. A. R. Hillman, D. C. Loveday, M. J. Swann, and S. Bruckenstein, *J. Chem. Soc. Faraday Trans.* **86**, 233 (1990).
146. A. R. Hillman, D. C. Loveday, and S. Bruckenstein, *J. Electroanal. Chem.* **274**, 157 (1989).
147. S. Bruckenstein, C. P. Wilde, M. Shay, A. R. Hillman, and D. C. Loveday, *J. Electroanal. Chem.* **258**, 457 (1989).
148. A. R. Hillman, D. C. Loveday, S. Bruckenstein, and C. P. Wilde, *J. Chem. Soc. Faraday Trans.* **86**, 437 (1990).
149. M. D. Ward, *J. Electrochem. Soc.* **135**, 2747 (1988).
150. M. D. Ward, *J. Phys. Chem.* **92**, 2049 (1988).
151. P. T. Varineau and D. A. Buttry, *J. Phys. Chem.* **91**, 1292 (1987).
152. A. J. Kelly, T. Ohsaka, N. Oyama, R. J. Forster, and J. G. Vos, *J. Electroanal. Chem.* **287**, 185 (1990).
153. A. J. Kelly and N. Oyama, *J. Phys. Chem.* **95**, 9579 (1991).
154. A. R. Hillman, N. A. Hughes, and S. Bruckenstein, *J. Electrochem. Soc.* **139**, 74 (1992).
155. G. Inzelt, *J. Electroanal. Chem.* **287**, 171 (1990).
156. G. Inzelt and J. Bacskai, *J. Electroanal. Chem.* **308**, 255 (1991).
157. R. Borjas and D. A. Buttry, *J. Electroanal. Chem.* **280**, 73 (1990).
158. T. Kawai, T. Kuwabara, S. Wang, and K. Yoshino, *J. Electrochem. Soc.* **137**, 3793 (1990).
159. S. Gottesfeld, A. Redonda, I. Rubinstein, and S. W. Feldberg, *J. Electroanal. Chem.* **265**, 15 (1989).
160. A. P. Clarke, J. G. Vos, and A. R. Hillman, *J. Electroanal. Chem.* **356**, 287 (1993).
161. A. R. Hillman, D. C. Loveday, and S. Bruckenstein, *Langmuir* **7**, 191 (1991).
162. B. A. Martin and H. E. Hager, *J. Appl. Phys.* **65**, 2627 (1989).
163. S. Bruckenstein, C. P. Wilde, M. Shay, and A. R. Hillman, *J. Phys. Chem.* **94**, 787 (1990).
164. S. Bruckenstein, A. R. Hillman, and M. J. Swann, *J. Electrochem. Soc.* **137**, 1323 (1990).
165. G. E. Walrafen, *J. Chem. Phys.* **52**, 4176 (1970).
166. G. Inzelt, R. W. Day, J. F. Kinstle, and J. Q. Chambers, *J. Electroanal. Chem.* **161**, 147 (1984).
167. P. Joo and J. Q. Chambers, *J. Electrochem. Soc.* **132**, 1345 (1985).
168. H. P. Gregor, J. Belle, and R. A. Marcus, *J. Am. Chem. Soc.* **76**, 1984 (1954).
169. R. Naegali, J. Redepenning, and F. C. Anson, *J. Phys. Chem.* **90**, 6227 (1986).
170. J. Koryta and J. Dvorak, *Principles of Electrochemistry* (Wiley, New York, 1987).
171. S. B. Tuwiner, *Diffusion and Membrane Technology* (Chapman and Hall, New York, 1962).
172. R. Schumacher, G. Borges, and K. K. Kanazawa, *Surf. Sci.* **163**, L621 (1985).
173. R. Schumacher, J. G. Gordon, and O. Melroy, *J. Electroanal. Chem.* **216**, 127 (1987).
174. R. Beck, U. Pittermann, and K. G. Weil, *J. Electrochem. Soc.* **139**, 453 (1992).
175. Y. Marcus, *Ion Solvation* (Wiley-Interscience, New York, 1985).
176. F. Helfferich, *Ion Exchange* (McGraw-Hill, New York, 1962).
177. H. P. Gregor, *J. Am. Chem. Soc.* **73**, 642 (1951).
178. R. W. Murray, A. G. Ewing, and R. A. Durst, *Anal. Chem.* **59**, 379A (1987).
179. F. C. Anson, *J. Phys. Chem.* **84**, 3336 (1980).
180. D. A. Buttry and F. C. Anson, *J. Am. Chem. Soc.* **106**, 59 (1984).
181. D. C. Bookbinder and M. S. Wrighton, *J. Am. Chem. Soc.* **102**, 5123 (1980).
182. H. D. Abruna and A. J. Bard, *J. Am. Chem. Soc.* **103**, 6898 (1981).
183. H. Akahoshi, S. Toshima, and K. Itaya, *J. Phys. Chem.* **85**, 818 (1982).
184. Z. Zhang and W. R. Seitz, *Anal. Chim. Acta* **160**, 47 (1984).
185. C. K. Chiang, *Polymer* **22**, 1454 (1981).

186. T. Ohsaka, K. Naoi, S. Ogano, and S. Nakamura, *J. Electrochem. Soc.* **134**, 2096 (1987).
187. C. E. D. Chidsey and R. W. Murray, *J. Phys. Chem.* **90**, 1479 (1986).
188. R. J. Forster and J. G. Vos, in *Comprehensive Analytical Chemistry*, vol. 27 (M. R. Smyth and J. G. Vos, eds.) (Elsevier, Amsterdam, 1992), chap. 7.
189. L. R. Faulkner, *Chem. Eng. News* **62**, 28 (1984).
190. R. W. Murray, *Phil. Trans. R. Soc. Lond.* **302**, 253 (1981).
191. W. J. Albery and A. R. Hillman, *Ann. Rep. Prog. Chem.* **C78**, 377 (1981).
192. W. J. Albery, A. W. Foulds, K. J. Hall, and A. R. Hillman, *J. Electrochem. Soc.* **127**, 654 (1980).
193. W. J. Albery and A. R. Hillman, *J. Electroanal. Chem.* **170**, 27 (1984).
194. W. J. Albery, W. R. Bowen, F. S. Fisher, A. W. Foulds, K. J. Hall, A. R. Hillman, R. G. Edgell, and A. F. Orchard, *J. Electroanal. Chem.* **107**, 37 (1980).
195. C. P. Andrieux, J. M. Dumas-Bouchiat, and J. M. Saveant, *J. Electroanal. Chem.* **131**, 1 (1982).
196. C. P. Andrieux, J. M. Dumas-Bouchiat, and J. M. Saveant, *J. Electroanal. Chem.* **169**, 9 (1984).
197. C. P. Andrieux and J. M. Saveant, *J. Electroanal. Chem.* **171**, 65 (1984).
198. C. P. Andrieux and J. M. Saveant, *J. Electroanal. Chem.* **142**, 1 (1982).
199. C. P. Andrieux and J. M. Saveant, *J. Electroanal. Chem.* **134**, 163 (1982).
200. A. P. Doherty, R. J. Forster, M. R. Smyth, and J. G. Vos, *Anal. Chem.* **64**, 572 (1992).
201. R. J. Forster and J. G. Vos, *J. Chem. Soc. Faraday Trans.* **87**, 1863 (1991).
202. A. P. Doherty and J. G. Vos, *J. Chem. Soc. Faraday Trans.* **88**, 2903 (1992).
203. C. P. Andrieux, O. Haas, and J. M. Saveant, *J. Am. Chem. Soc.* **108**, 8175 (1986).
204. A. P. Doherty, T. Buckley, D. Kelly, and J. G. Vos, *Electroanalysis* **6**, 553 (1994).
205. H. Larsson and M. Sharp, *J. Electroanal. Chem.* **381**, 133 (1995).
206. M. Sharp and H. Larsson, *J. Electroanal. Chem.* **386**, 189 (1995).
207. M. F. Mathias and O. Haas, *J. Phys. Chem.* **97**, 9271 (1993).
208. A. P. Doherty and J. G. Vos, *Anal. Chem.* **85**, 3424 (1993).
209. A. P. Doherty, M. A. Stanley, G. Arana, C. E. Koning, R. H. G. Brinkhuis, and J. G. Vos, *Electroanalysis* **4**, 333 (1995).
210. D. M. Kelly and J. G. Vos, *Electrochim. Acta* in press (1996).
211. A. R. Hillman, D. C. Loveday, A. Glidle, J. G. Vos, A. P. Clarke, D. Kelly, and S. Bruckenstein, *Macromol. Symp.* **80**, 323 (1994).
212. A. P. Clarke, J. G. Vos, A. R. Hillman, and A. Glidle, *J. Electroanal. Chem.* **389**, 129 (1995).
213. A. P. Clarke, J. G. Vos, H. L. Bradley, and A. R. Hillman, *J. Phys. Chem.* **99**, 15973 (1995).
214. B. A. Gregg and A. Heller, *J. Phys. Chem.* **95**, 5970 (1991).
215. B. A. Gregg and A. Heller, *J. Phys. Chem.* **95**, 5976 (1991).
216. B. A. Gregg and A. Heller, *Anal. Chem.* **62**, 258 (1990).
217. A. Heller, *J. Phys. Chem.* **96**, 3579 (1992).
218. A. Aoki, R. Rajagopalan, and A. Heller, *J. Phys. Chem.* **99**, 5102 (1995).
219. A. P. Doherty, M. A. Stanley, and J. G. Vos, *Analyst* **120**, 2371 (1995).

9

Applications of Electroactive Polymers in Bioelectrochemistry and Bioelectronics

Philip N. Bartlett and John Cooper

1. INTRODUCTION

Interest in the electrochemistry of redox enzymes and redox cofactors is driven both by basic research into electron transport in biological systems and by the desire to use biological electron transport systems. These systems can be used to make sensors for biologically important analytes and in the longer term, to develop bioelectronic devices in which the biological component is closely integrated with the electronic system.(1) One of the areas in which significant progress has been made in this field is in using electroactive polymers to form the bridge (or interface) between the biological world of electron transfer and the world of electronics.

Biological electron transport systems have evolved to be specific and selective: To pass electrons from one member of the electron transport chain to the next in orderly fashion. To do this they must recognize and interact with their redox partners, in effect to mimic the natural redox partners and thus bind productively and exchange electrons with the biological redox species. This is not always possible at simple metallic surfaces.

In this chapter we first describe components and the organization of some important biological redox systems. We then discuss advantages of electroactive polymers for this type of application and review the work so far in different areas of redox protein electrochemistry, redox enzyme electrochemistry, and redox

Philip N. Bartlett • Department of Chemistry, University of Southampton, Highfield, Southampton, SO9 5NH, United Kingdom. *John Cooper* • Department of Electronics and Electrical Engineering, University of Glasgow, Glasgow, G12 8LT, United Kingdom.

Electroactive Polymer Electrochemistry, Part 2: Methods and Applications, edited by Michael E. G. Lyons. Plenum Press, New York, 1996.

cofactor electrochemistry. This is a rapidly progressing area, and we finish with a brief forward look identifying some problems that remain to be tackled and some of the possibilities for developments in this field.

1.1. Biological Electron Transfer

Electron transfer reactions are vital for living systems. Electron transfer plays a key role in processes by which energy is harvested in both respiration and photosynthesis. Electron transfer reactions are linked to essential processes in living systems that separate charge across biological membranes and thus establish ion and pH gradients and drive chemical reactions in the cell.[2] A characteristic of these electron transfer processes is their high degree of organization, so that electron transport chains are established in which the flow of electrons occurs with a high degree of specificity form one component to the next. Disruption of these chains by for example the addition of nonspecific electron transfer mediators that can short circuit the process frequently leads to destruction of the system. Understanding factors that are important in the organization of these electron transport chains and interactions between redox partners has attracted much attention.

The mitochondria are sites of respiration and oxidative phosphorylation in all animal and higher plant tissues. They vary in shape and size depending on their origin. In all mitochondria there are two membranes, an outer membrane and an inner highly invaginated membrane. These two membranes divide the mitochondrion into two separate compartments. The mitochondrial electron transport chain is comprised of a series of electron carriers arranged in close physical proximity on the inner mitochondrial membrane. These carriers are of two types, those transporting single electrons and those transporting hydride or hydrogen ions and electrons. The main components are β-nicotinamide adenine dinucleotide (NADH)-dependent dehydrogenases, flavoproteins, ubiquinone, iron-sulphur proteins, and cytochromes.[3] The generally accepted sequence of carriers in the mitochondrial electron transport chain is shown in Fig. 9.1. Bacteria do not have mitochondria; instead their respiratory chains are located on the cytoplasmic membrane, which comprises the cell envelope. Bacterial electron transport chains (Fig. 9.2) are similar to those in the mitochondria, although generally more complex.

Two factors appear important in establishing sequential selective electron transfer in biological chains of this type. One is the arrangement of free-energy differences ΔG between chain components, so that the flow of electrons is thermodynamically downhill (Fig. 9.3). The second factor involves the use of distance and the effect of distance on the rate of electron transfer in providing selectivity. This latter factor has attracted much attention, and there have been many studies of the effect of distance on the rate of electron transfer reactions in both biological systems and model systems.[4,5] We can write a general expression based on the Marcus treatment for the rate of electron transfer:[6]

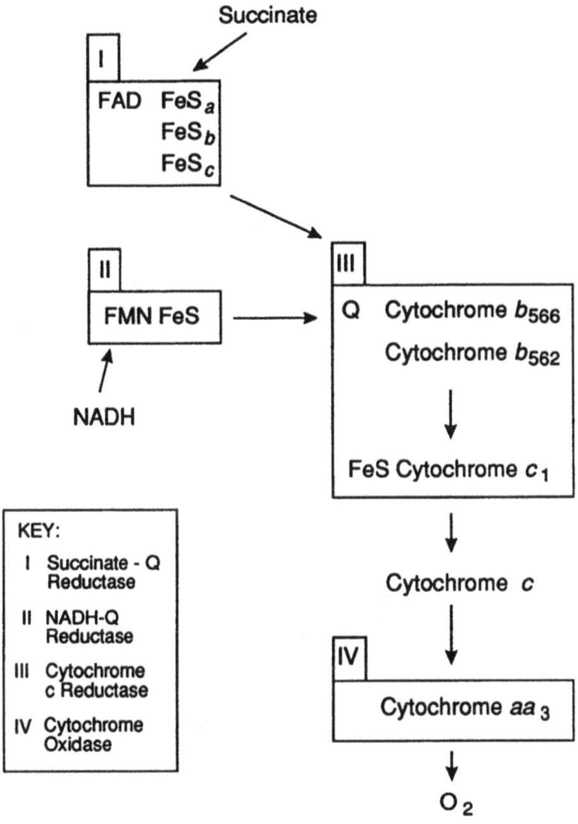

FIGURE 9.1. Sequence of components of the mitochondrial electron transport chain. NADH-Q reductase spans the mitochondrial membrane with the Q site within the membrane and the NADH site on the matrix side of the membrane. Succinate-Q reductase has a Q site within the membrane and a succinate site on the matrix side of the membrane. Cytochrome *c* reductase is a membrane-bound enzyme with cytochrome c_1 on the cystolic side of the membrane and cytochrome *b* in the membrane. Cytochrome *c* is soluble and found on the cystolic side of the membrane, while cytochrome oxidase translocates protons or electrons across the membrane. (Adapted from Ref. 3.)

$$k = K_E \, \nu \, \chi_{el} \exp\left(\frac{-\Delta G^{\ddagger}}{RT}\right) \tag{1}$$

where k is the rate constant for electron transfer, K_E describes the diffusion preequilibrium, ν is the effective frequency of motion along the reaction coordinate (of the order of a vibrational frequency, *ca.* 10^{13} s^{-1}), χ_{el} is the electronic transmission coefficient, and $\Delta G^{\ddagger}$ is the free energy of activation, which includes both inner- and outer-sphere contributions. In this expression the electron transmission coefficient χ_{el} depends on the distance r between the centers:

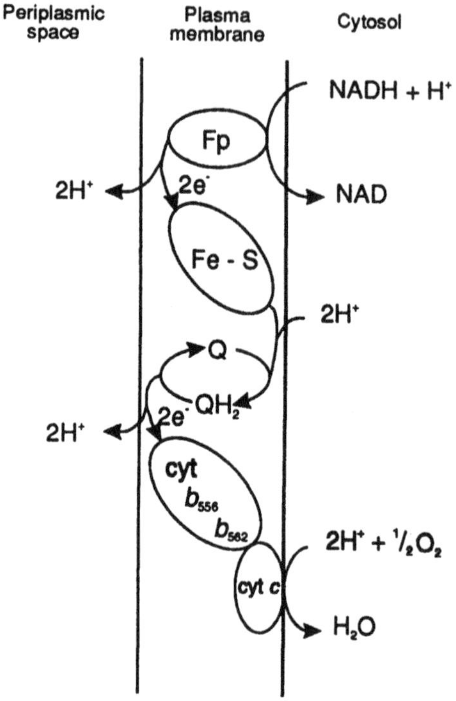

FIGURE 9.2. Example of a bacterial electron transport chain: Postulated mechanism for proton translocation in *Escherichia coli* during aerobic respiration. (Adapted from Ref. 3.)

$$\chi_{\mathrm{el}} = \chi_{\mathrm{el}(r=0)} \exp(-\beta r) \tag{2}$$

where $\chi_{\mathrm{el}(r=0)}$, the value at $r = 0$ corresponding to closest contact, is approximately 1, and β is a constant.

Various model systems were used to investigate the dependence of electron transfer rate on distance and except for some minor disagreements in detail, the broad picture that emerges for biological systems shows that χ_{el} falls exponentially as the distance increases and β is of the order of 12–14 nm^{-1}. Consequently electron transfer is possible only over distances less than about 2 nm.

When considering long-range electron transfer in proteins, there has been some discussion of the influence of the intervening medium and in particular of whether the electron passes along a particular path or makes significant use of particular protein side chains.[7] For instance it was postulated that aromatic side chains of proteins play an important role. However a recent study by Moser *et al.*[8] of electron transfer in the bacterial photoreaction center reveals no such effects in that case.

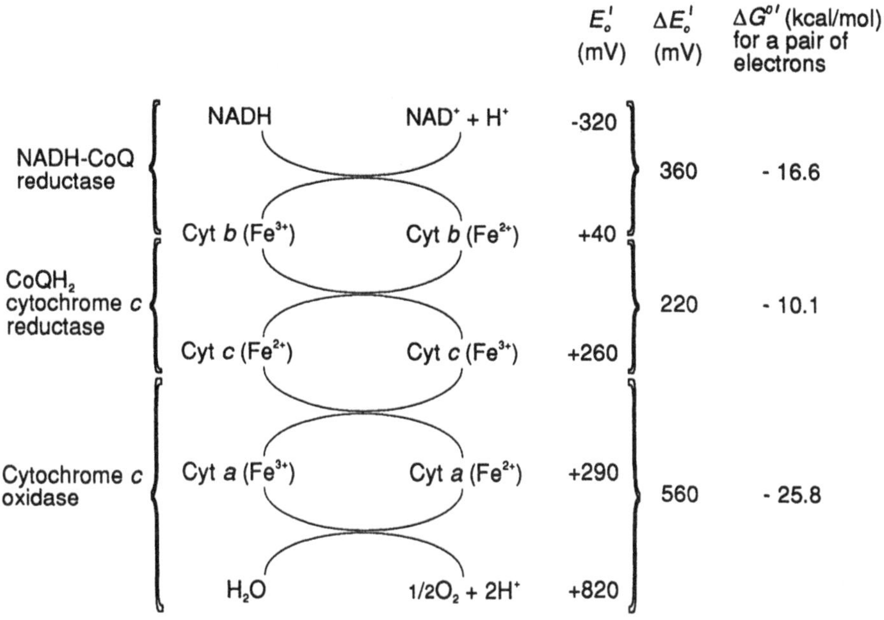

FIGURE 9.3. Energetics of the mitochondrial electron transport chain showing redox potentials and free-energy changes associated with each step. [Based on J. Darnell, H. Lodish, and D. Baltimore, *Molecular Cell Biology* (Scientific American Books, New York, 1990), p. 606.]

1.2. Components of Biological Electron Transport Chains

The various components of biological electron transport chains can be broadly divided into redox enzymes, redox proteins, and redox cofactors. This division is very convenient for our present discussion. In this section we briefly describe each of these component.

1.2.1. Redox Enzymes

By a redox enzyme we mean an enzyme containing a tightly bound redox-active prosthetic group, such as a flavin or a haem, that remains bound to the protein throughout the redox cycle. The term prosthetic group denotes a nonamino acid component of a conjugated protein. Redox enzymes can thus exist in both oxidized and reduced forms. Glucose oxidase is a typical example of such an enzyme and one that has often been studied. Fig. 9.4 shows in schematic form steps involved in the oxidation of glucose by molecular oxygen catalyzed by this enzyme. This type of enzyme mechanism is called a ping-pong mechanism.[9]

Redox enzymes can further be subdivided into terms of the redox centers present in the enzyme. Several types of redox center are known, including those

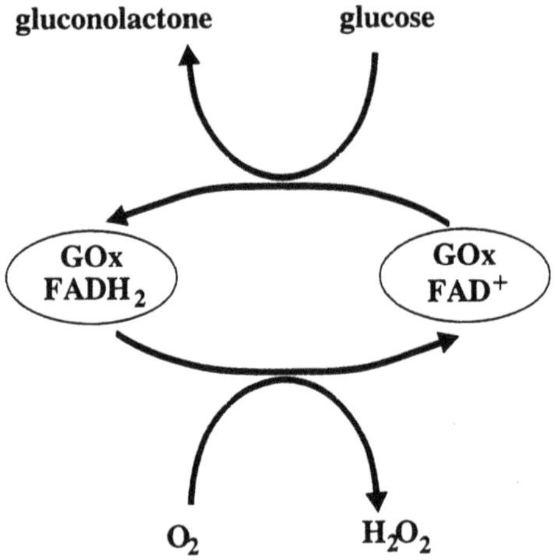

FIGURE 9.4. Ping-pong mechanism for the oxidation of glucose by oxygen catalyzed by glucose oxidase.

based on flavin, quinone, haem, and copper groups. Of these flavoproteins, which contain flavin prosthetic groups, have been studied most. Flavoproteins comprise a group of more than 80 enzymes containing either flavin adenine dinucleotide (FAD) or flavin mononucleotide (FMN) at the active site.(10) Of these the great majority contain FAD. The flavin unit is strongly associated to the protein structure, and in some cases, it is covalently bound to amino acid residues in the enzyme.(11) The flavin redox system is a two-electron couple with associated protonation equilibria (Fig.9.5).(12) Although redox properties of free flavin are readily studied and well-understood(13,14) results are not directly relevant to the behavior of the redox couple once it is bound within the protein. This is because the environment within the protein envelope is very different from that in solution and because this environment varies from enzyme to enzyme. Thus the redox potential of the flavin group at pH 7.0 in various proteins (not necessarily enzymes) has been shown to range from –730 – –50 mV versus SCE.(15) Not only does the protein envelope shift the overall thermodynamics of the two-electron oxidation/reduction of the flavin group, but it also affects the relative stabilities of the semiflavin form.

Flavoenzymes are further subdivided into oxidases and dehydrogenases on the basis of the mechanism of the enzyme-catalyzed reaction.(16) In oxidases the reduced flavin is reoxidized by molecular oxygen generating hydrogen peroxide.(17) In the dehydrogenases the reduced flavin does not react with oxygen.(18)

Quinone containing redox enzymes are a much less widely studied group, and to date fewer examples have been characterized.(19,20) The prosthetic group in this case is a quinone, pyrrolo-quinoline quinone (PQQ), and again it is a two-electron

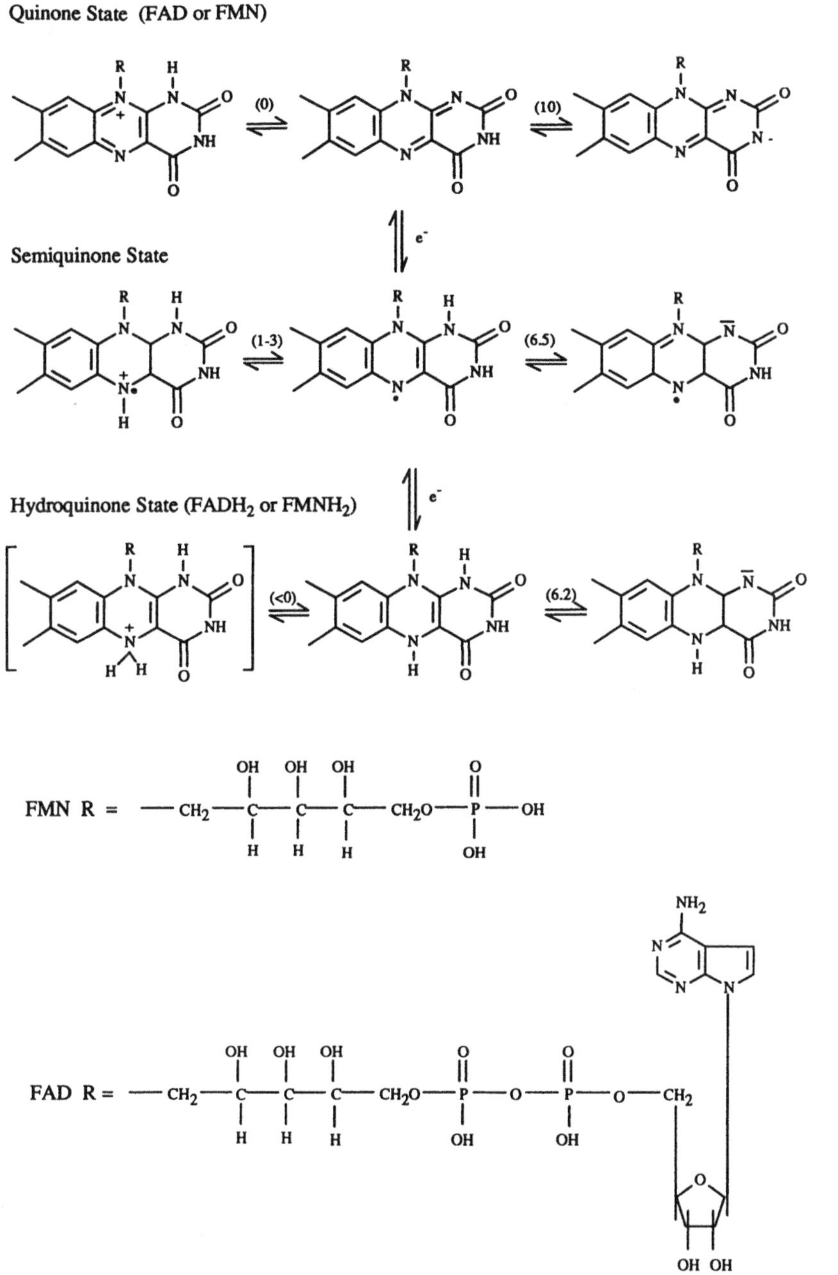

FIGURE 9.5. The flavin redox system showing associated protonation equilibria; pK$_a$ values are given in brackets.

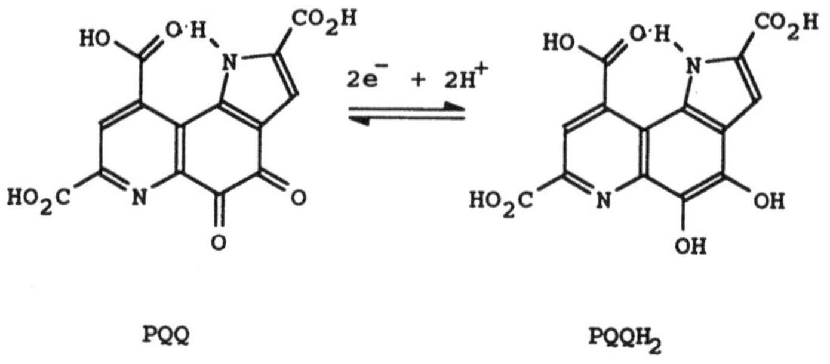

FIGURE 9.6. Structure of the pyrrolo-quinolone quinone (PQQ) prosthetic group.

redox couple with associated protonation equilibria (Fig. 9.6). The reduced PQQ is not reoxidized by molecular oxygen; hence these enzymes are referred to as dehydrogenases.

1.2.2. Redox Proteins

Redox proteins are a group of small electron transfer proteins containing one or more sites capable of undergoing redox reactions. They typically have molecular masses in the range of 5–15 kDa, and they are essentially globular in shape. The redox groups consist of metal-containing centers, typically iron, iron-sulphur, or copper located at or near the periphery of the protein. These properties distinguish small redox proteins from the redox enzymes, which are generally much larger and believed in many cases to have their catalytic centers embedded deep within the protein. In living systems redox proteins serve as electron donors or acceptors, reacting with other redox proteins or redox enzymes.

Cytochromes are found in all aerobic organisms. They act as electron carriers, and to achieve this function, they contain a haem site within the protein. Cytochromes are classified as *a, b*, or *c*, depending on the type of porphyrin present (Fig. 9.7). Most cytochromes are membrane bound, that is, water insoluble. An exception to this is cytochrome *c*, which is water soluble; as a consequence it has been particularly well studied.

Cytochrome *c* is a small haem protein with a molecular mass of around 12 kDa; the precise value depends on the origin of the protein. Its biological function is to accept an electron at Complex III of the respiratory chain and deliver it to Complex IV, the terminal oxidase, normally cytochrome *c* oxidase. In yeast and some bacteria, cytochrome *c* peroxidase (CcP, EC 1.11.1.5) may act as an alternative electron acceptor.[21] Cytochrome *c* has around 104 amino acid residues, 27 of which are known to be invariant between 50 different organisms.[22] The protein is about 3.4 nm in diameter, and crystal structures on a number of examples from

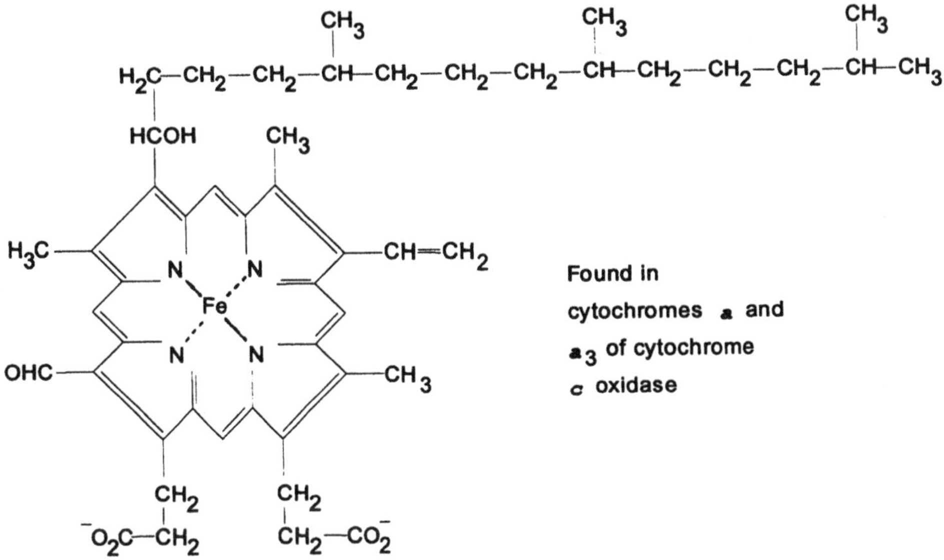

Found in cytochromes a and a3 of cytochrome c oxidase

a a - Type haem

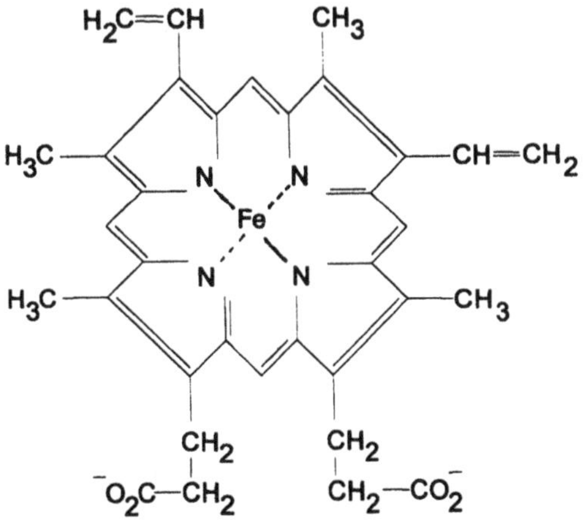

Found in cytochromes b562 and bc1 complex

b b - Type haem

FIGURE 9.7. Structures of the three haem prosthetic groups found in the mitochondrial respiratory chain: (a) a-type haem group found in cytochrome *a*, cytochrome a_3 and cytochrome oxidase; (b) b-type haem found in cytochrome b_{562} and cytochrome b_{566} of the cytochrome bc_1 complex; (c) c-type haem found in cytochrome *c* and cytochrome c_1.

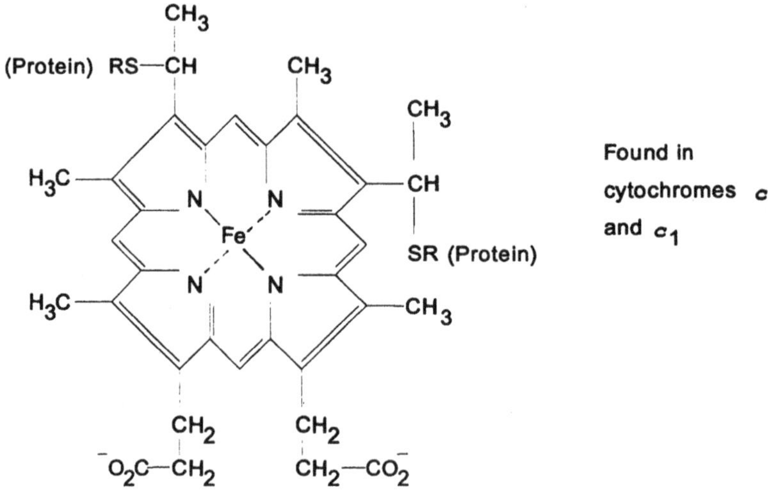

C c - Type haem

FIGURE 9.7. (Continued)

different species have been reported.[23] The haem group is covalently attached near the protein surface and surrounded by tightly packed hydrophobic side chains. This hydrophobic environment shifts the redox potential of the haem to values positive of that which it would have if it were solvated in aqueous solution.[24,25] In all examples studied to date, the haem is surrounded by clusters of lysine residues, positively charged at neutral pH. These lysines are believed to play an important role in recognizing and binding the cytochrome *c* to its natural redox partners, cytochrome oxidase or cytochrome *c* peroxidase, which have a complementary patch of negatively charged residues around their active sites. The study of the biological interface between prosthetic haem groups of cytochrome *c* and its natural redox partner CcP has provided the basis of much current knowledge of protein–protein redox interactions.[21] Processes are complex and involve charge matching, hydrogen bonding, and spatial fitting to bring the two protein molecules into close proximity. The essence of the molecular organization is a ring of basic lysines (Lys 8, 13, 27, and 72), positively charged at physiological pH, which extend away from the surface of cytochrome *c* to form salt bridges with carboxylate groups of aspartate 33, 34, 37, and 217 (negatively charged at physiological pH) present in CcP. This electrostatic interaction draws the two molecules together so that the two haem groups lie parallel, only 2.5 nm apart, thus facilitating electron transfer.

Another important group of redox proteins are the iron-sulphur proteins. These are proteins of relatively low molecular weight containing one or more iron atoms

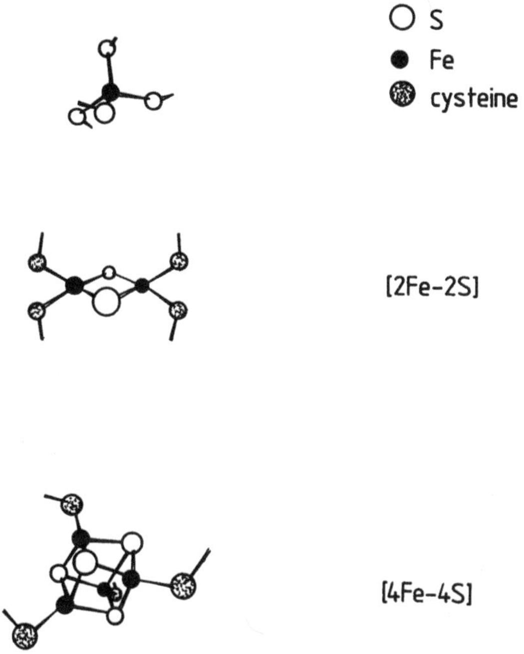

FIGURE 9.8. Structures of iron-sulphur centers of rubredoxin and ferredoxin. Rubredoxins are the simplest iron-sulphur proteins, and these contain one Fe-S center with the iron in a tetrahedral environment. Ferredoxins contain clusters of two [2Fe-2S] or four [4Fe-4S] iron atoms.

linked to sulphur atoms of cysteine residues and to inorganic sulphur. Rubredoxins are the simplest iron-sulphur proteins, and these contain one iron-sulphur center with the iron in a tetrahedral environment (Fig. 9.8). Ferredoxins contain either two- or four-iron atom clusters. For the two-iron atom cluster [2Fe-2S], the normal oxidation state is $[2Fe\text{-}2S]^{2+}$, and both iron atoms are involved in accepting an electron. In the four-iron cluster [4Fe-4S], a variety of oxidation states are possible with different numbers of Fe atoms reduced at any one time.

1.2.3. Redox Cofactors

In addition to redox enzymes and redox proteins, there are a number of low molecular weight redox cofactors that play a significant role in biological electron transport. Of particular interest are nicotinamide coenzymes and various quinones.

Two nicotinamide coenzymes, β-nicotinamide adenine dinucleotide (NAD^+) and β-nicotinamide adenine dinucleotide phosphate ($NADP^+$), have closely related structures (Fig. 9.9) and electrochemical properties. Although generally represented as cations, it is worth pointing out that in both cases the oxidized forms of coenzymes are anionic at neutral pH due to ionization of phosphate groups in the

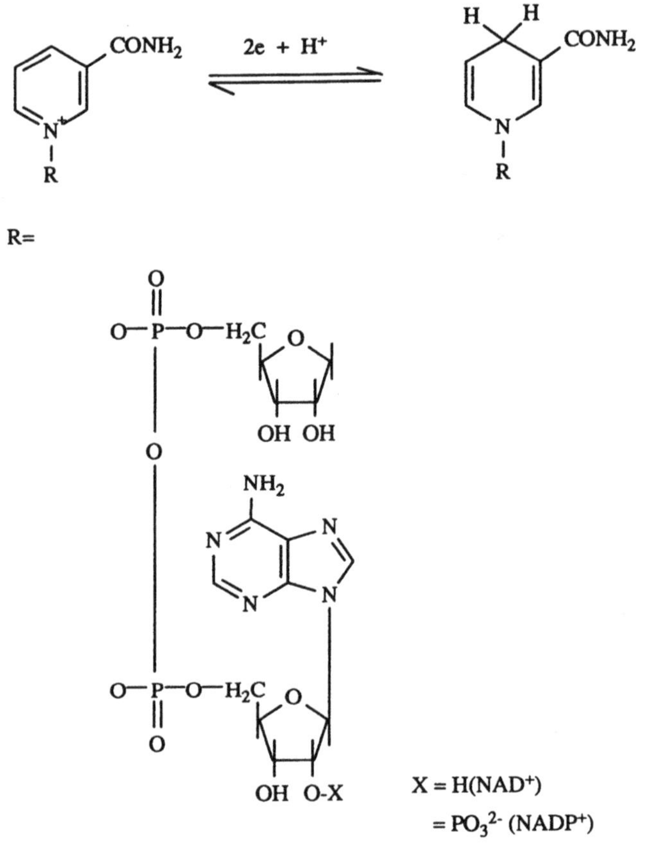

FIGURE 9.9. Structures of the NAD^+/NADH and $NADP^+$/NADPH coenzyme redox couples.

molecules. In both cases the oxidation reduction reaction involves two electrons and one proton

$$NAD(P)^+ + H^+ + 2e \rightarrow NAD(P)H$$

These two coenzymes act as carriers of reducing equivalents, and nicotinamide-dependent enzymes are found widely dispersed throughout living tissues. In animal cells NAD is usually present in larger amounts than NADP. The NAD-dependent enzymes are primarily involved in respiration, while NADP-dependent enzymes are involved in catabolism. The NAD is found in particularly large quantities in mitochondria, where the major route for its reoxidation is through reaction with ubiquinone.

Quinones are widely spread in living tissues, and two groups of compounds are particularly important, the ubiquinones (coenzyme Q) and the vitamins K. Quinones within the two groups vary in length *n* of the side chain present in the molecule (Fig. 9.10). For the ubiquinones, $n = 6-10$; for the vitamins K, $n = 6, 7$, or 9.[18] Ubiquinone is a lipid-soluble constituent of the mitochondrial respiratory chain, associated with the membrane, where it participates in the transport of electrons from organic substrates to oxygen. In plant tissues the same function is performed by closely related plastoquinones.

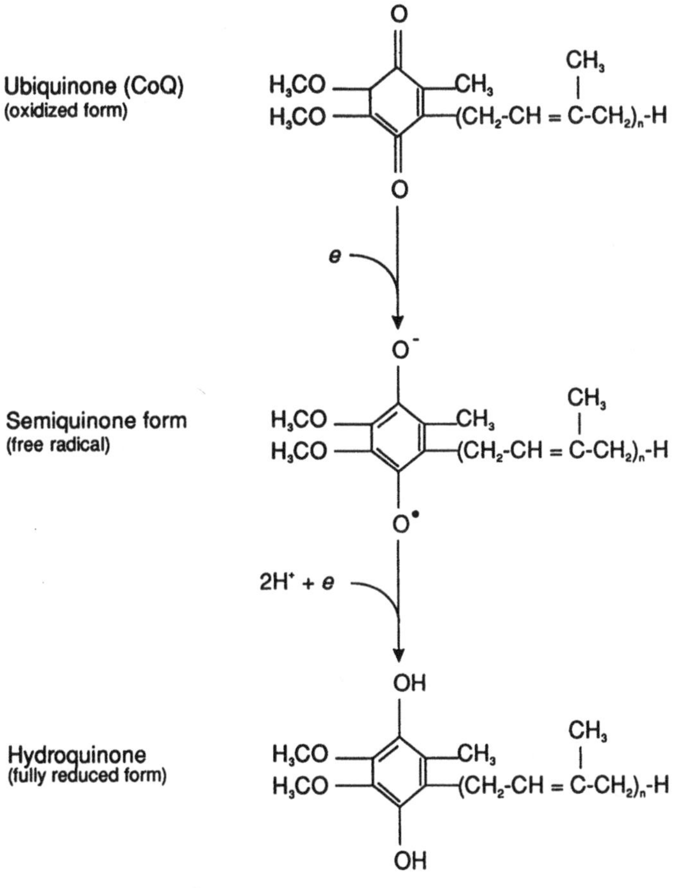

FIGURE 9.10. General structure of ubiquinone and related quinones. For ubiquinone (coenzyme Q) of human origin, $n = 10$. Quinones of this type are found in bacterial and mitochondrial membranes, and these are involved in electron and proton transport. Because of their long hydrophobic tail, these quinones are soluble in phospholipid bilayers.

2. ADVANTAGES OF ELECTROACTIVE POLYMERS

Since the first reports of electrode modification in the 1970s, electroactive polymers deposited at electrode surfaces have been a consistent and continuing area of research in electrochemistry. A wide variety of different types of electroactive polymer coating have been studied in greater or lesser detail, and much effort has been directed toward using modified electrode surfaces as electrocatalysts for many different types of application.

Electroactive polymers have a number of attractive features that account for this continuing interest. First they present a distributed array of catalytic sites. Thus in contrast to monolayer chemically modified electrodes, there are potentially a much greater number of reactive sites that can contribute to the catalytic current. Since these sites are distributed throughout the film, it is essential to consider the mass transport of reagents into the film and the mass transport of products out of these films when studying the overall kinetics of these processes. The coupled mass transport and kinetics in redox polymer films have been investigated in some detail, and good models exist for these processes.[26–28]

A further advantage of using electroactive polymers is that redox-active sites can be chosen or designed for the particular function they are to perform. Contrast this with the simple clean metallic electrode surface where there can be little control over interactions between the electrode surface and the redox couple on the molecular scale. In the context of biological redox couples, implications of this are very significant, because such systems have evolved to recognize their natural redox partners on the basis of molecular interactions between components. Using electroactive polymers allows the electrochemist to attempt to mimic these interactions so that the electrode is seen as a suitable partner for electron transfer to, or from, the biological molecule.

A final advantage of the electroactive polymer film particularly relevant to its use with redox enzymes is that the film can be used to entrap species at the electrode surface. This can confer several advantages. First it localizes the biological catalyst close to the electrochemical interface and in the right place for its efficient use. Second the electroactive film can be used to provide a favorable microenvironment in which an enzyme reaction can proceed. Third the polymer can be used to protect the electrode from fouling by nonspecific protein adsorption, and in certain circumstances, it can help exclude interferent species that might lead to undesirable side reactions or background currents.

When an enzyme is entrapped, it is necessary to consider the combined effects of mass transport, partition, and reaction within the film and enzyme kinetics to understand processes that determine the overall response. These interactions can be quite complex, although some progress has been made toward developing models for these systems.[29,30]

Electroactive polymers can be divided into three broad classes: electronically conducting polymers, such as poly(pyrrole), in which conduction is associated with motion of charge carriers along the chains; redox polymers, such as poly(vinylferrocene), in which conduction is associated with cross-exchange reactions between discrete redox sites; and polymer electrolytes, such as Nafion, in which conduction is associated with ion motions within the film. Examples of polymers from all three classes have been used in bioelectrochemical applications.

Films of these polymers can be deposited on the electrode surface by a variety of methods, including electropolymerization and dip, or drop, coating from solutions of the preformed polymer. The former technique has two advantages: Deposition can be controlled by adjusting the electrode potential, and the process is localized at the electrode surface. This has led many groups to study electropolymerized films as a method for immobilizing enzymes and other biological components. The use of dip or drop coating has the advantage that the polymer can be better characterized in terms of molecular weight and structure. However to use this technique, the polymer must be soluble in a suitable solvent and insoluble in the electrolyte solution ultimately used for electrochemical studies.

In the following sections, we review some applications of these different types of electroactive polymer film in bioelectrochemistry and bioelectronics. We begin by considering their uses with redox proteins.

3. APPLICATIONS IN REDOX PROTEIN ELECTROCHEMISTRY

The electrochemistry of redox proteins is characterized by a strong dependence on the nature of the electrode surface. Extensive studies by Hill *et al.* (cited in Refs. 31–33) show that provided the electrode surface is modified to be compatible with the redox protein, direct electrochemistry can be rapid. Their studies have emphasized the importance of the orientation of the protein at the electrode surface so that the distance over which the electron must transfer is not excessive. This is important because redox sites in these proteins are generally located toward one side of the protein, and the exponential dependence of the electron transmission coefficient χ_{el} on distance means that the rate of electron transfer drops rapidly as distance increases (Fig. 9.11). Most of this work has used low molecular weight modifiers adsorbed at the electrode surface,[34] although similar effects should be possible at polymer-modified electrode surfaces.

To date studies using electroactive polymers with redox proteins have been much less numerous despite the fact that it should be possible to extend the general principles elucidated by Hill and colleagues using adsorbed monolayers at electrode surfaces to design suitable electroactive polymers for this application. For example studies with poly(5-carboxyindole), a conducting polymer formed by electropolymerization of 5-carboxyindole[35] have shown that it can be used for the direct

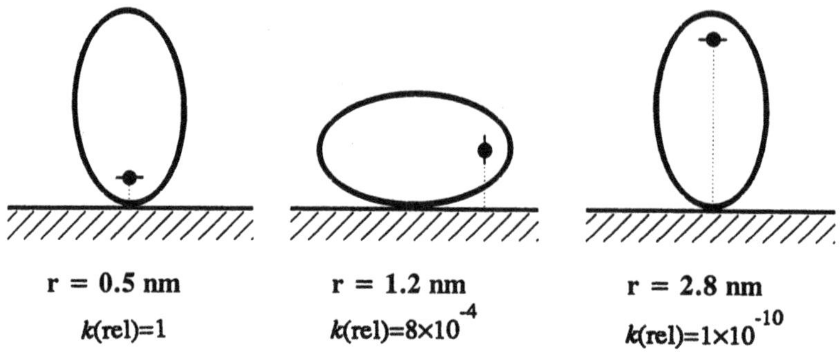

FIGURE 9.11. Effect of distance on the rate of electron transfer. The diagram is drawn schematically to represent cytochrome *c*, where the redox center is located close to the one side of the protein. Depending on the orientation of the protein at the electrode surface, the distance over which the electron must transfer, and hence the relative rate of electron transfer k_{rel}, varies.

oxidation and reduction of cytochrome *c*. Films were grown electrochemically from the monomer in acetonitrile, then transferred to an aqueous solution for redox protein studies. Electrochemistry of the protein at the conducting polymer electrode is well-behaved, and there is evidence of absorption of the enzyme onto the surface of the polymer.[36] This has been rationalized in terms of interactions between carboxylic acid groups attached to the polymer backbone (Fig. 9.12) and protonated lysine residues disposed about the heme redox site of the protein.

Redox polymers have been used as electrodes for various redox proteins, including ferredoxin, haemoglobin, and cytochrome *c*. Polymeric films of viologen (1,1′-dimethyl-4,4′-dipyridyl dichloride) deposited on gold minigrid electrodes were found to carry out the heterogeneous reduction of spinach ferredoxin, a 2Fe-2S protein with a molecular weight of about 11 kDa, although the precise nature of the polymeric material formed is unclear.[37,38] Poly(N-vinyl-4,4′-bipyridinium) films, formed by electropolymerization in moderately concentrated acid solution, have been shown to mediate reduction of horse heart cytochrome *c*.[39] In this case reduction of the cytochrome occurs near the reduction potential of the film, and not

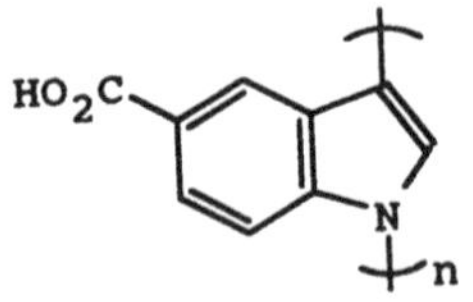

FIGURE 9.12. Proposed structure for poly(5-carboxyindole). (From Ref. 35.) *Note added in proof:* Subsequent work by Mackintosh and Mount [J.G. Mackintosh and A.R. Mount, *J. Chem. Soc., Faraday Trans.* **90**, 1121 (1994)] has shown that the polymerization of 5-carboxyindole leads to the formation of a polymer of asymmetric trimers of 5-carboxyindole rather than the linear polymer postulated by Bartlett *et al.* in their original work.

near the reduction potential of the protein; when the polymer film is oxidized, reduction of cytochrome *c* is blocked.

The electrochemistry of redox proteins incorporated into polymer electrolyte films has been reported. Oliver *et al.*[40] investigated the electrochemistry of cytochrome *c* in semirigid polyacrylamide and polyethyleneoxide gels at edge-plane pyrolytic graphite electrodes. The protein was incorporated into the films by sorption of microdroplets of aqueous solution into the nitrogen dried gels or by premixing with the polymer prior to casting on the electrode. The authors found that both the formal potential and heterogeneous electron transfer rate constant were identical in the gel and aqueous solution. Nafion, a commercial polyanionic, perfluorinated sulphonated polymer has also been used to entrap redox proteins.[41] In a preliminary study proteins were dissolved in buffered solutions of Nafion, the solution spread on a cleaned basal plane graphite surface, then air dried. The authors looked at the electrochemistry of cytochrome c_{551}, cytochrome b_5, cytochrome *c*, and azurin. They found that cytochrome c_{551}, cytochrome b_5, and azurin all gave appropriate responses but no electrochemistry was observed for cytochrome *c* under these conditions. They suggested that the highly positively charged cytochrome *c* was too tightly bound to the Nafion to interact with the electrode surface. This area of redox protein electrochemistry at electroactive polymer-coated electrodes is worthy of further study.

4. APPLICATIONS IN REDOX ENZYME ELECTROCHEMISTRY

In contrast to the area of redox protein electrochemistry, redox enzyme electrochemistry has received much greater attention, driven in many cases by the desire to construct practical, self-contained enzyme electrodes for commercial applications. Redox enzyme electrochemistry is also easier to study in many ways because the substrate or product is often detected electrochemically rather than the enzyme itself. Various types of electroactive polymers have been used with redox enzymes, including redox polymers, redox-active hydrogels, and electropolymerized films of conducting and nonconducting, polymers. We discuss each type of polymer in turn, starting with electropolymerized films.

4.1. Electropolymerized Films

Electropolymerization is an attractive approach to enzyme immobilization, a fact already recognized in the earliest studies in this area.[42,43] The use of electropolymerized films to immobilize enzymes has recently been reviewed,[44] so we provide an abridged discussion here.

Electrochemical polymerization is simple to carry out, and it can be controlled by manipulating the electrode potential. In principle this allows the polymer film thickness, and hence the amount of enzyme entrapped in close proximity to the electrode surface, to be controlled. Since electropolymerization occurs locally at the electrode surface, it can be used to confine an enzyme precisely at an electrode, without cross-immobilizing it on a neighboring electrode—an attractive property for fabricating arrays of enzyme microelectrodes.[45] Electropolymerization can also be used to build multilayer structures, either with one or more enzymes layered within a single polymer or with one enzyme within a multilayered copolymer.

A significant proportion of the published work has been concerned with the study of glucose oxidase (GOx: EC 1.1.3.4) immobilized in poly(pyrrole). This is in part due to interest in amperometric enzyme electrodes for glucose measurement and in part to the fact that GOx is a very robust enzyme and poly(pyrrole) is very easy to polymerize from aqueous solutions; thus conditions required to immobilize the enzyme are not demanding. In addition to the GOx/poly(pyrrole) system, a range of other enzymes and other electropolymerized films have been investigated.

Pyrrole can be electropolymerized to form electrically conducting poly(pyrrole) films at potentials greater than +0.6 V versus SCE. Films are generally grown either at a fixed potential or by cyclic voltammetry. Adding GOx to the polymerization solution has little effect on the polymerization process itself, but it generally leads to an increase (*ca.* $\times 2$) in the oxidative current, all other conditions remaining the same. A similar effect has been observed in the presence of other polyanions.[46] Since GOx has an isoelectric point pI of 4.2, the enzyme is negatively charged at pH 7.0, and it carries an overall charge from −9 to −10.[47] Poly(pyrrole) is deposited as a polycation; hence electrostatic interactions between polymer and enzyme play an important part in incorporating the enzyme into the growing film. Support for this view is provided by three pieces of evidence: (1) GOx can be adsorbed onto preformed poly(pyrrole) films;[48] (2) when poly(pyrrole) films containing glucose oxidase are grown in the presence of sulphonated hydroquinone[49] or ferrocyanide,[50] the amount of incorporated enzyme and the enzymic activity of the films decreases as the concentration of anion increases; (3) it is very difficult to incorporate positively charged proteins into poly(pyrrole) films.

Many redox enzymes can be denatured if the pH of the solution is taken too far from neutral or if nonaqueous solvents are used. It is therefore desirable, although not always essential, to use neutral aqueous solutions in the electropolymerization process, thus restricting the choice of monomers that can be used. One way to overcome this problem is to incorporate, or immobilize, the enzyme after deposition of the polymer film. This allows a much wider choice of deposition conditions. In addition to adsorbing the enzyme onto preformed poly(pyrrole) films,[48] GOx has also been covalently immobilized onto a wide range of preformed 3- and N-derivatized pyrrole carboxylates[51] or onto N-amino-substituted pyrroles.[52] This technique may increase current densities at a given glucose concentration by as much as twentyfold when compared to the best results obtained elsewhere. The authors make no claims for direct electron transfer between enzyme

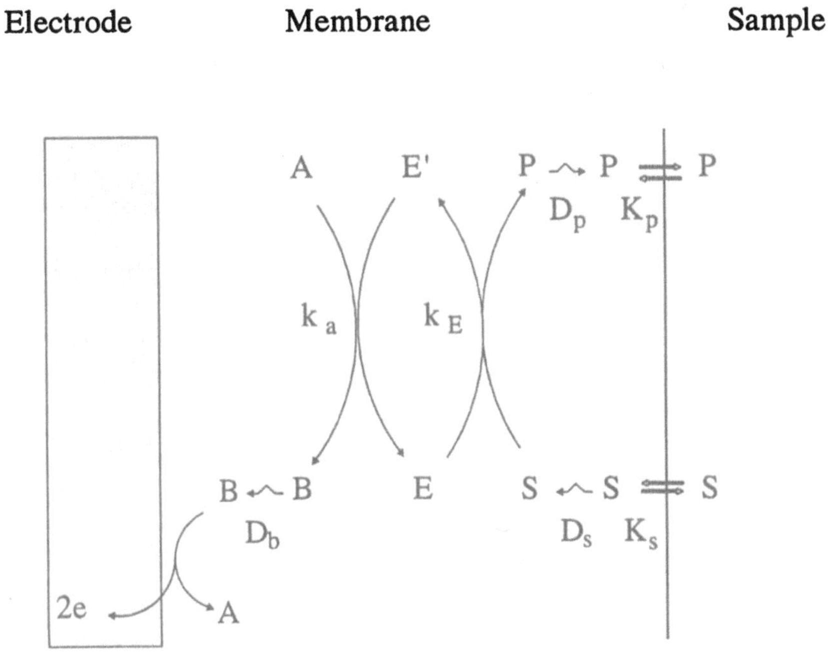

FIGURE 9.13. Model of an immobilzed enzyme layer at the electrode surface showing the various partition, diffusion, and reaction steps involved in the overall enzyme-catalyzed electrochemical reaction.

and polymer, and they cite the metallic anode surface as the position of electrocatalytic oxidation of H_2O_2. The N-carboxylate derivatized poly(pyrrole) films have also been shown actively to exclude ascorbate, bioactive amine proteins, and sulphydryl-containing peptides as interferents,[(51)] and this is an additional advantage if films are to be used for biosensor applications in such media as whole blood.

To understand the interplay of enzyme catalysis and mass transfer within polymer film, it is essential to develop models that take account of these effects, then compare the models' predictions with experiment. Fig. 9.13 illustrates the physicochemical processes involved in the enzymic turnover of substrate to product within a polymer film. Such processes include mass transport of substrate and product either to or from the film, partition of these species across the polymer–solution interface, transport of reactants and products within the film (by diffusion), and electrochemical reaction with enzymic products at the electrode surface. Effects of migration of charged species within the film are usually ignored.

It is important to recognize that enzyme kinetics for reactions within a polymer film are likely to differ from those for homogeneous solutions;[(53)] for example it is probable that immobilization of the enzyme decreases the turnover rate of the enzyme and increases the Michaelis constant K_M. It is also necessary to take into account effects of partition of both reactants and products into the film. It is quite possible that the concentration of substrate within the electropolymerized film is very different from the bulk solution concentration. Models of immobilized enzyme electrodes have been reviewed.[30,54,55]

To date there have been only a few detailed comparisons of these models with experimentation. One example of this approach is provided by work on oxygen mediation for the GOx-catalyzed oxidation of glucose in a poly(N-methylpyrrole) film.[56] In this case an investigation of the effect of enzyme loading and film thickness on the amperometric response showed that H_2O_2 generated by the enzyme reaction is detected by oxidation at the underlying platinum electrode, and not by direct oxidation on the polymer, as suggested elsewhere with the GOx/poly(pyrrole) enzyme electrode.

Evidence against a direct electrochemical mechanism for enzyme reoxidation was given in a detailed study of GOx immobilized in poly(pyrrole) films by Belanger *et al.*,[57] who showed that the conductivity of the film is destroyed during the reaction of the polymer with the enzymically produced H_2O_2. Similar results were found for poly(N-methylpyrrole). Further evidence that the response of GOx/poly(pyrrole) electrodes is due to H_2O_2 detection is provided by fact that when catalase is coimmobilized with GOx in a poly(pyrrole) film, no glucose-dependent response is obtained.[58] Catalase is an efficient scavenger of H_2O_2, and at equimolar concentrations with GOx, it removes peroxide before it reaches the anode.

On the other hand direct electron transfer between the poly(pyrrole) and GOx has been claimed.[59] Subsequent work on GOx immobilized in poly(N-methylpyrrole) suggests that the process is very inefficient unless the enzyme is partially denatured during immobilization.[60] In this study there was no evidence to support direct oxidation of the enzyme for films grown at 25° C. When films were grown at 50° C, the authors found evidence of direct oxidation of the enzyme, and they observed that selectivity of the enzyme was significantly reduced. They rationalize these observations by suggesting that at 50° C, the enzyme is partially denatured, so it is entrapped within the poly(N-methylpyrrole) film in a more open form, allowing access of polymeric chains to the active site of the enzyme.

Koopal *et al.* claim direct oxidation of GOx on poly(pyrrole) films grown in track-etch membranes.[61,62] This produces conducting microtubules[63] onto which the GOx is irreversibly adsorbed by subsequent adsorption and drying. These films show oxygen-independent responses to glucose at 0.35 V versus Ag/AgCl which the authors attribute to direct oxidation of the enzyme by the poly(pyrrole). They suggest that such direct electron transfer between the polymer and the enzyme is the result of the manner in which the poly(pyrrole) film is grown and subsequently treated with enzyme. However subsequent work by Kuwabata and Martin[64] casts doubt on this interpretation and indicates that the observed response is the result of direct oxidation of the glucose on the sputtered metal film used to contact the poly(pyrrole) microtubules rather than any enzyme catalyzed process.

One difficulty in testing for direct electrochemical oxidation of the enzyme by a polymer is that small amounts of mediator species can be very efficiently cycled within the film, giving the appearance of a direct electron transfer mechanism. It is therefore necessary to be very careful to consider all other possibilities before assuming that such a mechanism is the only explanation available.

The fact that for the most part, the mechanism for detecting oxidase enzyme activity within polymer films at electrodes is mediated by H_2O_2 has important consequences since H_2O_2 can be detected only at potentials greater than about 0.65 V versus Ag/AgCl. Enzyme electrodes fabricated in this manner as biosensors can suffer from significant interferences from species such as acsorbate, if used in biological fluids. A number of groups have tried to resolve this problem by using mediators. Ferrocene derivatives,[(48,65,66)] ferricyanide,[(50)] phenazine methosulphate,[(43)] benzoquinone,[(43,67)] or sulphonated benzoquinone,[(49)] and iodide in the presence of a molybdenum (VI) catalyst[(42)] have all been used. In addition poly(pyrrole) copolymerized with a ferrocene-substituted pyrrole has also been employed.[(68)] Using artificial mediators offers the advantage that electrodes can be operated at lower overpotentials, thus reducing effects of interferent species and eliminating, or greatly reducing, dependence on oxygen concentration.

Most of these studies have not addressed the question of where the reaction between the artificial mediator and the electrode occurs. Although it is often claimed that the mediator is reoxidized on the polymer, direct evidence to support this assertion is sparse. In many cases results can be equally well explained by diffusion of the mediator species to, and reoxidation on, the underlying electrode.

Detailed studies of glucose oxidase immobilized in poly(N-methypyrrole) films using both ferrocene monocarboxylic acid[(69)] and ferricyanide[(50)] as the mediator species indicate that in both cases poly(N-methylpyrrole) films are insulating and there is no evidence of direct reoxidation of the mediator on the polymer. This is an unexpected result, since these mediators are electroactive at poly(N-methylpyrrole) films that do not contain GOx. It appears that in addition to catalyzing the oxidation of glucose in the presence of a mediator, GOx can also catalyze the reduction of oxygen to H_2O_2 in the presence of mediators (M) as electron acceptors

$$O_2 + 2M + 2H^+ \rightarrow H_2O_2 + 2M^+$$

The H_2O_2 produced by this reaction can react with the poly(N-methylpyrrole) and destroy the conductivity of the film.

In addition to poly(pyrrole) and poly(N-methylpyrrole), other conducting polymers have been used for enzyme immobilization, including poly(thiophene),[(70)] poly(indole),[(71)] and poly(aniline).[(72–74)] Poly(thiophenes) and poly(indole) films have the possible disadvantage that the enzyme/polymer film must be grown from aprotic solvents, conditions that may denature the enzymes. The

GOx-poly(aniline) films have been grown from aqueous solutions, although acidic solutions (around pH 1) are necessary for electropolymerization, which can also adversely affect the enzyme. Although direct electron transfer between GOx and poly(aniline) has been postulated,[74] evidence to support this is not unequivocal. Based on experiments using poly(aniline) poly(aniline)/glucose oxidase bilayers, Hoa *et al.* have concluded that there is negligible direct electron transfer.[75] This difference may reflect different deposition conditions used in the two sets of experiments.

Other enzymes have been entrapped in electropolymerized films, including cholesterol oxidase (EC 1.1.3.6),[76] horseradish peroxidase (EC 1.11.1.7),[77] and uricase (EC 1.7.3.3).[78] Poly(pyrrole) has also been used to entrap a number of dehydrogenase enzymes, including alcohol dehydrogenase,[79] glucose dehydrogenase,[80] and lactate dehydrogenase.[81] These enzymes depend on the soluble cofactor, nicotinamide adenine dinucleotide, which can exist either in the oxidized (NAD^+) or reduced forms (NADH). Since NADH cannot be oxidized directly at poly(pyrrole) films, these enzyme electrodes depend on the catalytic reoxidation of NADH using a mediator either entrapped within the film or incorporated as a ligand. Examples of such mediators including ferricyanide[82] and napthoquinone-sulphonate.[83]

The direct oxidation of NADH at poly(pyrrole)-coated electrodes has been achieved at potentials as low as 0.1 V versus SCE by incorporating the enzyme NADH dehydrogenase into the conducting polymer film.[84] Results indicate direct but slow reoxidation of the enzyme by the conducting polymer. The rate of reoxidation of the enzyme is considerably enhanced by incorporating ferrocene or quinone derivatives as mediators within the films.

In addition to conducting polymer films, electropolymerized films of insulating polymers can also be used to entrap enzymes at electrode surfaces. The growth of nonconducting polymers is self-limiting; consequently films tend to be very thin (10–100 nm), generally resulting in fast diffusion of substrates and products to and from the enzyme. In addition to incorporating relatively large amounts of enzyme, these films were shown to be able to reject interferent species; hence they can improve selectivity.

Electropolymerization of phenols proceeds similarly to that for pyrrole; poly(phenol) itself is probably a mixture of para- and meta-linked units. Films are generally continuous and free from such defects as pinholes.[85] They have been used for corrosion protection,[86] as permselective films,[87–89] and as pH sensors.[90] Like poly(pyrrole) films, electropolymerized phenol films have a number of features that make them attractive for immobilizing an enzyme. First these films can be grown under electrochemical control from aqueous buffered solution at neutral pH. Second a wide variety of phenol derivatives are available that allow some control over the films' physical characteristics. Third these films are permselective, which could be useful in preventing interfering species from reaching the electrode surface or fouling the electrode if used as biosensors. Bartlett *et al.* investigated (1) immobilization of glucose oxidase at platinum electrodes in electropolymerized films using five different phenols[91] and (2) making microelectrodes for glucose detection.[92] Since these films are very thin, it is possible to assume that there is no concentration polarization within the film and to analyze the response to glucose according to a simple model that takes account of glucose/enzyme kinetics. In this way quantitative comparison between different films can be made.

Sasso *et al.*[93] used electropolymerized films of 1,2-diaminobenzene, a monomer that produces films with properties very similar to those of the phenols, to

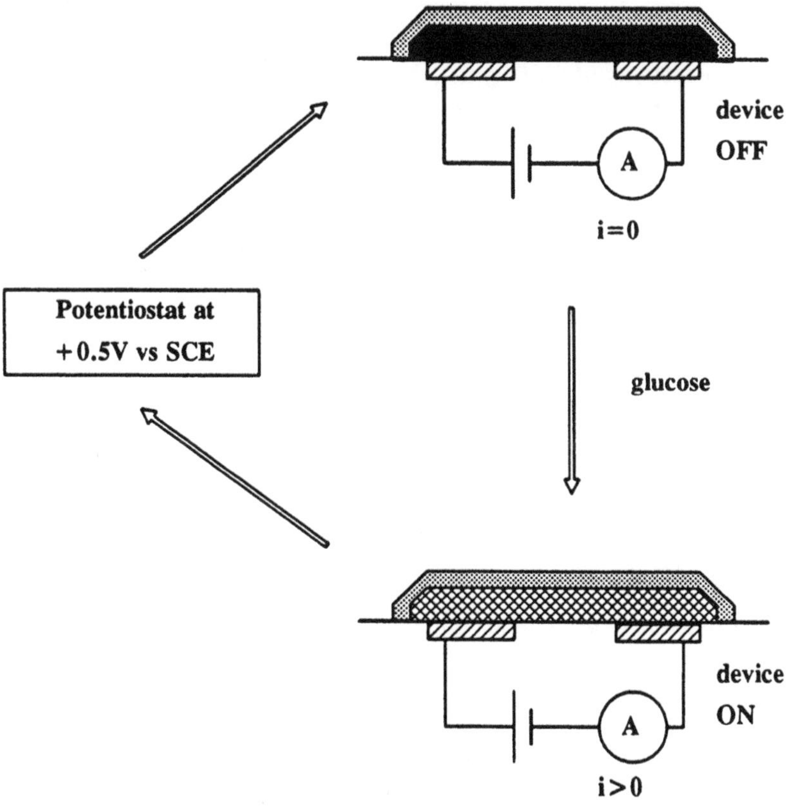

FIGURE 9.14. Schematic representation of an enzyme switch responsive to glucose based on glucose oxidase immobilized on a poly(aniline) film (see Ref. 96). At pH 5 and + 0.5 V versus SCE, the poly(aniline) is insulating. On adding glucose the polymer is reduced and switches to its conducting state. The device can be returned to the insulating state by electrochemical oxidation at 0.5 V.

reduce interferences in a glucose sensor. Malitesta *et al.*[94] also used the same monomer to immobilize glucose oxidase at a platinum electrode.

The technique of immobilizing enzymes in electropolymerized films has recently been used to make new devices in which an enzyme is immobilized on, or in, a conducting polymer film grown across a microelectrode array. Two types of device were described. In one type the enzyme-catalyzed reaction is linked to the redox chemistry of the conducting polymer so that the film is reduced by the enzyme-catalyzed reaction and thus changes its conductivity.[95,96] In the second type conductivity of the polymer is changed as a result of a local change in pH generated by the enzyme-catalyzed reaction.[75,97] In both cases conductivity of the polymer is monitored by applying a small (*ca.* 20-mV) potential across the polymer and measuring the current flowing through the film.

In devices of the first type, the conducting polymer starts in the oxidized state and is reduced as a result of the enzyme-catalyzed reaction of the substrate. This changes conductivity of the polymer, thus switching on or off the device. To reset the switch, it is necessary to connect the device to a potentiostat and reoxidize the film electrochemically (Fig. 9.14). Two devices have been reported to date, one based on poly(pyrrole)/poly(N-methylpyrrole) and responsive to NADH,[95] the second based on poly(aniline) and responsive to glucose.[96] In the latter case the device is switched from off to on, which appears to be advantageous, since it gives much faster response times.

In the second type of device, the response is generated by the local pH change (Fig. 9.15), so that when the substrate is removed, conductivity of the device returns to its original value. This approach suffers from the well-documented problems of all potentiometric pH-responsive enzyme devices.[98] In particular such sensors based on local pH changes are particularly sensitive to changes in the buffer capacity of the solution, so these do not work well in solutions of high buffer capacity for obvious reasons.

4.2. Redox Polymers

The attraction of using redox polymers as mediators for redox enzyme is that there are no freely diffusing species; hence it is possible to construct reagentless sensors. Such devices are therefore ideal for *in vivo* applications, and again much of the work has therefore concentrated on glucose oxidase. Early attempts to use redox polymers as mediators for glucose oxidase met with mixed success, but over the last 3 years, a clearer picture has emerged. One of the key insights that led to progress in this area was recognition of the need for flexibility in the polymer backbone to allow redox groups to reach an enzyme-active site[99] and optimization of the redox polymer–enzyme interaction, the so-called wiring of the enzyme.[100,101]

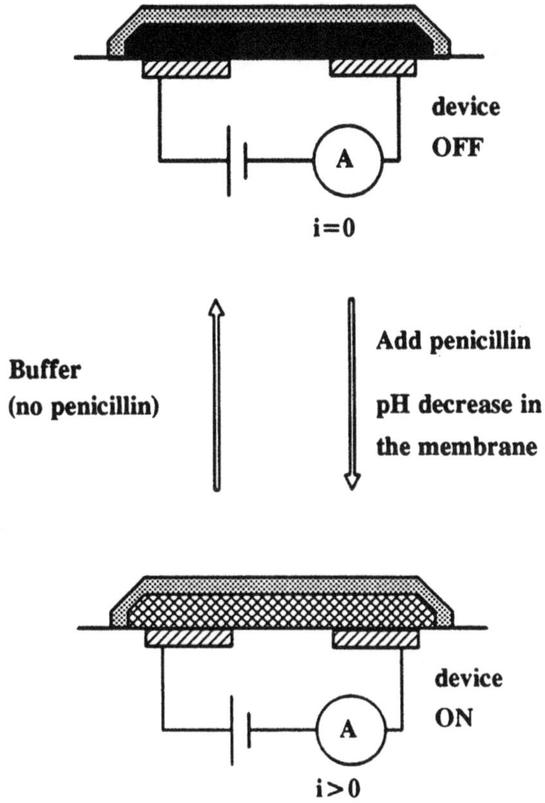

FIGURE 9.15. Schematic representation of an enzyme switch responsive to penicillin based on penicillinase immobilized on a poly(pyrrole) film (see Ref. 97). On adding penicillin protons are produced, which cause a local decrease in the solution pH and hence an increase in the conductivity of the polymer.

Initial studies using poly(vinylferrocene) as a mediator for glucose oxidase were disappointing,[102] apparently because of the rigidity of the polymer. Subsequent work using ferrocene-modified poly(siloxanes), in which the polymer backbone is highly flexible, clearly demonstrated efficient mediation for glucose oxidase. The importance of this flexibility was demonstrated by a study of a series of ferrocene-containing siloxane random block copolymers (Fig. 9.16),[103] in which the need for a compromise between increased polymer flexibility, favored by lower ferrocene content, and decreased spacing between redox sites, favored by higher ferrocene content, was demonstrated. Ferrocene-containing siloxane polymers have been used to make glucose electrodes[104] and mediate charge transfer to glycolate oxidase (EC 1.3.1.15).[105] In a recent study acrylamide-acrylic acid

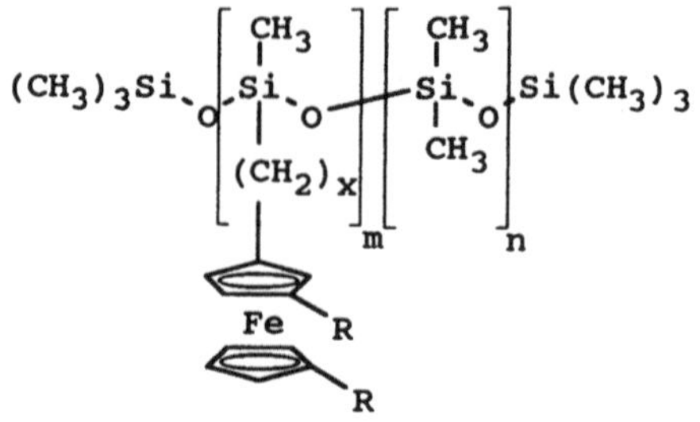

FIGURE 9.16. General structure of the ferrocene containing siloxane random block copolymers used by Hale *et al.* in Ref. 103).

ferrocene copolymers have also been shown to mediate glucose oxidase oxidation.[106]

A potential problem with using ferrocenes as mediators is the poor stability of some ferricinium ions in neutral aqueous solution.[107] As an alternative to the ferrocene-based system, Heller and his colleagues have made extensive use of osmium polymers as mediators.[101] By using a polycationic redox polymer (Fig. 9.17), he was able to form an electrostatic complex with the polyanionic enzyme, again glucose oxidase. At low ionic strength, the polycation and the enzyme are associated, while at high ionic strength, they dissociate unless the two are covalently bound together.[108] Furthermore these polycationic redox polymers are also strongly adsorbed at graphite, so they could be used to bind the enzyme to a graphite electrode to establish electrical communication between the active site of the enzyme and the electrode.[109]

By developing such systems and chemically cross-linking these polymers, Heller and colleagues were able to produce robust hydrophilic redox gels containing glucose oxidase. Initially this was achieved by using a poly(vinylpyridine) complex of $Os(bpy)_2Cl_2$ (MW 50 000) that was quaternized and then partially

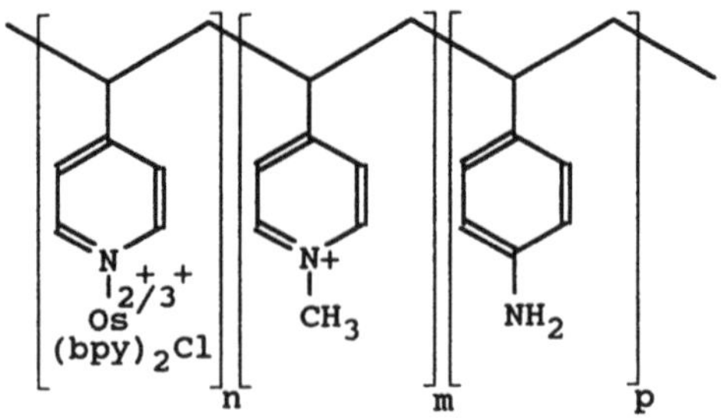

FIGURE 9.17. A quaternized redox polymer ($n = 0.14$, $m = 0.81$, $p = 0.05$) used to form electrostatic complexes with glucose oxidase (see Ref. 109).

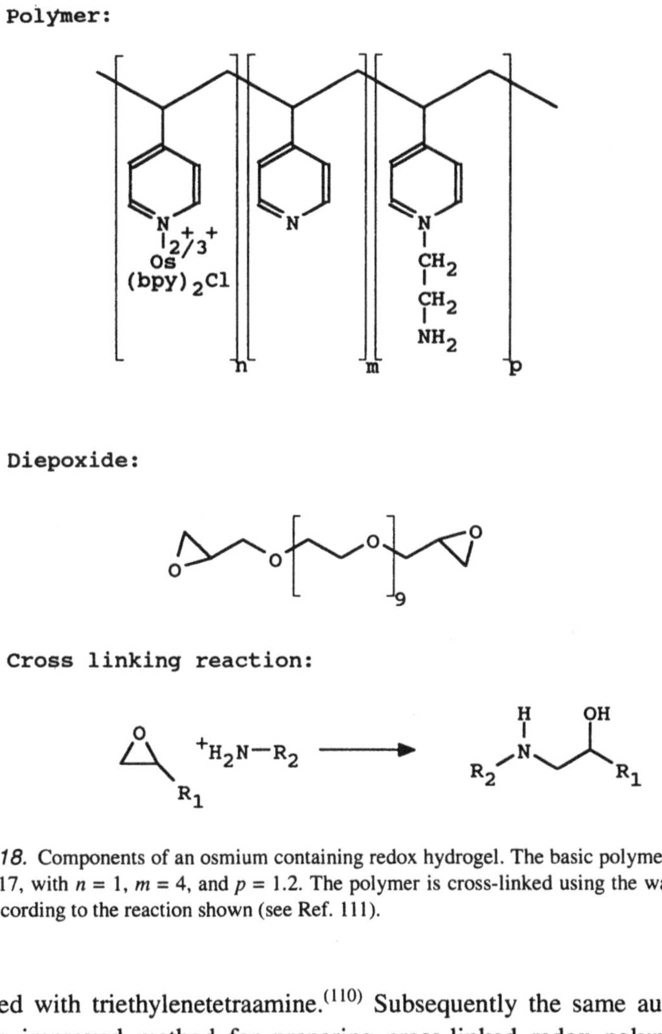

FIGURE 9.18. Components of an osmium containing redox hydrogel. The basic polymer is derived from Fig. 9.17, with $n = 1$, $m = 4$, and $p = 1.2$. The polymer is cross-linked using the water-soluble diepoxide according to the reaction shown (see Ref. 111).

cross-linked with triethylenetetraamine.(110) Subsequently the same authors described an improved method for preparing cross-linked redox polymer films containing glucose oxidase based around the use of a redox-conducting epoxy formed by cross-linking a poly(vinylpyridine) complex of $Os(bpy)_2Cl_2$ with a water soluble diepoxide (Fig. 9.18); this method is easier to carry out and gives more reproducible films.(111,112) Again there is efficient redox conduction through the film and excellent communication with the enzyme, so that these films can be used as reagentless biosensors; for example applying films of this type in glucose microsensors has been described(113) as well as a method for eliminating interferences based on oxidation of the interferant species by enzymatically generated hydrogen peroxide.(114,115) In addition to glucose oxidase, redox hydrogels of this type have been shown to work with lactate oxidase (EC 1.1.3.2), L-α-glycerol phosphate oxidase (EC 1.1.3.21),[116] cellbiose oxidase,[117] and horseradish peroxidase (EC 1.11.1.7).[118] It is clear that this flexible and successful approach is attracting increasing attention.

Viologen-acrylamide copolymers have been used in a similar manner to entrap enzymes and carry out direct reduction of the active site by bound viologen moieties.[119–121] In these studies the polymer was reduced chemically rather than electrochemically, but there appears no reason in principle why this same chemistry could not be carried out on an electrode. These viologen-acrylamide copolymers have been shown to reduce nitrate reductase (EC 1.9.6.1)[119] and glutathione reductase (EC 1.6.4.2).[120,121]

At present analysis of diffusion and kinetics in these redox polymer electrodes is at an early stage, although there have been some preliminary attempts to understand the behavior of the systems.[104,113] It is clear that more work on basic models for such electrodes, taking into account diffusion and enzyme kinetics is required.

5. APPLICATIONS IN REDOX COFACTOR ELECTROCHEMISTRY

In this section we review applications of electroactive polymers as electrode materials for use with redox cofactors, primarily NAD/NADH, and with components of redox proteins and redox enzymes, such as iron–sulphur clusters or flavins. We begin by considering the NAD/NADH couple.

The electrochemistry of NAD(P)/NAD(P)H has attracted much attention because of the many dehydrogenase enzymes (over 350) that use this cofactor. It was realized at an early stage that catalysis of NAD(P)/NAD(P)H electrochemistry represented a generic approach to designing amperometric enzyme electrodes. Problems associated with oxidation and reduction of this couple and with the design of modified electrodes for NADH oxidation or NAD reduction have recently been discussed.[54,122] Much of this work concentrated on using adsorbed redox species, such as meldola blue and its derivatives, rather than polymeric systems; more recently there has been increasing interest in applying electroactive polymers in this area. Some of the first studies using redox polymers concentrated on using orthoquinones covalently bound to a polymer backbone. These polymers suffered from poor charge transport thorough the film and rapid loss of activity.[123–125]

Oxidation of NADH at both poly(3-methylthiophene) and poly(pyrrole) has been studied. Using poly(3-methylthiophene) Atta *et al.*[126,127] observed some catalysis of the NADH oxidation, but they found that the effect was sensitive to the choice of background electrolyte. They suggested that adsorption of NADH onto the polymer was a prerequisite for oxidation. The best results were obtained in 0.01

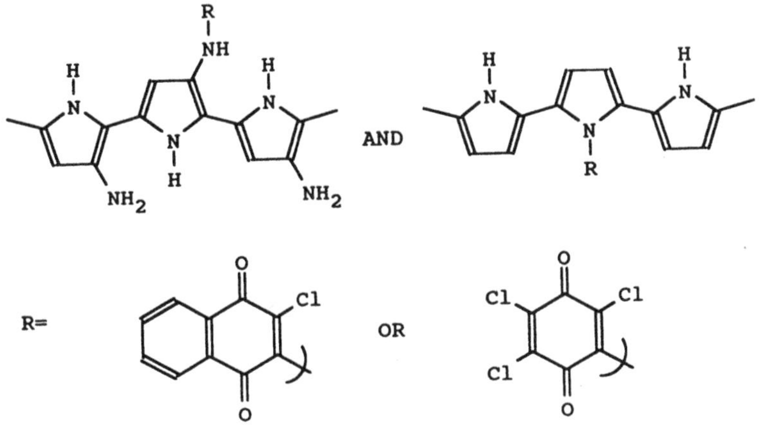

FIGURE 9.19. Structures of poly(pyrroles) modified by chloroanil and 2,3-dichloro-1,4-naphthoquinone (see Ref. 129).

mol dm^{-3} sulphuric acid; however this is not ideal because NADH undergoes hydrolysis in acidic solutions.[128] It is not clear how or why poly(3-methylthiophene) catalyzes the oxidation of NADH.

In the study of NADH oxidation at poly(pyrrole), Schuhmann and colleagues used poly(pyrrole) and modified poly(pyrrole) films.[129] For poly(pyrrole) itself and poly(pyrrole) containing electrostatically bound ferricyanide, vanadate, or molybdate anions, they observed no NADH oxidation, although these electroactive anions are known to oxidize the NADH in homogeneous solution. By using pyrrole modified by covalent attachment of chloranil or 2,3-dichloro-1,4-naphthoquinone (Fig. 9.19) structures, they were able to obtain catalytic oxidation of NADH and reasonable stability. It this case quinone functionalities are presumed to act as catalytic sites for oxidation, since they are known to be efficient homogeneous oxidants for NADH.[130]

Electropolymerized films of mediator species, such as mercaptoquinone,[131] N-methyl phenazinium ion,[132] and thionine (a redox dye),[133] were also shown to mediate NADH oxidation, although in these latter cases, the structure of the electropolymerized film is not well-established.

The reduction of NAD^+ at electroactive films has received less attention and represents a greater challenge because it must be achieved with regiospecificity if enzymatically active material is to be produced.[54] Rhodium (I) complexes are known to achieve this reduction in homogeneous solution,[134,135] and a preliminary note by Cosnier and Gunther has shown how this can be used in an electroactive polymer.[136] In experiments they used an electropolymerized film of

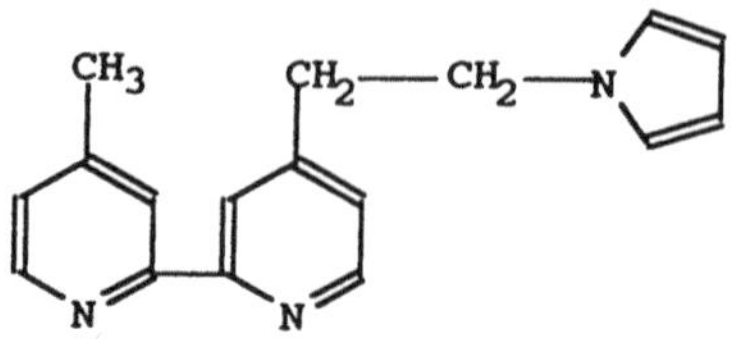

FIGURE 9.20. Structure of bis-4,4′-[bis-pyrrol-1-yl(methyl)methoxycarbonyl]-2,2′-bipyridyl.

$[Rh^{III}(C_5Me_5)(\mathbf{L})Cl]^+$, where **L** is bis-4,4′-[bis-pyrrol-1-yl(methyl)methoxycarbonyl]2,2′-bipyridyl (Fig. 9.20). On electropolymerization a poly(pyrrole) backbone with covalently attached rhodium redox centers is formed; these rhodium centers then act as catalytic sites for NAD^+ reduction. With this system the regiospecific reduction to give the 1,4 isomer of NADH (the enzymatically active isomer) is favored.

In addition to nicotinamide coenzymes, various other important biological redox centers have been studied. Although not so extensive, these studies illustrate some possible areas for future study. The flavin center is an important prosthetic group in many oxidases, and as such it has attracted study in its own right. Viologen-pendent poly(pyrrole) has been shown to catalyze the redox reaction of flavin mononucleotide,[(137)] with the bound viologen groups acting as catalytic redox centers for the process.

Flavin mononucleotide (FMN) is an anion at pH 7; therefore it can be doped into poly(pyrrole) by growing films in the presence of the ion.[(138)] Absorption spectra confirm the presence of FMN within films, and cyclic voltammery shows that entrapped flavin can be oxidized and reduced. It appears that entrapped reduced flavin is oxidized before the polymer backbone while the reduction of FMN and the backbone coincide. Electrochemistry of the entrapped material is shifted from that for the solution species, as we expect, due to the difference in the environment. The FMN has also been incorporated into poly(pyrrole-alkylammonium) films by ion exchange of the anion into the polymerized film.[(139)] Reversible voltammetry for incorporated FMN was observed with a slight shift in potential with respect to the solution species, and it was attributed to microenvironmental effects. These films were shown to catalyze oxygen reduction.

The electrochemistry of the PQQ (pyrrolo-quinoline quinone) prosthetic group has been investigated at poly(pyrrole)-coated electrodes, and PQQ has been entrapped within poly(pyrrole) films;[(140)] good electrochemistry was observed in both cases. Poly(pyrrole) has also been used to entrap adenosine triphosphate (ATP) anions, again by growing the film in the presence of the anion,[(141)] and as an electrode material for oxidizing ascorbate.[(142)] In the latter case the oxidation

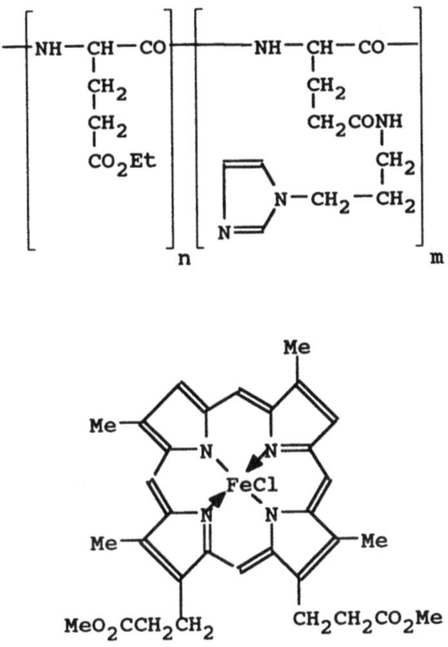

FIGURE 9.21. Structures of imidazole-modified poly(γ-ethyl L-glutamate) and deuteroporphyrin-IX-$Fe^{III}Cl$ used by Grimshaw and Trocha-Grimshaw to construct an analog of cytochrome *c* (see Ref. 143).

mechanism was studied, and the possible application in a sensor for ascorbate was demonstrated.

Finally we mention two pieces of work in which analogs of biological redox systems were constructed in membranes at an electrode surface. By coordinating a deuteroporphyrin-IX-$Fe^{III}Cl$ dimethyl ester to imidazole groups bound to poly(γ-ethyl L-glutamate (Fig. 9.21) deposited as a film at an electrode surface, Grimshaw and Trocha-Grimshaw constructed an analog of cytochrome *c*.[143] The coordinated iron centers can be reversibly oxidized and reduced in aqueous solution. In the second elegant example, electroactive $[4Fe\text{-}4S]^{2+}$ centers were bound into a designed and synthesized polymeric film of L-cystine derivatized pyrrole.[144] This produces electrode–polymer assemblies with electroactive cysteinyl S ligated, ferredoxinlike units that are a conducting polymer analog of the redox protein. The exciting aspect of this work is the ability to construct analogs of redox protein-active centers on an electrode surface and the potential for extending this for example by entrapping the FeMo cofactor of nitrogenase activity onto an electrode surface to achieve electrochemical nitrogen fixation.

6. CONCLUSION

The range and scope of applications of electroactive polymers in bioelectrochemistry is expanding rapidly. It is notable that much of the work reviewed here was published in the last 3 years. We are moving from the initial stages of these studies to a more mature phase in which detailed measurements and study are required to establish the mechanisms and nature of interactions within these systems and to design them for particular purposes.

REFERENCES

1. F. W. Scheller, F. Schubert, R. Renneberg, H -G. Müller, M. Jänchen, and H. Weise, *Biosensors* **1**, 135 (1985).
2. D. G. Nicholls, *Bioenergetics* (Academic Press, London, 1982).
3. C. Smith and E. J. Wood, *Energy in Biological Systems*. (Chapman and Hall, London, 1991).
4. G. McLendon, *Acc. Chem. Res.* **21**, 160 (1988).
5. S. S. Isied, *Prog. Inorg. Chem.* **32**, 443 (1984).
6. J. R. Bolton and M. D. Archer, in *Electron Transfer in Inorganic, Organic and Biological Systems* (J. R. Bolton, N. Mataga, and G. McLendon, eds.) (ACS, Washington, 1991), pp. 7–23.
7. P. Siddarth and R. A. Marcus, *J. Phys. Chem.* **94**, (1990) 8430–8434.
8. C. C. Moser, J. M. Keske, K. Warncke, R. S. Farid, and P. L. Dutton, *Nature* **355**, 796 (1992).
9. A. Fersht, *Enzyme Structure and Mechanism* (W. H. Freeman, Reading, PA, 1977).
10. M. Dixon and E. C. Webb, *Enzymes* (Longman, London, 1979), p. 479.
11. C. Walsh, *Acc. Chem. Res.* **13**, 148 (1980).
12. P. Hemmererich, C. Veeger, and H. C. S. Wood, *Angew. Chem.* **77**, 699 (1965).
13. B. Janík and P. J. Elving, *Chem. Rev.* **68**, 295 (1968).
14. G. Dryhurst, *Electrochemistry of Biological Molecules* (Academic Press, London, 1977), p. 365.
15. D. E. Metzler, *Biochemistry* (Academic Press, New York, 1977), p. 478.
16. S. Ghisla and V. Massey, *Eur. J. Biochem.* **181**, 1 (1989).
17. G. Dryhurst, K. M. Kadish, F. Scheller, and R. Renneberg, *Biological Electrochemistry*, vol. 1 (Academic Press, New York, 1982), p. 459.
18. M. Dixon and E. C. Webb, *Enzymes* (Longman, London, 1979), p. 483.
19. S. A. Salisbury, H. S. Forrest, W. B. T. Cruse, and O. Kennard, *Nature* **280**, 843 (1979).
20. J. A. Duine and J. J. Frank, *Trends in Biochemical Science* **6**, 278 (1981).
21. B. C. Finzel, T. L. Poulos, and J. Kraut, *J. Biol. Chem.* **259**, 3027 (1984).
22. E. Margoliash, in *Hemes and Hemoproteins* (B. Chance and R. Estabrook, eds.) (Academic Press, New York, 1986), p. 373.
23. T. Takano, O. B.-Kallai, R. Swanson, and R. E. Dickerson, *J. Biol. Chem.* **248**, 5234 (1973).
24. J. H. Wang, *Acc. Chem. Res.* **3**, 90 (1970).
25. A. L. Lehninger, *Biochemistry* (Worth, New York, 1975), p. 492.
26. C. P. Andrieux, J. M. Dumas-Bouchiat, and J. M. Saveant, *J. Electroanal. Chem.* **131**, 1 (1982).
27. W. J. Albery and A. R. Hillman, *J. Electroanal. Chem.* **170**, 27 (1984).
28. P. N. Bartlett, in *Biosensors: Fundamentals and Applications* (A. P. F. Turner, I. Karube, and G. S. Wilson, eds.) (Oxford University Press, Oxford, England, 1987), pp. 211–46.
29. J -M. Engasser and C. Horvath, in *Applied Biochemistry and Bioengineering. Vol. 1. Immobilized Enzyme Principles* (L. B. Wingard, E. Katchalski-Katzir, and L. Goldstein, eds.) (Academic Press, New York, 1976), p. 127.

30. P. N. Bartlett, P. Tebbutt, and R. G. Whitaker, *Progress in Reaction Kinetics* **16**, 55 (1991).
31. M. J. Eddowes and H. A. O. Hill, *J. Chem. Soc. Chem. Commun.* 771 (1977).
32. F. A. Armstrong, H. A. O. Hill, and N. J. Walton, *Acc. Chem. Res.* **21**, 407 (1988).
33. F. A. Armstrong, *Struct. Bonding* **73**, 137 (1990).
34. P. M. Allen, H. A. O. Hill, and N. J. Walton, *J. Electroanal. Chem.* **178**, 69 (1984).
35. P. N. Bartlett, D. H. Dawson, and J. Farrington, *J. Chem. Soc., Faraday Trans.* **88**, 2685 (1992).
36. P. N. Bartlett and J. Farington, *J. Electroanal. Chem.* **261**, 471 (1989).
37. H. L. Landrum, R. T. Salmon, and F. M. Hawkridge, *J. Amer. Chem. Soc.* **99**, 3154 (1977).
38. C. D. Crawley and F. M. Hawkridge, *Biochem. Biophys. Res. Commun.* **99**, 516 (1981).
39. C. M. Elliot and W. S. Martin, *J. Electroanal. Chem.* **137**, 377 (1982).
40. B. N. Oliver, J. O. Egekeze, and R. W. Murray, *J. Amer. Chem. Soc.* **110**, 2321 (1988).
41. C. E. W. Hahn, H. A. O. Hill, M. D. Ritchie, and J. W. Sears, *J. Chem. Soc., Chem. Commun.* 125 (1990).
42. M. Umaña and J. Waller, *Anal. Chem.* **58**, 2979 (1986).
43. N. C. Foulds and C. R. Lowe, *J. Chem. Soc., Faraday Trans. 1* **82**, 1259 (1986).
44. P. N. Bartlett and J. Cooper, *J. Electroanal. Chem.* **362**, 1 (1993).
45. B. F. Y. Yon Hin, R. S. Sethi, and C. R. Lowe, *Sensors and Actuators* **B1**, 550 (1990).
46. T. Shimidzu, A. Ohtani, T. Iyoda, and K. Honda, *J. Electroanal. Chem.* **224**, 123 (1987).
47. R. Bentley, in *Enzymes*, vol. 7 (P. D. Boyer, H. A. Hardy, and H. Myrback, eds.) (Academic Press, New York, 1973), p. 567.
48. J. M. Dicks, S. Hattori, I. Karube, A. P. F. Turner, and T. Yokozawa, *Ann. Biol. Clin.* **47**, 607 (1989).
49. Y. Kajiya, H. Sugai, C. Iwakura, and H. Yoneyama, *Anal. Chem.* **63**, 49 (1991).
50. P. N. Bartlett, Z. Ali, and V. Eastwick-Field, *J. C. S. Faraday Trans.* **88**, 2677 (1992).
51. T. Schalkhammer, E. Mann-Buxbaum, G. Urban, and F. Pittner, *J. Chromatography* **510**, 355 (1990).
52. W. Schuhmann and R. Kittsteiner-Eberle, *Biosensors and Bioelectronics* **6**, 263 (1991).
53. J-M. Engasser and C. Horvath, in *Applied Biochemistry and Bioengineering*, vol. 1: *Immobilized Enzyme Principles* (L. B. Wingard, E. Katchalski-Katzir, and L. Goldstein, eds.) (Academic Press, New York, 1976), p. 127.
54. M. J. Eddowes, in *Biosensors: A Practical Approach* (A. E. G. Cass, ed.) (IRL Press, Oxford, 1990), p. 211.
55. P. N. Bartlett and K. F. E. Pratt, *Biosensors and Bioelectronics* **8**, 451 (1993).
56. P. N. Bartlett and R. G. Whittaker, *J. Electroanal. Chem.* **224**, 37 (1991).
57. D. Belanger, J. Nadreau, and G. Fortier, *J. Electroanal. Chem.* **274**, 143 (1989).
58. J. M. Cooper and D. Bloor, *Electroanalysis* **5**, 883 (1993).
59. S-I. Yabuki, H. Shinohara, and M. Aizawa, *J. Chem. Soc., Chem. Commun.* 945 (1989).
60. P. de Taxis du Poet, S. Miyamoto, T. Murakami, J. Kimura, and I. Karube, *Anal. Chim. Acta* **235**, 255 (1990).
61. C. G. J. Koopal, M. C. Feiters, R. J. N. Nolte, B. de Ruiter, R. B. M. Schasfoort, *Biosensors and Bioelectronics* **7**, 461 (1992).
62. C. G. J. Koopal, B. de Ruiter, and R. J. M. Nolte, *J. Chem. Soc., Chem. Commun.* 1691 (1991).
63. C. G. J. Koopal, M. C. Feiters, R. J. M. Nolte, B. de Ruiter, R. B. M. Schasfoot, R. Czajka, and H. Van Kempen, *Synthetic Metals* **51**, 397 (1992).
64. S. Kuwabata and C. R. Martin, *Anal. Chem.* **66**, 2757 (1994).
65. C. Iwakura, Y. Kajiya, and H. Yoneyama, *J. Chem. Soc., Chem. Commun.* 1019 (1988).
66. F. Mizutani and M. Asai, *Bull. Chem. Soc. Jpn.* **61**, 4458 (1988).
67. P. Janda and J. Weber, *J. Electroanal. Chem.* **300**, 119 (1991).
68. N. C. Foulds and C. R. Lowe, *Anal. Chem.* **60**, 2473 (1988).
69. R. G. Whitaker, Ph. D. diss., University of Warwick, 1989.
70. J. C. Scott, P. Pfluger, M. T. Krounbi, and G. B. Street, *J. Phys. B.* **28**, 2140 (1983).

71. P. C. Pandey, *J. Chem. Soc., Faraday Trans. 1* **84**, 2259 (1988).
72. P. N. Bartlett and R. G. Whitaker, *Biosensors* **3**, 359 (1987/88).
73. H. Shinohara, T. Chiba, and M. Aizawa, *Sensors and Actuators* **13**, 79 (1988).
74. J. C. Cooper and E. A. H. Hall, *Biosensors and Bioelectronics* **7**, 473 (1992).
75. D. T. Hoa, T. N. Suresh Kumar, N. S. Punekar, R. S. Srinivasa, R. Lal, and A. Q. Contractor, *Anal. Chem.* **64**, 2645 (1992).
76. Y. K. Kajiya, R. Tsuda, and H. Yoneyama, *J. Electroanal. Chem.* **301**, 155 (1991).
77. T. Tatsuma, M. Gordaira, and T. Watanabe, *Anal. Chem.* **64**, 1183 (1992).
78. M. Shaolin, K. Jinquing, and Z. Jianbing, *J. Electroanal. Chem.* **334**, 121 (1992).
79. S. Yabuki, H. Shinohara, Y. Ilariyama, and M. Aizawa, *J. Electroanal. Chem.* **227**, 179 (1990).
80. S. Yabuki, F. Mizutani, and M. Asai, *Biosensors and Bioelectronics* **6**, 311 (1991).
81. A. G. Dubinin, F. C. Li, Y. R. Li, and J. T. Yu, *Bioelectrochemistry and Bioenergetics* **25**, 131 (1991).
82. T. Matsue, H. Yamada, H.-C. Chang, I. Uchida, K. Nagata, and K. Tomita, *Biochim. Biophys. Acta* **1038**, 29 (1990).
83. Y. Kajiya, H. Matsumoto,, and H. Yoneyama, *J. Electroanal. Chem.* **319**, 185 (1991).
84. T. Matsue, N. Kasai, M. Narumi, M. Nishizawa, H. Yamada, and I. Uchida, *J. Electroanal. Chem.* **300**, 111 (1991).
85. F. Bruno, M. C. Pham, and J. E. Dubois, *Electrochim. Acta* **22**, 451 (1977).
86. G. Mengoli and M. M. Musiani, *J. Electrochem. Soc.* **134**, 643C (1987).
87. Y. Ohnuki, T. Oshaka, T. Matsuda, and N. Oyama, *J. Electroanal. Chem.* **158**, 55 (1983).
88. T. Oshaka, T. Hirokawa, H. Miyamoto, and N. Oyama, *Anal. Chem.* **59**, 1758 (1987).
89. J. Wang, S-P. Chen, and M. S. Lin, *J. Electroanal. Chem.* **273**, 231 (1989).
90. G. Cheek, C. P. Wales, and R. J. Nowak, *Anal. Chem.* **55**, 380 (1983).
91. P. N. Bartlett, P. Tebbutt, and C. Tyrrell, *Anal. Chem.* **64**, 138 (1992).
92. P. N. Bartlett and D. J. Caruana, *Analyst* **117**, 1287 (1992).
93. S. V. Sasso, R. J. Pierce, R. Walla, and A. M. Yacynych, *Anal. Chem.* **62**, 1111 (1990).
94. C. Malitesta, F. Palmisano, L. Torsi, and P. G. Zambonin, *Anal. Chem.* **62**, 2735 (1990).
95. T. Matsue, M. Nishizaura, T. Sawaguchi, and I. Uchida, *J. Chem. Soc., Chem. Comm.* 1029 (1991).
96. P. N. Bartlett and P. R. Birkin, *Anal. Chem.* **66**, 1552 (1994).
97. M. Nishizawa, T. Matsue, and I. Uchida, *Anal. Chem.* **64**, 2642 (1992).
98. M. J. Eddowes, D. G. Pedley, and B. C. Webb, *Sensors and Actuators* **7**, 233 (1985).
99. P. D. Hale, T. Inagaki, H. I. Karan, Y. Okamoto, and T. J. Skotheim, *J. Amer. Chem. Soc.* **111**, 3482 (1989).
100. A. Heller, *Acc. Chem. Res.* **23**, 128 (1990).
101. A. Heller, *J. Phys. Chem.* **96**, 3579 (1992).
102. J. A. Chambers and N. J. Walton, *J. Electroanal. Chem.* **250**, 417 (1988).
103. P. D. Hale. L. I. Boguslavsky, T. Inagaki, H. I. Karan, H. S. Lee, T. J. Skotheim, and Y. Okamoto, *Anal. Chem.* **63**, 677 (1991).
104. L. Gorton, H. I. Karan, P. D. Hale, I. Y. Okamoto, and T. A. Skotheim, *Anal. Chim. Acta* **228**, 23 (1990).
105. P. D. Hale, T. Inagaki, H. S. Lee, H. I. Karan, Y. Okamoto, and T. A. Skotheim, *Anal. Chim. Acta* **228**, 31 (1990).
106. E. J. Calvo, C. Danilowicz, and L. Diaz, *J. Chem. Soc., Faraday Trans.* **89**, 377 (1993).
107. R. Szentrimay, P. Yeh, and T. Kuwana, in *Electrochemical Studies of Biological Systems* (D. Sawyer, ed.) (ACS, Washington DC, 1977), p. 143.
108. Y. Degani and A. Heller, *J. Amer. Chem. Soc.* **111**, 2357 (1989).
109. M. V. Pishko, I. Katakis, S.-E. Lindquist, L. Ye, B. A. Gregg, and A. Heller, *Angew. Chem. Int. Ed. Engl.* **29**, 82 (1990).
110. B. A. Gregg and A. Heller, *Anal. Chem.* **62**, 258 (1990).

111. B. A. Gregg and A. Heller, *J. Phys. Chem.* **95**, 5970 (1991).
112. B. A. Gregg and A. Heller, *J. Phys. Chem.* **95**, 5976 (1991).
113. M. V. Pishko, A. C. Michael, and A. Heller, *Anal. Chem.* **63**, 2268 (1991).
114. R. Maidan and A. Heller, *J. Amer. Chem. Soc.* **113**, 9003 (1991).
115. R. Maidan and A. Heller, *J. Amer. Chem. Soc.* **64**, 2889 (1992).
116. I. Katakis and A. Heller, *Anal. Chem.* **64**, 1008 (1992).
117. M. Elmgren, S -E. Lindquist, and G. Henriksson, *J. Electroanal. Chem.* **341**, 257 (1992).
118. M. Vreeke, R. Maidan, and A. Heller, *Anal. Chem.* **64**, 3084 (1992).
119. I. Willner, A. Riklin, and N. Lapidot, *J. Amer. Chem. Soc.* **112**, 6438 (1990).
120. I. Willner and N. Lapidot, *J. Amer. Chem. Soc.* **113**, 3625 (1991).
121. I. Willner, R. Kasher, E. Zahavy, and N. Lapidot, *J. Amer. Chem. Soc.* **114**, 10963 (1992).
122. L. Gorton, *J. Chem. Soc., Faraday Trans. 1* **82**, 1245 (1986).
123. C. Degrand and L. L. Miller, *J. Amer. Chem. Soc.* **102**, 5728 (1980).
124. M. Fukui, A. Kitani, C. Degrand, and L. L. Miller, *J. Amer. Chem. Soc.* **104**, 28 (1982).
125. A. N. K. Lau and L. L. Miller, *J. Amer. Chem. Soc.* **105**, 5271 (1983).
126. N. F. Atta, A. Galal, A. E. Karagözler, H. Zimmer, J. F. Rubinson, and H. B. Mark, *J. Chem. Soc., Chem. Commun.* 1347 (1990).
127. N. F. Atta, A. Galal, A. E. Karagözler, G. C. Russell, H. Zimmer, and H. B. Mark, *Biosensors and Bioelectronics* **6**, 333 (1991).
128. R. D. Braun, K. S. V. Santhanam, and P. J. Elving, *J. Amer. Chem. Soc.* **97**, 2591 (1975).
129. W. Schuhmann, R. Lammert, M. Hämmerle, and H -L. Schmidt, *Biosensors and Bioelectronics* **6**, 689 (1991).
130. B. W. Carlson and L. L. Miller, *J. Amer. Chem. Soc.* **107**, 479 (1985).
131. G. Arai, M. Matsushita, and I. Yasumori, *Nippon Kagaku Kaishi* **5**, 894 (1985).
132. A. Torstensson and L. Gorton, *J. Electroanal. Chem.* **130**, 199 (1981).
133. T. Ohsaka, K. Tanaka, and K. Tokuda, *J. Chem. Soc., Chem. Commun.* 222 (1993).
134. R. Wienkamp and E. Steckhan, *Agnew. Chem., Int. Ed. Engl.* **21**, 782 (1982).
135. R. Ruppert, S. Herrmann, and E. Steckhan, *Tetrahedron Lett.* **28**, 6583 (1987).
136. S. Cosnier and H. Gunther, *J. Electroanal. Chem.* **315**, 307 (1991).
137. M. Kijima, A. Sakawaki, and T. Sato, *Chem. Lett.* 499 (1991).
138. F. Battaglini, C. Bonazzola, and E. J. Calvo, *J. Electroanal. Chem.* **309**, 347 (1991).
139. S. Cosnier and C. Innocent, *J. Electroanal. Chem.* **338**, 339 (1992).
140. H. Shinohara, G. F. Kahn, Y. Ikariyama, and M. Aizawa, *J. Electroanal. Chem.* **304**, 75 (1991).
141. A. Boyle, E. Geniès, and M. Fouletier, *J. Electroanal. Chem.* **279**, 179 (1990).
142. M. E. G. Lyons, W. Breen, and J. Cassidy, *J. Chem. Soc., Faraday Trans.* **87**, 115 (1991).
143. J. Grimshaw and J. Trocha-Grimshaw, *J. Chem. Soc., Chem. Commun.* 157 (1990).
144. C. J. Pickett, K. S. Ryder, and J.-C. Moutet, *J. Chem. Soc., Chem. Commun.* 694 (1992).

10

Analytical Applications of Polymer-Modified Electrodes

Dónal Leech

1. INTRODUCTION

Before the advent of modified electrodes, electrochemistry was restricted by the range of electrode materials available and burdened with problems associated with the adsorption of a wide variety of compounds at their surfaces. In the early 1970s several groups realized that the deliberate modification of electrode surfaces could be used to tailor the electrode for specific applications. The pioneering work of Lane and Hubbard[(1)] on the modification of platinum electrodes with adsorbed unsaturated monomers was followed by the development of general schemes for the covalent attachment of monolayers to electrode surfaces via silanization reactions[(2–4)] or cyanuric chloride derivatization of electrodes.[(5)] For many applications it was realized that films thicker than monolayers would be beneficial. This led to the modification of electrodes with polymer films, first demonstrated by Miller's group,[(6,7)] who prepared polymers containing such redox groups as quinones and nitroaromatics.

This chapter describes progress in the application of polymer-modified electrodes (PMEs) to voltammetric chemical sensing. Several excellent reviews on the general preparation, properties, theory, and applications of modified electrodes have appeared over the past decade.[(8–12)] Literature surveys and reviews of advances in the field of chemical sensors[(13)] and electroanalysis[(14,15)] are also available. The area of polymeric potentiometric electrodes and the vast research field of biosensors are not dealt with in this chapter, which introduces the reader to selected representative examples of the diverse modification strategies available for prepar-

---

Dónal Leech • Département de Chimie, Université de Montréal, Montréal, Québec, Canada.

Electroactive Polymer Electrochemistry, Part 2: Methods and Applications, edited by Michael E. G. Lyons. Plenum Press, New York, 1996.

ing PMEs and their application to electroanalysis. An introduction to the preparation of polymer films on electrodes is presented, followed by examples of their analytical application.

2. PREPARING AND COATING POLYMER LAYERS

2.1. Preformed Polymers

Preformed polymers coated on a variety of electrode surfaces have found application in electroanalysis. Polymeric films containing coordinated redox groups were among the first PMEs developed because of their obvious potential as electrocatalysts.[16] Preformed polymers are normally adsorbed on the surface of the electrodes in a mechanism assumed to involve a combination of adsorption processes and the insolubility of the polymer in the electrolyte being used.

The simplest method for fabricating PMEs from preformed polymers is based on adsorption from solution by dipping the electrode in a polymer solution. The reproducibility of this method is questionable however because of the need to shake or wipe excess solution from the surface before the solvent evaporates.[6,17] More reproducible films can be prepared by dropping a measured aliquot of a polymer solution onto the electrode surface, followed by the slow evaporation, preferably in a solvent-saturated chamber,[11] of the solvent. For thin polymer films, the technique of spin coating can produce more homogeneous and reproducible layers. This technique involves the drop coating and solvent evaporation of the polymer solution on a rapidly rotating electrode.

Using preformed redox polymers as electrode modifiers has the advantage of completely characterizing polymers prior to modifying the electrode surface. Control of the layer thickness of the deposited films can be achieved by carefully controlling the amount deposited by using drop- or spin-coating procedures.

Coating electrodes with preformed ion-exchange polymers can also be achieved by drop-, dip-, or spin-coating methods. These layers can be used for electrostatic preconcentration of analytes or as permselective barriers to prevent electrode fouling and interferences. In addition equilibration of deposited ion-exchange polymers with solutions containing redox-active ions can lead to electrocatalytic polymeric layers, as first demonstrated for the incorporation of $Fe(CN)_6^{4-/3-}$ in protonated poly(4-vinylpyridine) (PVP) films by Oyama and Anson.[18] The codeposition of both the ion-exchanger and redox ion has been reported to yield better behaved films.[19] One of the most widely used ion-exchange materials is the perfluorosulfonated Nafion polymer developed by Dupont, which possesses both anionic and hydrophobic domains for accumulation of metal complexes and hydrophobic analytes.[20–22] Other ion-exchange materials, such as clays[23] and zeolites,[24] have been coated on electrodes and used to immobilize redox-active groups.

A general approach for incorporating electrocatalysts into p-chlorosulfonated polystyrene films, devised by Ellis and Meyer,[25] is based on the reactivity of the chlorosulfonate group attached to the polymer film deposited on the electrode surface toward compounds, which are selected to be redox-active, containing amino, hydroxy, and carboxylate groups. Various metal redox complexes have been incorporated into polystyrene films in this way.[25]

2.2. Preparation from Monomers

The direct formation of polymer films on electrode surfaces from their monomeric constituents is an attractive method for the preparation of PMEs. One such method is the polymerization of vinyl monomers from the gas phase using a radiofrequency plasma discharge. However note that polyvinylferrocene (PVF) films formed by plasma polymerization have been found to be cross-linked to a greater degree than electrodeposited PVF films, thus limiting penetration of electrolyte ions.[26]

Chemical coupling via silanization reactions of electroactive silanes to electrode surfaces is a useful procedure for preparing monolayer coatings.[4,8] This principle has been extended to the preparation of polymeric films by using bis-trialkoxysilylated monomer derivatives of viologen[27] and cobaltocenium[28] that can couple to the electrode surface and form polymeric films by subsequent cross-linking.

A convenient method for *in situ* formation of polymeric films on electrode surfaces is electropolymerization. For the formation of three-dimensional electropolymerized films, it is essential for the layers formed to be redox-active or conducting to allow subsequent oxidation or reduction of the monomer at the layer or for the layer formed to be permeable to the monomer to allow it to react at the electrode surface. Reductive electropolymerization of vinyl or acetylene containing bipyridine ligands attached to metal complexes of ruthenium, osmium, and iron have received the most attention. The mechanism for the formation of polymer films from these monomers involves radical coupling of the bipyridine ligands, even though the anionic polymerization is initiated at the vinyl group.[29] Electropolymerization of a $[Ru(v\text{-}bpy)_3]^{2+}$ polymer film, where v-bpy represents a 4-vinyl-4′-methyl-2,2′-bipyridine ligand, and the subsequent electrochemistry of the film is depicted in Fig. 10.1. The reductive electropolymerization of aldehyde-substituted bipyridine ligands complexed to iron and zinc to form 1,2-diol linkages between bipyridine ligands has recently been reported.[30] Oxidative electropolymerization of aromatic amines,[31] phenols,[32] and vinyl-substituted complexes[33] can be achieved. Amino-substituted ligands, such as phenanthroline metal com-

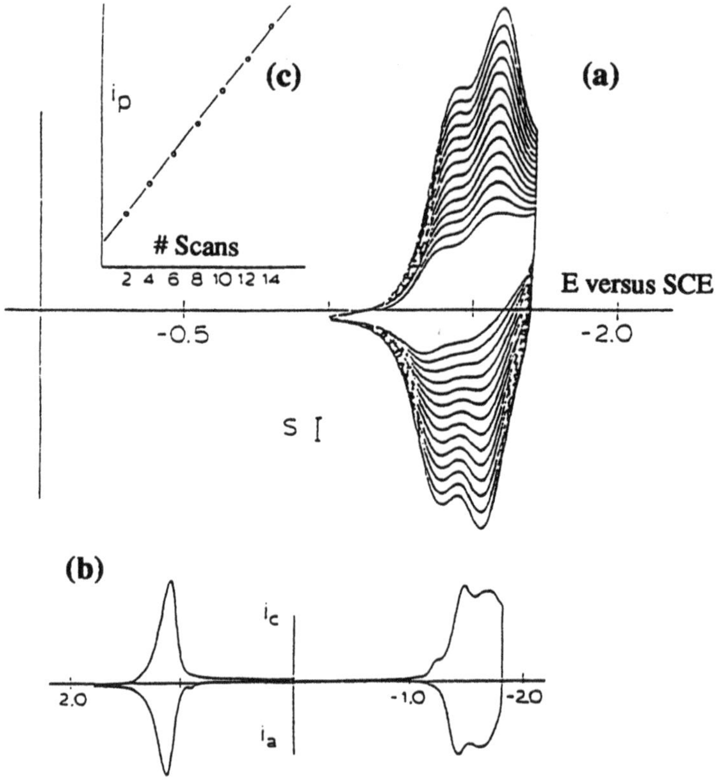

FIGURE 10.1. (a) Repetitive cyclic voltammograms for the electropolymerization of films of $[Ru(v\text{-}bpy)_3]^{2+}$ on glassy carbon electrodes. (b) Cyclic voltammograms of the modified electrode in supporting electrolyte only. (c) The correlation between the growth and the scan number. (From Ref. 29a with permission of the American Chemical Society)

plexes[31] and porphyrins,[33] can undergo oxidative electropolymerization to yield electroactive films.

Perhaps the most important development in electropolymerization is the discovery that conductive polymer films can be prepared by electrooxidation of such monomers as pyrrole, thiophene, and aniline.[34,35] These polymers, which incorporate counterions (dopants) into the polymer matrix during polymerization, have proved to be convenient stable supports for incorporating such reagents as electrocatalytic centers[35] or complexing agents.[36] Films of conducting polymers on electrode surfaces can also function as an electrocatalytically active material.[37] The preparation, conduction mechanism, and applications of conducting polymers have been reviewed,[34,35,38] and a reference handbook on conducting polymers is available to the interested reader.[39]

Purely inorganic films on electrode surfaces for electrocatalytic and ion-sensing applications have been produced by the electrochemical precipitation of Prussian Blue (PB) and its ruthenium analog on electrode surfaces.[40] The nickel analog of PB can be prepared by dissolution of a nickel electrode in the presence of ferricyanide.[41]

Electrodeposition and electropolymerization methods are convenient direct procedures for preparing modifying layers. The thickness and frequently the permeation and charge transport characteristics of the polymer films can be controlled by the applied potential. Lack of knowledge of the precise structure of the electropolymerized or electrodeposited polymer films however represents a problem.

3. ANALYTICAL APPLICATIONS OF POLYMER-MODIFIED ELECTRODES

The application of modified electrodes can be exploited in such technologies as energy storage, microelectrochemical devices, supramolecular chemistry, electrochromic displays, electrocatalysis, solar energy conversion and electroanalysis.[8–12]

Attachment of polymeric films to electrode surfaces for electroanalysis is usually undertaken to achieve at least one or a combination of the following goals. To achieve these objectives modifiers with electrocatalytic, molecular recognition, preconcentration, or permselective properties are applied to electrode surfaces.

Enhanced signals over unmodified electrodes or generation of a signal where none was previously detected
Increased stability of the electrode and reproducibility toward detection of the analyte of interest
Selective recognition of the target analyte
Reduction or elimination of interferences

3.1. Electrocatalysis

Initial research into the application of PMEs focused on their potential use in electrocatalysis. Much of this work centered on preformed redox polymers containing coordinated electroactive metal complexes because of the synthetic flexibility and the ability to control loading of the electrocatalytic center in the modifying layer. Electrostatic binding of electroactive ions into ionomeric polymer films is a convenient procedure for preparing electrocatalytic layers, although care must be taken to minimize leaching of the electroactive center.[8–12]

Voltammetric detection of a solution species normally depends on that species being reduced or oxidized at the electrode. Frequently in electrochemistry electrode kinetics are such that the redox reaction occurs at more positive or negative potentials than predicted on the basis of thermodynamics. Modification of electrode surfaces with redox centers that can mediate charge transfer to the analyte, as depicted in Fig. 10.2, may reduce this overpotential to allow the analyte to be detected within the electrolyte potential window and to bring the redox potential into a region where electroactive interferences are reduced. In addition to shifting the operating formal potential of the analyte species, modified electrodes can also increase the rate of the reaction over that observed at unmodified electrode surfaces.

The use of a three-dimensional polymer film reaction layer for mediated electrocatalysis can lead to increased currents at the electrode surface compared to monolayer or submonolayer coverages of the redox mediator.[8–12] For optimum sensor performance in the mediated catalysis, electrocatalytic reaction should proceed throughout the entire polymer film, which implies that high permeability of the films to the analyte is essential and charge transport through the layers must be rapid. Theoretical approaches for predicting and optimizing the response of electrocatalytic redox PMEs have been developed to a considerable extent, as reviewed elsewhere in this volume. Unfortunately the exploitation of the theory has been limited in the field of electroanalysis, with most researchers favoring a haphazard approach to optimization of the performance of electrocatalytic PMEs.

$$O + ne^- \Leftrightarrow R$$

$$R + Y \Rightarrow O + Z$$

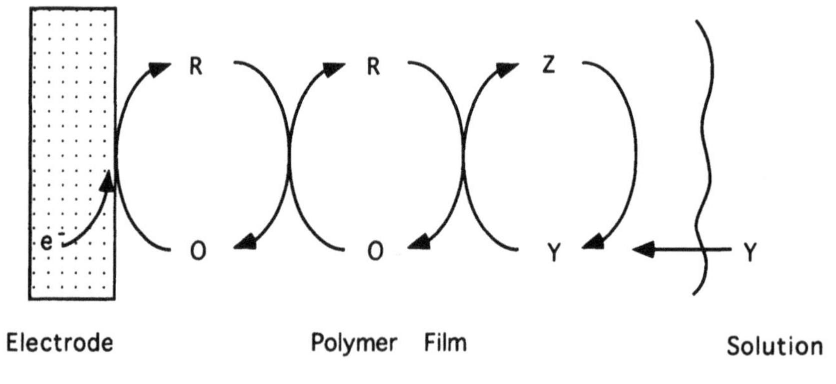

FIGURE 10.2. A model for the mediated reduction of species Y by the O/R redox couple immobilized in a polymeric film at the electrode surface. The governing process may be either substrate diffusion to the film; substrate partition into and diffusion within the layer; charge transport within the PME film (governed by relative rates of electron hopping, redox site diffusion, counterion transport, or polymer motions) and electrode; or mediation kinetics.

Selected PMEs shown to be useful in the electrocatalytically detecting analytes are discussed in the following paragraphs.

Many researchers have focused on the preparation and catalytic properties of polymer-bound ruthenium and osmium species because of their proven ability to catalyze homogeneous reactions and the vast synthetic chemistry available for their preparation. A series of preformed polymers of $[M(bpy)_2(pol)_nCl]Cl$, where M can be a Ru(II) or Os(II) metal center coordinated to 2,2′-bipyridine ligands (bpy) and anchored to a pyridine or imidazole nitrogen of a PVP or poly(N-vinylimidazole) polymer (pol), have been prepared and characterized with respect to charge transport rates and mechanisms in drop-coated films on electrode surfaces.[42,43] Electrodes coated with films of the ruthenium polymer have been shown to mediate the oxidation of Fe(II),[42b] nitrite, and nickel bis(2-hydroxyethyl)dithiocarbamate.[44] Improvements in the stability of this system in flowing streams was achieved by overcoating the film with a layer of poly(3-methylthiophene) or poly(N-ethyltyramine).[45] Macroelectrodes[46] and microelectrodes[47] modified with the osmium polymer have been shown to mediate the electroreduction of the nitrite ion. A flow-through system using the osmium polymer coated onto a carbon fiber microelectrode has been used to determine accurately the nitrite content in processed meats.[47] The mediated reduction of Fe(III) was postulated to occur in significant portions of films of the osmium polymer coated on rotating disk electrodes.[48] A dual-electrode sensor for the speciation of Fe(II) and Fe(III) has subsequently been developed by modifying one electrode with the ruthenium polymer and the other with the osmium polymer.[49] Operation in flowing streams in a parallel configuration permitted the simultaneous detection of 5×10^{-8} mol dm^{-3} of Fe(II) and Fe (III) with a linear dynamic range extending up to 1×10^{-2} mol dm^{-3}. Typical flow injection analysis (FIA) traces for the repeated injection of 1-mM levels of Fe(II) and Fe(III) at the dual-polymer-modified electrodes are shown in Fig. 10.3. Electronic cross talk between the two parallel electrodes limited the sensing system to samples with Fe(III)/Fe(II) concentration ratios from 100–0.01. Of several inorganic cations tested for both Fe(II) and Fe(III) interference, only Cu(II) yielded an interfering reduction current at the osmium electrode. Optimum selectivity for Fe(III) over Cu(II) was obtained at thicker films and higher flow rates in FIA.

Other ruthenium- and osmium-containing polymers with catalytic properties have been investigated,[8–12] although few have been developed as practical sensors. Vinning and Meyer exchanged an oxobridged dimer of ruthenium ($[(bpy)_2(H_2O)Ru^{III}]_2O^{4+}$) into polymer films of Nafion and polystyrene sulfonate coated on glassy carbon electrodes.[50] The Ru^V-Ru^V form of the dimer was found to catalytically oxidize Cl^- to Cl_2, although the catalyst was not stable because of deactivation, thought to occur by oxidative anation involving binding of polymer sulfonate groups to the ruthenium complex.

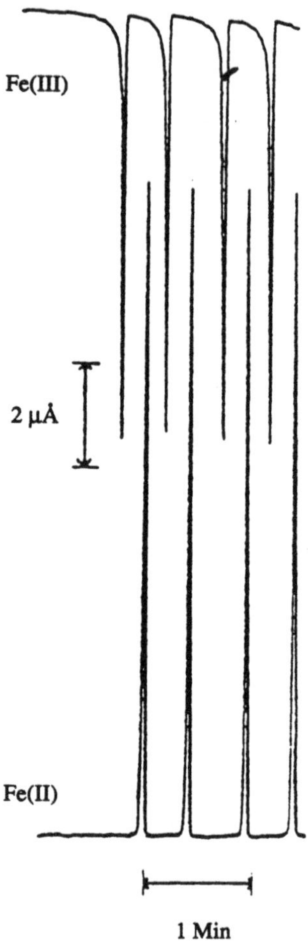

FIGURE 10.3. The FIA traces obtained at a dual-electrode thin layer cell for detecting 1 mM of Fe(II) (*bottom*) and Fe(III) (*top*) in a pH 1.0 0.2-M Na_2SO_4 background electrolyte. The electrodes were modified with 1×10^{-9} mol cm^{-2} surface coverages of $[Ru(bpy)_2(PVP)_{10}Cl]Cl$ and $[Os(bpy)_2(PVP)_{10}Cl]Cl$ films and held at constant potentials of 0.85 V and 0.12 V (versus SCE), respectively. (From Ref. 49 with permission of the American Chemical Society)

Several studies concentrate on immobilizing catalytic metal particles in polymeric films. Kao and Kuwana prepared films of poly(vinylacetic acid) on glassy carbon electrodes by refluxing them in neat monomer solutions.[51] Electrodeposition of Pt spherical particles in these films resulted in electrocatalytic electrodes for O_2 reduction and H_2 generation. Electrodeposited Pt in conducting polyaniline films yield extremely stable layers that are electrocatalytic toward methanol oxidation,[52] while glassy carbon electrodes modified with Pt-containing Nafion films were used in flow injection analysis of H_2O_2.[53] Liu and Wang[54] deposited Ni onto glassy carbon electrodes modified with ionomeric poly(ester sulfonic acid) films of Eastman Kodak AQ polymers. The modified electrodes exhibited increased electrocatalytic activity toward the oxidation of carbohydrates and amino acids in comparison to a metallic nickel electrode. Modified electrodes containing cobalt phthalocyanine (CoPC) electrocatalytic particles for postseparation detection of carbohydrates were prepared by electropolymerization of a cobalt-4,4′,4″,4‴-tetraaminophthalocyanine monomer on glassy carbon electrodes.[55] This electrode offered features similar to the carbon-paste-modified electrodes[56] with the added advantages of increased stability and the capability of operating in organic media.

The analytical application of particle-dispersed-modified electrodes to the selective detection of a single analyte is limited because of broad catalytic activities; thus their use as electrochemical detectors following chromatographic separation of carbohydrates is often suggested. Similar nonspecific catalytic PMEs consisting of electrocatalytic RuO_2 particles[57] and $Ru(OH_2)_6^{2+}$ in Nafion[58] have been shown to catalyze the oxidation of catechol and of alcohols, respectively; these could presumably be used in place of the carbon paste RuO_2-modified electrode developed for postseparation detection of carbohydrates and alcohols.[59] Other electrocatalytic particle electrodes were prepared from lead dioxide[60] in polypyrrole and CoPc entrapped behind a permselective cellulose acetate film.[61]

The mixed-valent inorganic film of PB on platinum electrode surfaces was shown to mediate the oxidation of ascorbic acid.[62] A ruthenium analog of PB formed by depositing an inorganic film onto carbon electrodes from a solution containing $RuCl_3$ and $K_4Ru(CN)_6$ can electrocatalytically oxidize As(III)[63] with a high stability. A typical response at these films to As(III) ions is shown in Fig. 10.4. Ruthenium films can also be used for electrocatalytic oxidation and detection of thiocyanate levels in flowing solutions with a linear dynamic range of 0.2–4 μM.[64] These films on electrodes also have the capability of catalyzing the oxidation of insulin, and they can be used to detect levels as low as 4.1 ng in FIA.[65] An inorganic film consisting of a nickel-hexacyanoferrate-modified glassy carbon electrode was used as a catalytic layer for detecting Fe(III) following HPLC separation without having to remove oxygen from the mobile phase.[66] A detection limit of 2.3 ng cm^{-3} of Fe(III) was determined, and a linear working range up to 0.9 μM was established.

The PMEs were investigated for their ability to mediate oxidation of such biologically important analytes as ascorbic acid (AA) and dihydronicotinamide adenine dinucleotide (NADH). Van Koppenhagen and Majda[67] incorporated ferricyanide ions into a film of polyacrylamide and acrylamide/vinylpyridine copolymers on electrode surfaces by ion exchange, and they examined the kinetics of the mediated oxidation of AA by ferricyanide. The mediated reaction occurred at lower anodic potentials (200 mV versus Ag/AgCl) than the corresponding

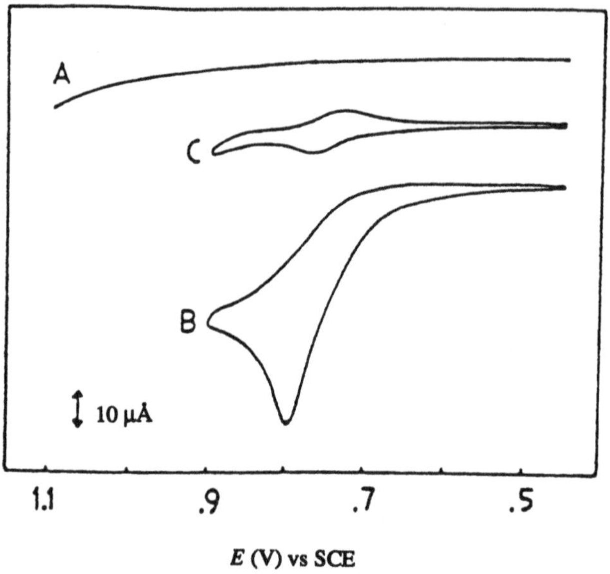

FIGURE 10.4. Cyclic voltammograms for 2-mM $NaAsO_2$ at (A) an unmodified glassy carbon electrode and (B) a glassy carbon electrode modified with a $RuCl_3/K_4Ru(CN)_6$ film. The electrolyte was 0.5-M NaCl, pH 2. (C) A cyclic voltammogram of the modified electrode in blank electrolyte solution. (From Ref. 63 with permission of the American Chemical Society)

oxidation at unmodified electrode surfaces (600 mV versus Ag/AgCl). Ferrocyanide ions electrostatically incorporated into PVP and subsequently admixed into a carbon paste matrix also yielded modified electrodes for the electrocatalytic oxidation of AA.[68] Another approach to AA detection involves mediation by electronically conducting films of polypyrrole.[35] The kinetics and mechanism of the mediated oxidation of AA at electrochemically prepared polypyrrole films doped with either the chloride or dodecylbenzene sulfonate ions was investigated by Lyons *et al.*[69] The mediated reaction was shown to occur at the polymer solution interface at a rapid rate, with a response time of 1–2 seconds for oxidation at rotating disk electrodes held at a constant 500-mV operating potential (versus Ag/AgCl). A linear response to AA from submicromolar to millimolar concentrations was maintained during 15 days in contact with the polypyrrole/dodecylbenzene sulfonate polymer, although the slope decayed to less than 50% of the original value during this period. Mao and Pickup[70] investigated the mediated oxidation of AA at poly-[1-methyl-3-(pyrrol-1-ylmethyl)pyridinium]-modified electrodes that contained electrostatically bound ferrocyanide. These films are capable of preconcentrating AA in the polymer layer, and oxidation of AA is mediated by both the polymer and the ferricyanide redox couple.

A modified electrode capable of mediating the electrooxidation of NADH would be extremely useful in constructing enzyme electrodes for substrates of the NAD^+-dependent dehydrogenases. The NADH oxidation at unmodified electrode surfaces is kinetically slow and limited by strong adsorption of oxidized products. A systematic investigation of the kinetics of homogeneous mediation of NADH oxidation by Tse and Kuwana[71] identified o-quinone as a catalytic mediator. The quinone moiety confined to a glassy carbon electrode surface by condensation with 3,4-dihydroxybenzylamine mediated the oxidation of NADH but exhibited short lifetimes because of electrode passiviation. The use of adsorbed redox dyes, such as the phenoxazine Meldola Blue,[72] and conducting organic salt electrodes of N-methylpyridinium/tetracyanoquinodimethane (NMP^+TCNQ)[73] for the mediated oxidation of NADH has also received considerable attention, but these systems involve electrocatalysis by monolayer coverages of mediator and the dissolved electrode/mediator, respectively.

Polymeric films active in the mediated oxidation of NADH have been prepared and studied. Films of the electropolymerized conducting polymer, poly(3-methylthionine) (P3MT) on platinum electrodes displayed a catalytic activity toward NADH oxidation and yielded extremely stable films with high resistance toward surface passivation.[74] Hajizadeh *et al.*[75] immobilized the electron transfer mediator thionine in a polymeric film on graphite electrodes by cross-linking with triisocyanate. This PME mediated the oxidation of NADH at a potential of 117 mV (versus AgAgCl), as shown in Fig. 10.5, representing a reduction of 300 mV in the oxidation potential for NADH at an unmodified graphite electrode. An applied potential of 200 mV at the modified electrode yielded a detection limit of 0.5 μM at a signal-to-noise ratio greater than 2. The current for the oxidation of the interferent ascorbic acid at the modified electrode was diminished to 13% of that observed at the unmodified electrode at the same potential. Ohsaka *et al.*[76] have electropolymerized thionine to yield stable films of poly(thionine) on basal plane pyrolytic graphite electrodes. Electroactive polymer films reduce the potential required for NADH oxidation to approximately 100 mV (versus Ag/AgCl) and yield a stable response and a linear dynamic range from 1 μM–1 mM.

Cox *et al.*[77] developed a modified electrode for the catalytic oxidation and detection of the nitrite ion by electrostatic binding of $IrCl_6^{2-}$ into quaternized PVP (QPVP) films. A working linear range from 8 μM–4 mM for determining nitrite was established. The modifying layer also prevented surface passivation of the underlying electrode by thiocyanate. A modified electrode containing cobalt-cyclam (cobaltic complex of 1,4,8,11-tetraazacyclotetradecane) incorporated into a Nafion film on a gold disk electrode has been prepared and characterized.[78] The PME was employed to investigate the kinetics of catalytic reduction of the nitrite ion in strongly alkaline electrolyte. Dong *et al.* electropolymerized 4-vinylpyridine from a PVP film on a glassy carbon electrode.[79] Incorporating an isomolybdic acid catalyst ($H_4Mo_8O_{26}$) by ion exchange into the protonated PVP in acidic media

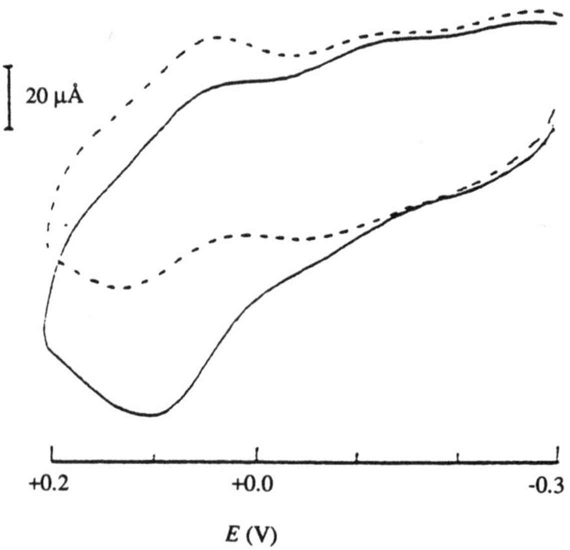

FIGURE 10.5. Cyclic voltammogram for the oxidation of 2-mM NADH in a 0.1-M phosphate buffer, pH 7 at a cross-linked thionin-modified electrode (*solid line*). The dotted line represents the response obtained at the modified electrode in the absence of NADH. (From Ref. 75 with permission from Marcel-Dekker)

yielded a film that could be used to detect (by electrocatalytic reduction) chlorate and bromate ions.

3.2. Preconcentration and Permselectivity

The determination of heavy metals, proteins, and inorganic complexes is frequently achieved by preconcentrating these species as amalgams or surface-adsorbed species on mercury electrode surfaces followed by voltammetric stripping.[80] Stripping voltammetry has a parallel in the field of modified electrodes where PMEs can be exploited for their ability to preconcentrate analytes at the electrode surface followed by voltammetric stripping and detection. In addition permselective films on electrodes can separate the response of the analyte of interest from that of coexisting interferences. Permselective and preconcentrating barrier films can also provide a means of preventing the fouling of electrode surfaces by large molecules or passivating electrochemical reactions. These preconcentrating and permselective polymeric coatings can achieve high sensitivities by favorably partitioning the analyte within the film on the basis of charge or polarity or by preventing interfering species from reaching the underlying electrode surface on the basis of size, charge, or polarity.

An early example of permselective barrier films involved using an overlayer of thick polyethylene membranes on a gold electrode to block surface fouling by proteins.[81] Another successful size-exclusion membrane barrier is the cellulose acetate (CA) film popularized by Wang and other researchers.[82–84] Base hydrolysis of CA films results in the formation of uniform pinholes through the film; the size of these pinholes can be controlled by the hydrolysis time.[83] The CA films have been demonstrated to prevent surface fouling of electrodes used for anodic stripping voltammetry[84] and to minimize the effect of the protein matrix in detecting H_2O_2.[82] Kuhn *et al.*[85] attempted to extend the molecular weight cutoff range the CA films by using the phase inversion process. These CA films were cast from solvent/nonsolvent solutions to yield size exclusion membranes consisting of a thin permselective outer layer and a more porous sublayer. These membranes permitted the rapid permeation of a 1500-dalton poly(ethylene glycol) ester of ferrocene;[85a] however the reproducibility of results presents a problem with these CA membranes.[85] Christie *et al.*[86] demonstrated that thin films of plasticized polyvinylchloride (PVC), normally used for potentiometric ion-selective electrode applications, applied to electrodes over a polycarbonate dialysis membrane offered improved selectivity ratios for the amperometric detection of phenolic compounds and H_2O_2 in the presence of the common biological interferents, ascorbic acid and uric acid, over those observed at the dialysis membrane alone or at a composite dialysis/membrane.

Electropolymerized films of conducting and nonconducting polymers on electrodes were investigated for use as permselective membranes. Films of electropolymerized poly(2,6-dimethyl-1,4-phenylene oxide)[87] and polyphenol[88] on platinum electrodes were developed as pH-responsive films because of their permselectivity to hydrogen ions. Wang *et al.*[32] controlled the permselectivity of electropolymerized films of polyaniline, polypyrrole, and polyphenol toward such biologically important analytes as H_2O_2, catechol, and acetaminophen in the presence of ascorbic acid and uric acid by varying the electropolymerization time or monomer concentration. Representative FIA traces for the application of these polymer films in analyte discrimination are shown in Fig. 10.6. Films of polyaniline were shown to exclude the protein albumin effectively from the platinum electrode surface, thereby preventing surface fouling.[32] Glassy carbon electrodes modified with electropolymerized films of the conducting polymer poly(3-methylthiophene) were reported to allow the stable oxidation and detection of phenol, which would normally passivate unmodified electrode surfaces.[89] Overoxidation of electropolymerized films of polypyrrole yields electronically nonconducting but ionically conducting ion-selective films that can exclude anionic species.[90] Permeability and selectivity of overoxidized polypyrrole depend not only on the doping; anions used to prepare the film[90–92] but also on the electrode substrate, this is attributed to different mechanisms in polymer growth at various substrates.[90] Overoxidized polypyrrole films prepared in the presence of dodecyl sulphate[91] and indigo

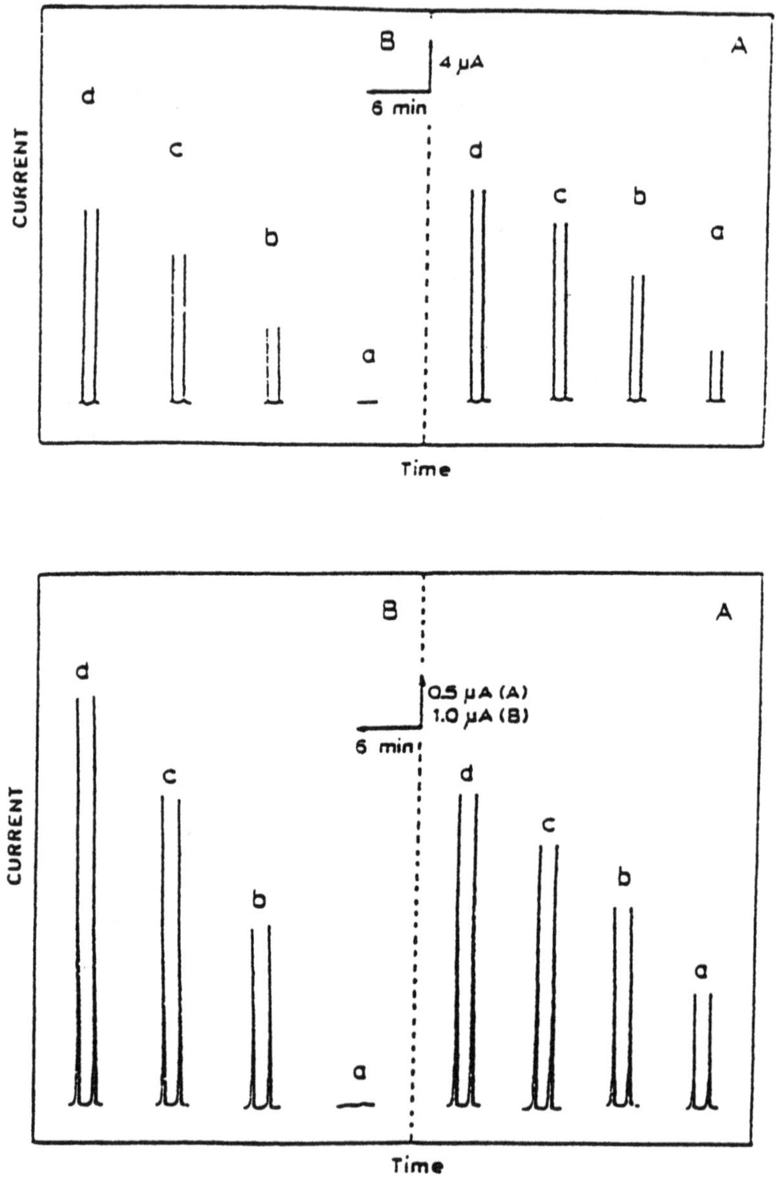

FIGURE 10.6. The FIA traces at polyaniline (7-min anodization, *top*) and polyphenol (0.5-mM phenol concentration, *bottom*) modified electrodes (B) to determine catechol (*top*) and acetaminophen (*bottom*). Corresponding traces for the unmodified electrodes are shown in (B). Traces in (a) show the response to the injection of 0.1-mM uric acid (*top*) or NADH (*bottom*), with traces (b)–(d) representing the response on adding 50-μM catechol (*top*) or acetaminophen (*bottom*). (From Ref. 32 with permission from Elsevier)

carmine[92] dopant anions effectively discriminated against ascorbic acid in detecting dopamine levels, resulting in an improved detection limit of 40 nM for dopamine determination.

While previously discussed films offer permselective discrimination from interferences based on the screening ability of membranes formed, improved detection limits can be achieved by preferentially partitioning and accumulating the analyte in films based on electrostatic or hydrophobic interactions. Ionomeric films offer the dual capability of preconcentration and discrimination from interferences. One of the most widely used ionomeric coatings is the perfluorosulfonated cation-exchange polymer Nafion. The ability of Nafion to inhibit anionic interferences and prevent surface passiviation has led to the development of a modified mercury–Nafion film for the anodic stripping voltammetry of heavy metals.[93] Whitely and Martin[94] examined the effect of film properties on the accumulation and detection of the model analyte $Ru(NH_3)_6^{3+}$ in Nafion films. They reported a procedure for regenerating the modified electrode following voltammetric determination of the analyte.[94,95] Nafion has been extensively studied for the selective accumulation and subsequent detection of cationic neurotransmitters in the presence of anionic ascorbate and urate interferences in solution and for the *in situ* monitoring of a neurotransmitter release.[95–98] The ability of cationic analytes to partition into Nafion polymer films was shown also to be a function of their hydrophobicity, with more hydrophobic alkylpyridinium,[95] inorganic,[99] and neurotransmitter[96] cations yielding higher partition coefficients. Toniolo *et al.*[100] used a Nafion film in a poorly conducting medium (low electrolyte concentration) to accumulate Tl^+ and Pb^{2+} cations with subsequent electrochemical detection in a solid-state cell, thus circumventing the dissociation of the cation from the Nafion film due to competing electrolyte cations normally seen in a conventional cell. The high stability of the Nafion ionomer on electrodes following a thermal-curing procedure[101] resulted in its use as a protective coating to prevent degradation of subcutaneously implanted Ag/AgCl reference electrodes in rats.[102] Nafion is also the coating of choice for the performance of voltammetry in such highly resistive media as supercritical fluids,[103] since it provides a conductive connection between the working microelectrode and a quasi-reference electrode, and it can also partition electroactive cations for subsequent voltammetric quantification.

The cation-exchange poly(ester sulfonic acid) AQ polymers have also been studied for their ability to preconcentrate metal ions[104] and cationic neurotransmitters.[15,105] These polymers are not so effective as the Nafion coating in excluding anionic interferences from ascorbic acid and uric acid.[105,106] Niwa *et al.*[106] modified interdigitated array electrodes (IDAs) with both an outer-Nafion and inner-AQ29D layer for the selective detection of dopamine at nanomolar levels in the presence of a 100-fold excess of acsorbic acid. The outer layer of Nafion acts as an effective barrier to ascorbic acid permeation to IDA electrodes. Redox cycling of dopamine in the AQ29D layer at the IDA electrodes increases sensitivity, while the ascorbic acid that penetrates this layer is destroyed by irreversible oxidation. Other polymeric cation-exchange materials used to modify electrode surfaces include poly(styrene sulfonate),[107] natural ionic polysaccharides,[108] chelating resins,[109] clays,[23,110] and carrageenan hydrogels.[111] The molecular sieve size selectivity and cation-exchange ability of zeolite layers on electrode surface has led to their use in entrapping electrocatalytic particles,[24,112] detecting accumulated metals,[113] and as sensors for the alkali-metal cations.[112]

The most frequently used anion-exchange material is protonated or quaternized poly(4-vinylpyridine) (QPVP), which has been used for the electrostatic incorporation of anionic metal complexes[114] and for the preconcentration and detection of low levels of these metal complexes in solution.[115] Cassidy and Tokuda[116] reported the use of a QPVP layer to detect reversibly low levels of Cu(II) in chloride electrolytes. The Cu(II) is accumulated in the polycationic layer by reduction of the metal at a potential of –0.2V (versus SCE) to the $CuCl_2^-$ complex. Subsequent oxidation strips the metal from the layer while allowing sensitive detection of submicromolar concentrations of the accumulated species. Riley and Wallace[117] investigated the accumulation and reductive stripping determination of gold chloride and cyanide complexes in the presence of Cu(II) at both drop-coated and electropolymerized films of protonated PVP. Electropolymerized films were found to be more mechanically stable than preformed polymer coatings on glassy carbon electrode surfaces, and these films allowed the detection of ppb levels of $AuCl_4^-$ in the presence of a 50-fold excess of Cu(II). Recently a perfluorinated polycationic analog of Nafion became available (Tosoh Soda Manufacturing) that can be used for the preconcentration of anionic species. Oyama *et al.*[118] reported the high stability and ion-exchange capacity of these perfluorinated polymers using ferricyanide and other multiply charged anionic redox complexes as model analyses. The accumulation of ferricyanide in a Tosflex® perfluorinated polymer manufactured by Tosoh was investigated.[119] The application of electrodes modified with the Tosflex polymer to determine low levels of Cu(II) in a scheme similar to that developed by Cassidy and Tokuda[116] for QPVP-modified electrodes yields micromolar detection limits.[120] Other anion-exchange materials used to modify electrodes include protonated poly(L-lysine);[121] electropolymerized films of N,N-dimethylaniline,[122] and liquid anion exchangers, such as Amberlite LA2, in carbon paste matrices.[123]

Using two or more films on electrode surfaces can increase the discrimination from coexisting interference[106] and also lead to bifunctional PMEs that possess both electrocatalytic sites and permselective discrimination of such interferences as the CoPC/CA films described by Wang *et al.*[61] A second approach for selectively detecting an analyte involves using arrays of modified electrodes and applying chemometric pattern recognition methods to the resulting fingerprint for the analyte of interest.[124] Wang *et al.*[124] used an electrode array of partially selective films of Nafion, hydrolyzed CA, PVP, AQ polymer, and lipid to detect neurotransmitters in the presence of catechol and ascorbic acid. The response for selected analytes at an array of coated electrodes is shown in Fig. 10.7, demonstrating fingerprint patterns obtainable in these arrays.

Conducting polymer films, such as polypyrrole, polythiophene, and polyaniline, on electrode surfaces can be switched between the oxidized conducting state and the reduced insulating state. These redox-switching processes are depicted in Fig. 10.8. Ingress and egress (termed doping/undoping) of counteranions accompanies this switching behavior. The ability of conducting polymer films of polypyrrole to incorporate and expel the counteranion on switching between oxidized and reduced states was exploited by Ikariyama and Heineman for detecting the counteranion.(125) In the FIA system developed, polypyrrole was electropolymerized onto a platinum electrode, and the modified electrode was held at an oxidizing potential in the flow system with a background mobile phase of glycine that does not dope the polymer. Introducing a sample containing a dopant anion, such as carbonate, phosphate, or acetate, resulted in an anodic current peak at the modified electrode, thus allowing detection of electroinactive ions. Polarization of the

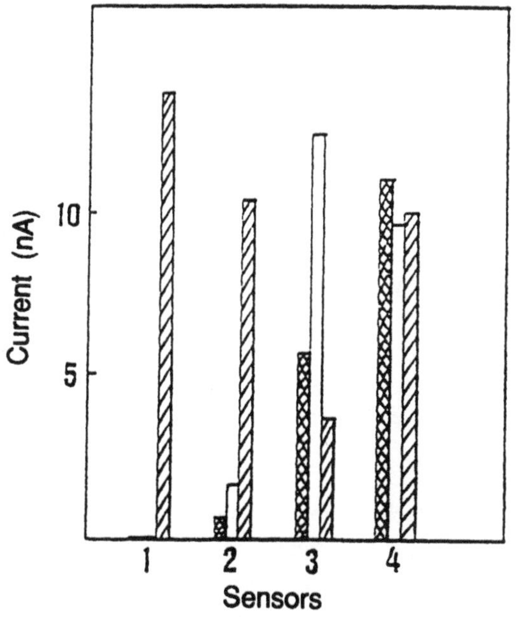

FIGURE 10.7. Array response pattern obtained for flow injection of 2-μM promethazine (*crosshatch*), dopamine (*open*), and catechol (*hatch*) at electrodes modified with 20-min hydrolyzed CA (1), PVP (2), Nafion (3), and lipid (4) films. The operating potential was set at 0.8 V, and the background electrolyte consisted of a 0.05-M phosphate buffer at pH 5.5. (From Ref. 124 with permission for the American Chemical Society)

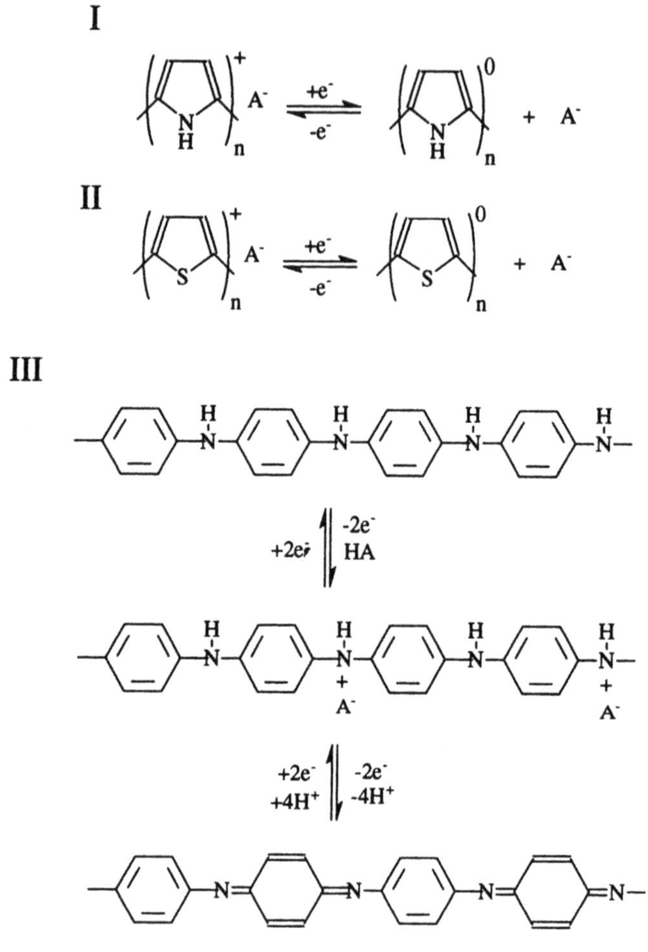

FIGURE 10.8. Postulated oxidation-reduction processes for (I) polypyrrole, (II) polythiophene, and (III) polyaniline. (Adapted from Ref. 156 with permission from the Royal Society of Chemistry)

modified electrode at a reducing potential resulted in undoping and the regeneration of the sensing surface for the next sample injection. Ye and Baldwin[(126)] reported that films of polyaniline, which can also be used for electroinactive ion detection, offer higher stability than polypyrrole films. Wang and Liu[(127)] demonstrated that selectivity of the polyanailine-modified electrode can be varied by changing the mobile phase. Sadik and Wallace[(128)] investigated the effect several variables have on the doping/undoping process at polypyrrole-modified electrodes on detecting electroinactive ions. They concluded that the observed signal can be a combination of changes in conductivity of both the solution and the polymer film, and the electrochemically induced ion-exchange process. The ability to control electrostatic doping/undoping of conducting polymers electrochemically has prompted investigations into using conducting polymers as stationary phase packings on carbon particles in liquid chromatography.[129]

Films of the insoluble mixed-valent transition metal hexacyanoferrate polymers on electrodes can also be used in an analogous manner to conducting polymers to detect electroinactive cations.[130–133] The zeolitic structure of films determines cation selectivity, with both voltammetric and mass[130] responses of the films being affected by cation type and concentration. Amos *et al.*[131] showed that 10^{-8} M levels of the Cs^+ ion could be detected on nickel hexacyanoferrate films in the presence of a large excess of Na^+ ions. Using an electrochemical quartz crystal microbalance allowed detection of both mass and charge changes in PB films on electrodes by the dual measurement of frequency and current.[130] Thomsen and Baldwin[132] investigated the application of films of cupric hexacyanoferrate on glassy carbon electrodes for detecting submicromolar levels of the K^+, NH_4^+, Rb^+, and Cs^+ ions following liquid chromatographic separation. Specific selectivity of the Cs^+ ion over the others cations was obtained by carefully controlling the electrode potential. Li *et al.*[133] used colloidal particles of cupric hexacyanoferrate as a counteranion dopant of the conducting polymer polypyrrole to form a stable film on electrode surfaces capable of showing cation selectivity.

3.3. Molecular Recognition

Modification of electrode surfaces with functional groups that can complex the target analyte offers an alternative to accumulation based on electrostatic or polarity alone. Polymeric materials containing ligand-binding sites are most frequently used in metal ion analysis. Preconcentration into functionalized polymer films depends to a large extent on the stability constant of the metal–ligand complex. This allows the researcher to choose an appropriate ligand for the particular application, for example polymeric films modified with ligands known to form solution complexes with stability constants as high as 10^{50} for selected metal[134] can be used as sensitive metal sensors. Alternatively a ligand can be selected on the basis of its specific binding to a particular metal, such as the 2,9-disubstituted phenanthrolines used to selectively determine copper in the presence of iron because of their diminished affinity for iron.[135]

Immobilization of complexing agents on electrode surfaces was achieved by several procedures. Lane and Hubbard[1] first demonstrated the principle by adsorbing a salicylate ligand, which could complex Fe(III) solution species, onto platinum electrodes. Cox and Majda adsorbed adenosine monophosphate onto platinum electrodes to detect Fe(II) from solution using cyclic voltammetry.[136] Price and Baldwin[137] used ferrocene carboxaldehyde as a model analyte to investigate a modified electrode developed to determine aldehydes and ketones. The formation of imine linkages between the analyte and a coating of allylamine on platinum electrode surfaces with subsequent voltammetric detection yielded a detection limit of 0.1 μM for the ferrocene compound. Other surface-bound ligands that were investigated include a [(carboxyyamido)propyl]trioxysilylidine complex bound to a glassy carbon electrode for Cu(II) detection,[138] iminodiacetic acid groups bound to glassy carbon electrodes using carbodiimide coupling to detect Ag(I) ions[139] and a trioctylphosphine oxide (TOPO) surface-modified electrode to detect U(VI).[140] Simple adsorption of complexing ligands onto electrode surfaces for subsequent metal analysis was used to detect trace amounts of Au(III) and Al(III) using adsorbed aza crown ether[141] and alizarin[142] ligands, respectively.

Incorporating complexing ligands into a carbon paste matrix is a convenient method of preparing modified electrodes, although the lack of stability of these systems and their inability to function in nonaqueous media are problematic. Carbon paste electrodes modified with the TOPO ligand were used to extract and detect Au(III).[143] This ligand immobilized in a PVC matrix on a mercury film electrode was also used for the sensitive detection of several metals, including Bi(III) in copper alloys.[144]

Carbon paste electrodes modified with a 2,9-dimethyl-1,10-phenanthroline ligand were used for the complexation and detection of Cu(I) levels using anodic differential pulse voltammetry (DPV).[145] Quantitative results up to 10 μM were possible with a detection limit of 0.3 μM attainable. Of the metal ions studied, including Cu(II), Pb(II), Ni(II), Fe(III), Co(II), and Zn(II), only Ag(I) interfered in determining Cu(I). Other ligands, such as calixarenes,[146] dimethylglyoxime,[147] thiacrown compounds,[148] and dithizone,[149] were incorporated into the carbon paste matrix to detect heavy metals, Ni(II), Ag(I), and Au(III), respectively.

Incorporating chelating complexes into polymer-modified surfaces for the subsequent detection of their metal ligands was achieved. The most common procedure is to incorporate the complex into an ion-exchange polymer film on the electrode surface, frequently Nafion. Crown ether complexes have been ion-exchanged into Nafion films on carbon fiber microelectrodes to yield a modified electrode with increased sensitivity toward the detection of the neurotransmittters dopamine and serotonin.[150] The combination of preconcentration into the Nafion film and the interaction of crown ether oxygen atoms with the positively charged amino group of neurotransmitters allowed the simultaneous *in vivo* detection of less than a 10-nM concentration of dopamine and serotonin. A regioselective electrode that can be used to determine *o*-nitrophenol in the presence of *p*-nitrophenol with a selectivity ratio of 33 was reported based on coating a Nafion-coated glassy carbon electrode with α-cyclodextrin.[151] Zen and Lee[152] used a Nafion film containing dimethylglyoxime and subsequently coated with a mercury film to detect sub-ppb levels of Ni(II). The electrode response was presumed to arise from a combination of preconcentration at the Nafion ionomer, amalgam formation at the mercury film, and complexation with the ligand. The Nafion film also functioned as an effective barrier to interference effects from such organic surface-active compounds as Triton-X-100. A glassy carbon electrode coated with a film formed by the codeposition of the lead ionophore methylenebis(N,N′-diisobutyldithiocarbamate) and Nafion was used to accumulate and detect by subsequent voltammetry of Pb(II) to levels of 10^{-7}M.[153] Labuda *et al.*[154] deposited derivatives of 4-acylpyrazolone ligands and the oxine ligand on graphite electrodes and subsequently coated the modified electrodes with a Nafion film. The ability of these modified electrodes to complex Cu(I) was investigated and the effect of competing solution ligands on the response evaluated.

These modified electrodes were used to determine nanomolar to micromolar levels of labile Cu(II) in river samples. Electrodes for the detection of Ag(I) concentrations were prepared in a similar fashion using macrocyclic thia ligands.[155] The stripping peak current was shown to be proportional to the stability constant of the immobilized ligand, and detection limits as low as 10^{-10} M could be obtained. Incorporating ligands into conducting polymer films on electrode surfaces during electropolymerization is possible by allowing the ligand to act as a dopant counterion.[156] Such ligands as dithiocarbamates,[157] ethylenediaminetetraacetic acid,[158] and bathocuproine disulfonate and bathophenanthroline disulfonate (BPS)[159] were incorporated into polypyrrole films in this way to detect metal ions. Square-wave voltammograms of a polypyrrole-coated electrode containing the BPS ligand as a counteranion in the presence and absence of Cu(II) are shown in Fig. 10.9.

Chemically attaching the ligand to the polymer or forming polymers from functionlized ligands can offer increased stabilities in comparison to ion-exchange immobilization strategies. Gehron and Brajter-Toth[160] formed complexation polymers to detect Fe(II) by cross-linking QPVP films with a bathophenanthrolinedisulfonic acid ligand. A linear response was obtained from 10^{-8}–10^{-7} M for detecting Fe(II) by accumulation followed by DPV at this modified electrode. The Ni(II)-sensitive modified electrodes were prepared by forming films from tetrakis(3-methoxy-4-hydroxyphenyl)porphyrin with Ni(II) as a central metal.[161] Demetalation is possible in acidic media, with resulting porphyrin polymers capable of detecting 8×10^{-8} M concentrations of Ni(II) in solution by accumulation followed by DPV of the Ni(II)/(III) oxidation. Exposure to acid regenerated films for subsequent analyses.

An innovative procedure for the reproducible detection of metal ions using complexation schemes at PMEs was devised by Abruña and coworkers.[162–167] The systems developed rely on the presence of a redox couple in the polymer film in addition to immobilized complexing ligand sites. The redox couple, selected to have a formal potential removed from the metal-ligand complex to be detected, acts as an internal standard to allow facile evaluation of saturation effects, normally problematic in complexation studies with PMEs. The redox couple also increases the between- and within-film reproducibility by using ratiometric methods to

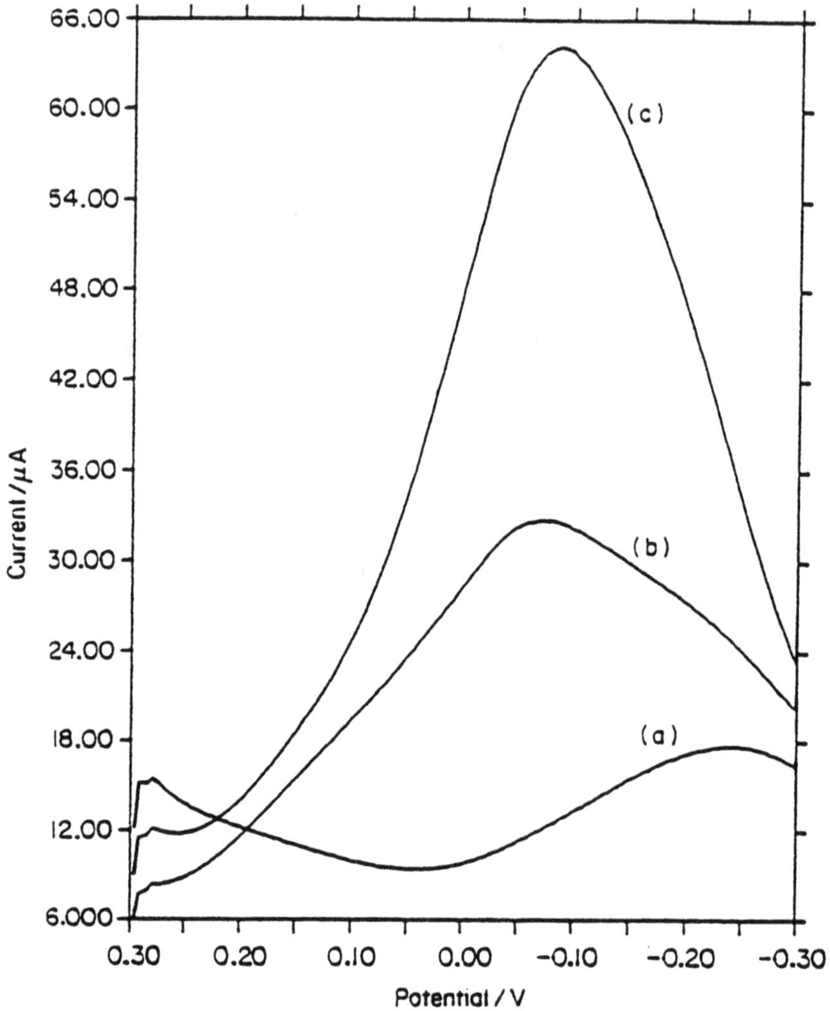

FIGURE 10.9. Square-wave voltammograms of a polypyrrole/bathophenanthroline disulfonate modified electrode (a) before and (b) after exposure to a 5-mM solution of Cu(II) (c) or Cu(I). (From Ref. 159 with permission from Elsevier)

determine metal concentrations. The initial report[(162)] demonstrated the utility of this technique using electropolymerized copolymers of the vinylferrocene and 4-vinyl-4′-methyl-2,2′-bipyridine (v-bpy) monomers to detect Fe(II) levels. Ligands, such as sulfonated bathophenanthroline, sulfonated bathocuproine, and diethyldithiocarbamate, were also ion-exchanged into electropolymerized QPVP/vinylferrocene copolymers to detect Fe(II) and Cu(I) levels.[(162)] General schemes for preparing polymeric films containing complexation sites and a redox couple were reported based on electropolymerization of vinylferrocene or $[M(v\text{-}bpy)_3]^{2+}$ polymers, where M can be Ru or Os, with ligand-containing monomers, such as vinylpyridine, acrylonitrile, or v-bpy, or by ion-exchange of such ligands as sulfonated bathophenanthroline and diethyldithiocarbamate into electropolymerized films of $[M(v\text{-}bpy)_3]^{2+}$ on electrode surfaces.(163) Modified microelectrodes for detecting Fe(II) and Ca(II) levels were subsequently developed using a copolymer of v-bpy/vinylferrocene and a viologen polymer containing ion-exchanged antipyrylazo III, respectively.(164) These PMEs permitted the detection of 5×10^{-8} M levels of Fe(II) and submicromolar levels of Ca(II) in the presence of excess Mg(II). Cha and Abruña(165) studied a series of Cu(I)-binding ligands incorporated into $[Ru(v\text{-}bpy)_3]^{2+}$ films by ion-exchange; they found that the log of the detected current at constant Cu(I) concentration correlates with the log of the ligand-metal formation constant, indicating the retention of relative binding strength on immobilization. In a subsequent study the effect of competing metals and ligands and of decreasing electrode size on detecting Cu(I) using these films was investigated.(166) Modified microelectrodes and ultramicroelectrodes exhibited increased sensitivities in comparison to a macroelectrode, and the presence of either oxalate or phosphate ions in the solution resulted in a steadily increasing current at the copper oxidation potential. Cha *et al.*(167) also developed modified electrodes for detecting 10^{-8} M mercury levels using a $[Ru(v\text{-}bpy)_3]^{2+}$ film containing ion-exchanged mercury ligand, such as picolinic acid, tryptophan, or nicotinic acid.

4. CONCLUSION

The future development of polymer-modified electrodes for application to analysis will continue. This research area has yielded dividends in recent years, as is apparent from the preceding sections. Problems still exist however in applying polymer layers to electroanalysis. The definition of film stability and methods for improving the stability of current systems needs to be addressed. Increased attention to elucidating underlying mechanisms involved in the sensing process and factors that can influence them is essential for a complete understanding of polymer-modified electrodes. Determining and optimizing the rates of charge transport, analyte permeation, and mediation or complexation kinetics for each system are recommended for characterizing polymer layers.

Promising approaches for the development of easy-to-prepare sensitive and selective modifying layers include the self-assembly of monolayers and polymeric films on metal electrode surfaces(168) and the coordination of macrocylic ligands to conducting polymer monomers to electropolymerize polymer films possessing molecular recognition properties.(169) Molecularly imprinting templates into polymer systems[170] to design modified electrode surfaces for specific analysis also shows great promise for electroanalytical applications.

REFERENCES

1. R. F. Lane and A. T. Hubbard, *J. Phys. Chem.* **77**, 1401 (1973).
2. P. R. Moses, L. M. Wier, and R. W. Murray, *Anal. Chem.* **47**, 1882 (1975).
3. J. R. Lenhard and R. W. Murray, *J. Electroanal. Chem.* **78**, 195 (1977).
4. M. S. Wrighton, R. G. Austin, A. B. Bocarsly, J. M. Bolts, O. Haas, K. D. Legg, L. Nadjo, and M. C. Palazzotto, *J. Electroanal. Chem.* **87**, 429 (1978).
5. A. W. C. Lin, P. Yeh, A. M. Yacynych, and T. Kuwana, *J. Electroanal. Chem.* **84**, 411 (1977).
6. L. L. Miller and M. R. Van De Mark, *J. Am. Chem. Soc.* **100**, 639 (1978).
7. L. L. Miller and M. R. Van De Mark, *J. Am. Chem. Soc.* **100**, 3223 (1978).
8. R. W. Murray in *Electroanalytical Chemistry*, vol. 13, (A. J. Bard, ed.), Marcel Dekker, New York, 1984), p. 191.
9. H. D. Abruña, *Coord. Chem. Rev.* **86**, 135 (1988).
10. A. R. Hillman in *Electrochemical Science and Technology of Polymers*, vol. 1, (R. G. Linford, ed.) (Elsevier, Amsterdam, 1987), pp. 103, 241.
11. R. J. Forster and J. G. Vos in *Comprehensive Analytical Chemistry*, vol. 27 (G. Svehla, ed.) M. R. Smyth and J. G. Vos, (volume eds.) (Elsevier, Amsterdam, 1992), p.465.
12. A. Merz, *Topics Curr. Chem.* **152**, 49 (1990).
13. J. Janata, *Anal. Chem.* **64**, 196R (1992).
14. P. Edelman and J. Wang, eds., *Biosensors and Chemical Sensors* (American Chemical Society, Washington, DC, 1992).
15. J. Wang, *Electroanalysis* **3**, 255 (1991).
16. L. R. Faulkner, *Chem. Eng. News* **62**, 28 (1984).
17. A. Merz and A. J. Bard, *J. Am. Chem. Soc.* **100**, 3222 (1978).
18. N. Oyama and F. C. Anson, *J. Electrochem. Soc.* **127**, 247 (1980).
19. K. Shigehara, N. Oyama, and F. C. Anson, *J. Am. Chem. Soc.* **103**, 2552 (1981).
20. M. N. Szentirmay and C. R. Martin, *Anal. Chem.* **56**, 1898 (1984).
21. I. Rubinstein and A. Bard, *Anal. Chem.* **53**, 102 (1981).
22. D. A. Buttry and F. C. Anson, *J. Am. Chem. Soc.* **106**, 59 (1984).
23. A. J. Bard and P. K. Ghosh, *J. Am. Chem. Soc.* **105**, 5691 (1983).
24. C. G. Murray, R. J. Nowak, and D. R. Rolison, *J. Electroanal. Chem.* **164**, 205 (1984).
25. C. D. Ellis and T. J. Meyer, *Inorg. Chem.* **23**, 1748 (1984).
26. (a) E. F. Bowden, M. F. Dautartas, and J. F. Evans, *J. Electroanal. Chem.* **219**, 49 (1987). (b) E. F. Bowden, M. F. Dautartas, and J. F. Evans, *J. Electroanal. Chem.* **219**, 71 (1987). (c) E. F. Bowden, M. F. Dautartas, and J. F. Evans, *J. Electroanal. Chem.* **219**, 91 (1987).
27. R. N. Dominey, N. S. Lewis, J. A. Bruce, D. C. Bookbinder, and M. S. Wrighton, *J. Am. Chem. Soc.* **104**, 4671 (1982).
28. R. A. Simon, T. E. Mallouk, K. A. Daube, and M. S. Wrighton, *Inorg. Chem.* **23**, 3119 (1985).
29. (a) P. Denisevich, H. D. Abruña, C. R. Leidner, T. J. Meyer, and R. W. Murray, *Inorg. Chem.* **21**, 2153 (1982). (b) J. M. Calvert, R. H. Schmehl, B. P. Sullivan, J. S. Facci, T. J. Meyer, and R. W. Murray, *Inorg. Chem.* **22**, 2151 (1983).
30. S. Gould, C. Earl, B. P. Sullivan, and T. J. Meyer, *J. Electroanal. Chem.* **350**, 143 (1993).
31. C. D. Ellis, L. D. Margerum, R. W. Murray, and T. J. Meyer, *Inorg. Chem.* **22**, 1283 (1983).
32. J. Wang, S. P. Chen, and M. S. Lin, *J. Electroanal. Chem.* **273**, 231 (1989).
33. R. M. Kellet and T. G. Spiro, *Inorg. Chem.* **24**, 2378 (1985).

34. A. Deronzier and J-C. Moutet, *Acc. Chem. Res.* **22**, 249 (1989).
35. M. D. Imisides, R. John, P. J. Riley, and G. G. Wallace, *Electroanalysis* **3**, 879 (1991).
36. Y. P. Lin and G. G. Wallace, *J. Electroanal. Chem.* **247**, 145 (1988).
37. R. A. Saraceno, J. G. Pack, and A. G. Ewing, *J. Electroanal. Chem.* **197**, 265 (1986).
38. J. Heinze, *Topics Curr. Chem.* **152**, 1 (1990).
39. T. A. Skotheim, ed., *Handbook of Conducting Polymers*, vols. 1 and 2 (Marcel Dekker, New York, 1986).
40. K. Itaya, I. Uchida, and V. D. Neff, *Acc. Chem. Res.* **19**, 162 (1986).
41. A. B. Bocarsly and S. Sinha, *J. Electroanal. Chem.* **137**, 157 (1982).
42. (a) J. M. Kelly, J. M. Clear, G. M. O'Connell, and J. G. Vos, *J. Chem. Res.* **M**, 3039 (1981). (b) C. P. Andrieux, O. Haas, and J. M. Saveant, *J. Am. Chem Soc.* **108**, 8175 (1986). (c) M. E. G. Lyons, H. G. Fay, J. G. Vos, and A. J. Kelly, *J. Electroanal. Chem.* **250**, 207 (1988). (d) D. Leech, R. J. Forster, M. R. Smyth, and J. G. Vos, *J. Mater. Chem.* **1**, 629 (1991).
43. (a) R. J. Forster and J. G. Vos, *Macromol.* **21**, 2432 (1988). (b) A. J. Kelly, T. Ohsaka, N. Oyama, R. J. Forster, and J. G. Vos, *J. Electroanal. Chem.* **287**, 185 (1990). (c) R. J. Forster, J. G. Vos, and M. E. G. Lyons, *J. Chem. Soc. Faraday Trans.* **87**, 3761 (1991).
44. J. N. Barisci, G. G. Wallace, E. A. Wilke, M. Meaney, M. R. Smyth, and J. G. Vos, *Electroanalysis* **1**, 245 (1989).
45. G. G. Wallace, M. Meaney, M. R. Smyth, and J. G. Vos, *Electroanalysis* **1**, 357 (1989).
46. A. P. Doherty, R. J. Forster, M. R. Smyth, and J. G. Vos, *Anal. Chim. Acta* **255**, 45 (1991).
47. M. M. Malone, A. P. Doherty, M. R. Smyth, and J. G. Vos, *Analyst* **117**, 1259 (1992).
48. R. J. Forster and J. G. Vos, *J. Chem. Soc. Faraday Trans.* **87**, 1863 (1991).
49. A. P. Doherty, R. J. Forster, M. R. Smyth, and J. G. Vos, *Anal. Chem.* **64**, 572 (1992).
50. W. J. Vinning and T. J. Meyer, *Inorg. Chem.* **25**, 2023 (1986).
51. W. H. Kao and T. Kuwana, *J. Am. Chem. Soc.* **106**, 473 (1984).
52. K. M. Kost, D. E. Bartak, B. Kazee, and T. Kuwana, *Anal. Chem.* **60**, 2379 (1988).
53. B. T. Tay, K. P. Ang, and H. Gunasingham, *Analyst* **113**, 617 (1988).
54. A. Liu and E. Wang, *Anal. Chim. Acta* **280**, 223 (1993).
55. X. Qi, R. P. Baldwin, H. Li, and T. F. Guarr, *Electroanalysis* **3**, 119 (1991).
56. (a) L. M. Santos and R. P. Baldwin, *Anal. Chem.* **58**, 848 (1986). (b) A. M. Tolbert, R. P. Baldwin, and L. M. Santos, *Anal. Lett.* **22**, 683 (1989).
57. M. E. G. Lyons, C. H. Lyons, A. Michas, and P. N. Bartlett, *Analyst* **117**, 1271 (1992).
58. C. Shi and F. C. Anson, *J. Electroanal. Chem.* **362**, 273 (1993).
59. (a) J. Wang and Z. Taha, *Anal. Chem.* **62**, 1413 (1990). (b) D. Leech, J. Wang, and M. R. Smyth, *Electroanalysis* **3**, 37 (1991).
60. D. Velayutham and M. Noel, *Talanta* **39**, 481 (1992).
61. J. Wang, T. Golden, and R. Li, *Anal. Chem.* **60**, 1642 (1988).
62. F. Li and S. Dong, *Electrochim. Acta* **32**, 1511 (1987).
63. J. A. Cox and P. J. Kulesza, *Anal. Chem.* **56**, 1021 (1984).
64. J. A. Cox, T. Gray, and K. R. Kulkarni, *Anal. Chem.* **60**, 1710 (1988).
65. J. A. Cox and T. Gray, *Anal. Chem.* **61**, 2462 (1989).
66. P. J. Kulesza, K. Bratjer, and E. Dabek-Zlotorzynska, *Anal. Chem.* **59**, 2776 (1987).
67. (a) J. E. van Koppenhagen and M. Majda, *J. Electroanal. Chem.* **189**, 379 (1985). (b) J. E. van Koppenhagen and M. Majda, *J. Electroanal. Chem.* **238**, 113 (1987).
68. P. W. Geno, K. Ravichandran, R. P. Baldwin, *J. Electroanal. Chem.* **183**, 155 (1985).
69. M. E. G. Lyons, W. Breen, and J. Cassidy, *J. Chem. Soc. Faraday Trans.* **87**, 115 (1991).
70. H. Mao and P. G. Pickup, *J. Electroanal. Chem.* **265**, 127 (1989).
71. (a) D. C. Tse and T. Kuwana, *Anal. Chem.* **50**, 1315 (1978). (b) C. Ueda, D. C. Tse, and T. Kuwana, *Anal. Chem.* **54**, 850 (1982).
72. L. Gorton, A. Tortensson, H. Jaegfeldt, and G. Johansson, *J. Electroanal. Chem.* **161**, 103 (1984).

73. W. J. Albely and P. N. Bartlett, *J. Chem. Soc. Chem. Commun.* 234 (1984).
74. N. F. Atta, A. Galal, E. Karagözler, H. Zimmer, J. F. Robinson, and H. B. Mark, Jr., *J. Chem. Soc. Chem. Commun.* 1347 (1990).
75. K. Hajizadeh, H. T. Tang, H. B. Halsall, and W. R. Heineman, *Anal. Lett.* **24**, 1453 (1991).
76. T. Ohsaka, K. Tanaka, and K. Tokuda, *J. Chem. Soc. Chem. Commun.* 222 (1993).
77. (a) J. A. Cox and P. J. Kulesza, *J. Electroanal. Chem.* **175**, 105 (1984). (b) J. A. Cox and K. R. Kulkarni, *Analyst* **111**, 1219 (1986).
78. L. Ma, B.-Y. Zhang, H.-L. Li, and J. Q. Chambers, *J. Electroanal. Chem.* **362**, 201 (1993).
79. S. Dong, F. Song, B. Wang, and B. Liu, *Electroanalysis* **4**, 613 (1992).
80. (a) J. Wang, *Am. Lab.* **19**, 41 (1985). (b) K. Z. Brainina, *Talanta* **18**, 513 (1971).
81. A. J. Bard, *J. Chem. Ed.* **60**, 302 (1983).
82. G. Sittampalam and G. S. Wilson, *Anal. Chem.* **55**, 1608 (1983).
83. J. Wang and L. D. Hutchins, *Anal. Chem.* **57**, 1536 (1985).
84. J. Wang and L. D. Hutchins, *Anal. Chem.* **58**, 402 (1986).
85. (a) L. S. Kuhn, S. G. Weber, and K. Z. Ismail, *Anal. Chem.* **61**, 303 (1988). (b) L. S. Kuhn and S. G. Weber, *Electroanalysis* **3**, 941 (1991).
86. I. M. Christie, P. H. Treloar, and P. Vadgama, *Anal. Chim. Acta* **269**, 65 (1992).
87. T. Ohsaka, T. Hirokawa, H. Migamoto, and N. Oyama, *Anal. Chem.* **59**, 1758 (1987).
88. G. Cheek, C. P. Wales, and R. J. Nowak, *Anal. Chem.* **55**, 380 (1983).
89. J. Wang and R. Li, *Anal. Chem.* **61**, 2809 (1989).
90. (a) M. Freund, L. Bodalbhai, and A. Brajter-Toth, *Talanta* **38**, 95 (1991). (b) A. Witowski and A. Brajter-Toth, *Anal. Chem.* **64**, 635 (1992). (c) Z. Gao, M. Zi, and B. Chen, *Anal. Chim. Acta* **286**, 213 (1994).
91. Z. Gao and A. Ivaska, *Anal. Chim. Acta* **284**, 393 (1993).
92. Z. Goa, B. Chen, and M. Zi, *Analyst* **119**, 459 (1994).
93. (a) B. Hoyer, T. M. Florence, and G. E. Batly, *Anal. Chem.* **59**, 1608 (1987). (b) B. Hoyer and T. M. Florence, *Anal. Chem.* **59**, 2839 (1987).
94. L. D. Whitely and C. R. Martin, *Anal. Chem.* **59**, 1746 (1987).
95. M. W. Epenscheid, A. R. Ghatak-Roy, R. B. Moore III, R. M. Penner, M. N. Szentirmay, and C. R. Martin, *J. Chem. Soc. Faraday Trans. 1* **82**, 1051 (1986).
96. (a) G. A. Gerhadt, A. F. Oke, F. Nagy, B. Moghaddam, and R. N. Adams, *Brain Res.* **290**, 390 (1984). (b) G. Nagy, G. A. Gerhadt, A. F. Oke, M. E. Rice, R. N. Adams, R. B. Moore III, M. N. Szentirmay, and C. R. Martin, *J. Electroanal. Chem.* **188**, 85 (1985). (c) P. Capella, B. Ghasemzadeh, K. Mitchell, and R. N. Adams, *Electroanalysis* **2**, 175 (1990).
97. (a) E. W. Kristensen, W. G. Kuhr, and R. M. Wightman, *Anal. Chem.* **59**, 1752 (1987). (b) R. M. Wightman, in *Biosensor Technology: Fundamentals and Applications*, (R. P. Buck, W. E. Hatfield, M. Umaña, and E. F. Bowden, eds.) (Marcel Dekker, New York, 1990), p. 39.
98. H. Ji and E. Wang, *J. Chromatog.* **410**, 111 (1987).
99. (a) H. L. Yeager and A. Steck, *Anal. Chem.* **51**, 862 (1979). (b) H. L. Yeager and A. Steck, *J. Electrochem. Soc.* **128**, 1880 (1981).
100. R. Toniolo, G. Bontempelli, G. Schiavon, and G. Zotti, *J. Electroanal. Chem.* **356**, 67 (1993).
101. Z. Fan and D. J. Harrison, *Anal. Chem.* **64**, 1304 (1992).
102. F. Moussy and D. J. Harrison, *Anal. Chem.* **66**, 674 (1994).
103. (a) A. C. Michael and R. M. Wightman, *Anal. Chem.* **61**, 270 (1989). (b) A. C. Michael and R. M. Wightman, *Anal. Chem.* **61**, 2193 (1989).
104. G. Che and S. Dong, *J. Electroanal. Chem.* **346**, 239 (1993).
105. (a) J. Wang and T. Golden, *Anal. Chem.* **58**, 2679 (1989). (b) J. Wang, *Electroanalysis* **3**, 255 (1991).
106. Q. Niwa, M. Morita, and H. Tabei, *Electroanalysis* **6**, 237 (1994).

107. (a) M. Majda and L. R. Faulkner, *J. Electroanal. Chem.* **137**, 149 (1982). (b) X. Chen, P. He, and L. R. Faulkner, *J. Electroanal Chem.* **222**, 223 (1987).
108. J. Wang, Z. Taha, and N. Naser, *Talanta* **38**, 81 (1991).
109. R. Agraz, M. T. Sevilla, and L. Hernández, *Anal. Chim. Acta* **273**, 205 (1993).
110. (a) K. Itaya and A. J. Bard, *J. Phys. Chem.* **89**, 5565 (1985). (b) G. Villemure and A. J. Bard, *J. Electroanal. Chem.* **283**, 403 (1990).
111. A. L. Crumbliss, S. C. Perine, A. K. Edwards, and D. P. Rillema, *J. Phys. Chem.* **96**, 1388 (1992).
112. (a) D. R. Rolison, R. J. Nowak, T. A. Welsh, and C. A. Murray, *Talanta* **38**, 27 (1991). (b) M. D. Baker and C. Senaratne, *Anal. Chem.* **64**, 697 (1992).
113. P. Hernandez, E. Alda, and L. Hernandez, *F. Z. Anal. Chem.* **327**, 676 (1987).
114. (a) N. Oyama and F. C. Anson, *Anal. Chem.* **52**, 1192 (1980). (b) K. Doblhofer and R. Lange, *J. Electroanal. Chem.* **237**, 13 (1987). (c) B. Keita, L. Nadjo, and J. P. Haeussler, *J. Electroanal. Chem.* **243**, 481 (1988).
115. A. Sugii, N. Oqawa, K. Harada, and K. Nishimura, *Anal. Sci.* 399 (1988).
116. J. F. Cassidy and K. Tokuda, *J. Electroanal. Chem.* **285**, 287 (1990).
117. P. J. Riley and G. G. Wallace, *Electroanalysis* **3**, 191 (1991).
118. N. Oyama, T. Ohsaka, and T. Okajima, *Anal. Chem.* **58**, 979 (1986).
119. L. Dunsch, L. Kavan, and J. Weber, *J. Electroanal. Chem.* **280**, 313 (1990).
120. P. Ugo, L. M. Moretto, and G. A. Mazzocchin, *Anal. Chim. Acta* **273**, 229 (1993).
121. (a) F. C. Anson, T. Ohsaka, and J. M. Savéant, *J. Phys. Chem.* **87**, 640 (1983). (b) F. C. Anson, J. M. Savéant, and K. Shigehara, *J. Am. Chem. Soc.* **105**, 1096 (1983).
122. N. Oyama, T. Ohsaka, and T. Shimizu, *Anal. Chem.* **57**, 1526 (1985).
123. (a) K. Kalcher, *Talanta* **33**, 489 (1986). (b) X. Cai, K. Kalcher, W. Diewald, C. Neuhold, and R. J. Magee, *F. J. Anal. Chem.* **345**, 25 (1993). (c) W. Diewald, K. Kalcher, X. Cai, and R. J. Magee, *Anal. Chim. Acta* **273**, 237 (1993).
124. J. Wang, G. D. Rayson, Z. Lu, and H. Wu, *Anal. Chem.* **62**, 1924 (1990).
125. Y. Ikariyama and W. R. Heineman, *Anal. Chem.* **58**, 1803 (1986).
126. J. Ye and R. P. Baldwin, *Anal. Chem.* **60**, 1982 (1988).
127. E. Wang and A. Liu, *Anal. Chim. Acta* **252**, 53 (1991).
128. O. A. Sadik and G. G. Wallace, *Electroanalysis* **5**, 555 (1993).
129. (a) H. Ge and G. G. Wallace, *Anal. Chem.* **61**, 198 (1989). (b) H. Ge, P. R. Teasdale, and G. G. Wallace, *J. Chromatog.* **544**, 305 (1991). (c) T. Nagaoka, M. Fujimoto, H. Nakao, K. Kakuno, J. Yano, and K. Ogura, *J. Electroanal. Chem.* **364**, 179 (1994).
130. M. R. Deakin and H. Byrd, *Anal. Chem.* **61**, 290 (1989).
131. L. J. Amos, A. Duggal, E. J. Mirsky, P. Ragonesi, A. B. Bocarsly, and P. A. Fitzgerald-Bocarsly, *Anal. Chem.* **60**, 396 (1988).
132. (a) K. N. Thomsen and R. P. Baldwin, *Anal. Chem.* **61**, 2594 (1989). (b) K. N. Thomsen and R. P. Baldwin, *Electroanalysis* **2**, 263 (1990).
133. Y. Li, W. Zhang, and S. Dong, *Electroanalysis* **5**, 431 (1993).
134. IUPAC Chemical Data Series no. 22, *Stability Constants of Metal-Ion Complexes, Part B, Organic Ligands* (compiled by D. D. Perrin) (Pergamon Press, Oxford, England, 1979).
135. A. A. Schilt, *Analytical Applications of 1,10-phenanthroline and Related Compounds,* (Pergamon Press, Oxford, England 1969).
136. J. A. Cox and M. Majda, *Anal. Chim. Acta* **118**, 271 (1981).
137. J. P. Price and R. P. Baldwin, *Anal. Chem.* **52**, 1940 (1980).
138. T. J. Miwa, L. T. Jin, and A. Mizuiki, *Anal. Chim Acta* **160**, 135 (1984).
139. J. Wang, P. Teng, and T. Golden, *Anal. Chim. Acta* **194**, 129 (1987).
140. K. H. Lubert and M. Schnurrbusch, *Anal. Chim. Acta* **186**, 57 (1986).
141. I. Turyan and D. Mandler, *Anal. Chem.* **65**, 2089 (1993).
142. A. J. Downard, H. Kipton, J. Powell, and S. Xu, *Anal. Chim. Acta* **251**, 157 (1991).

143. K. Kalcher, H. Greschonig, and R. Pietsch, *F. Z. Anal. Chem.* **327**, 513 (1987).
144. J. Lexa and K. Stulik, *Talanta* **32**, 1027 (1985).
145. S. V. Prahbu, R. P. Baldwin, and L. Kryger, *Anal. Chem.* **59**, 1074 (1987).
146. D. W. M. Arrigan, G. Svehla, S. J. Harris, and M. A. McKervey, *Electroanalysis* **6**, 97 (1994).
147. (a) R. P. Baldwin, J. K. Christensen, and L. Kryger, *Anal. Chem.* **58**, 1790 (1986). (b) K. N. Thomsen, L. Kryger, and R. P. Baldwin, *Anal. Chem.* **60**, 151 (1988).
148. S. Tanaka and H. Yoshida, *Talanta* **36**, 1044 (1989).
149. K. Kalcher, *F. Z. Anal. Chem.* **325**, 181 (1986).
150. T. Baumeyer, J. Dittrich, and F. Crespi, *Electroanalysis* **5**, 565 (1993).
151. T. Matsue, U. Akiba, and T. Osa, *Anal. Chem.* **58**, 2096 (1986).
152. J-M. Zen and M-L. Lee, *Anal. Chem.* **65**, 3238 (1993).
153. D. W. M. Arrigan, G. Svehla, J. Alderman, and W. A. Lane, *Analyst* **119**, 287 (1994).
154. J. Labuda, M. Vanícková, E. Uhlemann, and W. Mickler, *Anal. Chim. Acta* **284**, 517 (1994).
155. J. Labuda and M. Vanícková, *Electroanalysis* **5**, 141 (1993).
156. P. R. Teasdale and G. G. Wallace, *Analyst* **118**, 329 (1993).
157. (a) D. M. T. O'Riordan and G. G. Wallace, *Anal. Chem.* **58**, 128 (1986). (b) L. Yuping and G. G. Wallace, *Anal. Lett.* **22**, 669 (1989).
158. L. Yuping and G. G. Wallace, *J. Electroanal. Chem.* **247**, 145 (1988).
159. K-K. Shiu, S-K. Pang, and H-K. Cheung, *J. Electroanal. Chem.* **367**, 115 (1994).
160. M. J. Gehron and A. Brajter-Toth, *Anal. Chem.* **58**, 1488 (1986).
161. T. Malinski, A. Ciszewski, J. R. Fish, and L. Czuchajowski, *Anal. Chem.* **62**, 909 (1990).
162. A. R. Guadalupe and H. D. Abruña, *Anal. Chem.* **57**, 142 (1985).
163. A. R. Guadalupe, L. M. Wier, and H. D. Abruña, *Am. Lab.* **18**, 102 (1986).
164. H. C. Hurrell and H. D. Abruña, *Anal. Chem.* **60**, 254 (1988).
165. S. K. Cha and H. D. Abruña, *Anal. Chem.* **62**, 274 (1990).
166. E. Lorenzo, J. Fernández, F. Pariente, K. Trible, B. Pendley, and H. D. Abruña, *J. Electroanal. Chem.* **356**, 43 (1993).
167. S. K. Cha, B. K. Ahn, J-U. Hwang, and H. D. Abruña, *Anal. Chem.* **65**, 1564 (1993).
168. (a) S. E. Creager, D. M. Collard, and M. A. Fox, *Langmuir* **6**, 1617 (1990). (b) H. O. Finklea and D. D. Hanshew, *J. Electroanal. Chem.* **347**, 327 (1993). (c) C. E. D. Chidsey and D. N. Loiacono, *Langmuir* **6**, 682 (1990).
169. P. D. Beer, O. Kocian, R. J. Mortimer, and C. Ridgway, *Analyst* **117**, 1247 (1992).
170. (a) R. Arshady and K. Mosbach, *Makromol. Chem.* **182**, 687 (1981). (b) G. Vlatakis, L. I. Anderson, R. Müller, and K. Mosbach, *Nature* **361**, 645 (1993).

11

Polymer-Based Amperometric Biosensors

Emmanuel I. Iwuoha and Malcolm R. Smyth

1. INTRODUCTION

Electrochemical detection of analytes can give rise to very elegant methods in analytical chemistry. Such electroanalytical applications are characterized by excellent sensitivity, good selectivity toward the analyte, and simplicity in the design and use of instrumentation. However, the limited number of electrode/conductor surfaces available and the problem of electrode instability caused by passivation of electrode and slow electrode kinetics mean that there is a restricted number of analytes suitable for electrochemical detection. To circumvent this and other related problems, electrode surfaces can be modified to change chemical and physical properties of the electrode material. The added impetus of these chemically modified electrodes (CMEs) is tailoring the electrode surface for improved quantitative and qualitative detection of a particular analyte. This is due to the high selectivity of the electrode surface for catalytic, ion-exchange or complexation reactions, improvements in permselectivity, or high specificity due to incorporating various biocomponents.

Since the pioneering work of Updike and Hicks[1] in developing enzyme electrodes and Lane and Hubbard[2,3] in direct chemical modification of electrode surfaces, a great deal of attention has been paid to developing chemically modified electrodes. The result is that virtually every substance that is electroactive, or for which its chemical reaction can be coupled to the electrode-modifying matrix and/or electron transfer mediator, can now be detected electrochemically. Principles, techniques, and scope of application of CMEs as sensors in analytical

---

Emmanuel I. Iwuoha and Malcolm R. Smyth • School of Chemical Sciences, Dublin City University, Dublin 9, Ireland.

Electroactive Polymer Electrochemistry, Part 2: Methods and Applications, edited by Michael E. G. Lyons. Plenum Press, New York, 1996.

chemistry have appeared in several reviews.[4–13] The thrust of this chapter focuses on CMEs incorporating biological components, such as enzymes, in a polymer-modified matrix that couples bioreaction with electrochemical detection. Particular emphasis is on polymer-based amperometric biosensors. These are bioactive modified electrodes in which the specificity and selectivity of a bioreaction culminate in a current response proportional to the analyte concentration. Enzymes and other biological components of modified electrodes that catalyze biospecific redox reactions produce electroactive species that can be detected at the electrode. Through this approach many electroactive molecules of biological interest have been detected.

1.1. Properties of Polymer Films for Biosensors

Polymer-modified amperometric biosensors are popular for several reasons. The ease of entrapping an enzyme in preformed or *in situ* polymer coatings on electrode surfaces makes immobilization of enzymes in polymer films an attractive procedure for preparing biosensors. Unlike monolayers, which are not particularly stable and can hold only a limited amount of surface-bound species,[14] polymers can be applied to an electrode surface in any thickness. The result is that a concentration of modifying species, such as electron transfer mediators and enzymes, can be increased considerably. Also because the polymer interior is hydrophobic, high enzymatic activity is favored. A variety of polymeric materials are available; these include preformed polymers and those formed at electrode surfaces by electropolymerization or gamma irradiation. In addition, methods of preparation are easy to perform and controllable. The ease of improving electrochemical properties of some polymers by functionalizing with redox-active substances offers great opportunity for producing reagentless biosensors. This flexibility of polymer films is particularly important in mass producing biosensors. The three-dimensional structure of polymer films are stable under electrochemical conditions and permeable to ionic species, molecules, and electrons. In addition, most polymers are stable in not only the aqueous phase but also in organic phases. The membrane, ion-exchange, and size-exclusion properties of polymers can be harnessed to ensure permselectivity and eliminate electrode fouling in biosensors. It is possible to combine polymer films to enhance stability and response of biosensors. This is a method of producing multilayered film sensors that can be applied to real sample matrices without pretreatment.

1.2. Preparation of Polymer Film Biosensors

1.2.1. Preformed Polymer Films

Polymer film biosensors are formed by coating the electrode surface with preformed polymers or by *in situ* formation and deposition of polymer films on the electrode surface. Preformed polymers used in biosensors are either not functionalized or functionalized. The former is made up of only an organic moiety, such as poly(vinylpyridine) (PVP) and poly(styrene sulphonic acid) (PSSA).[16,17] Functionalized preformed polymers contain active sites that may be redox-active, as in poly(vinylferrocene),[18] poly[vinylpyridine Os(2,2′-bipyridyl)$_2$Cl]Cl,[19–21] and ferrocene-modified polysiloxane[22–24] or ion-exchange moieties, such as poly(ester sulphonic acid) ionomer (Eastman AQ polymer)[25–27] and Nafion perfluorinated-sulphonated ion-exchange polymer.[28–30] Polymer films are usually formed by spin coating or dip coating the electrode surface with a solution of the polymer with or without other sensor constituents, such as an enzyme. Where only the polymer solution is involved, the resulting film after evaporating the solvent can serve as a support for immobilizing the enzyme, mediator, or ion-exchange species to impart the requisite specificity and sensitivity to the biosensor.

Another method of casting preformed polymer films on electrode surface involves using droplet evaporation in a desiccator, open air, or with a heat gun; cross-linking reagents, such as glutaraldehyde, are usually needed in this procedure. The advantage of this method is that the thickness of the film can be controlled through the amount of biosensor film solution deposited on the electrode surface.

1.2.2. Electropolymerized Films

The *in situ* oxidative electropolymerization at the electrode surface is achieved with a suitable monomer. The mode of electrochemical formation of polymers on the electrode surface can be oxidative or reductive.[31–34] However, polymer films applied in biosensor preparations are usually formed by oxidative electropolymerization from solutions containing the monomer, the enzyme, and other additives. The procedure and mechanism of oxidative polymerization are reviewed in several articles.[9–13, 35–37] Oxidative electropolymerization leads to the formation of insoluble cross-linked conducting or nonconducting polymer films in very intimate contact with the electrode. If carried out in aqueous solutions of film-modifying species, such as enzymes and mediators, electrochemical polymerization results in a simple (one-step), rapid, and flexible method for biosensor preparation. The electropolymerization technique is also a reliable method for casting a new polymer film on a precoated electrode to enhance stability, sensitivity, and permselectivity. For example, poly(pyrrole) film may be electrochemically deposited on an Os-polymer/GOx-modified electrode.[21] In addition, a stable composite, reagentless biosensors can be formed by depositing or entrapping both enzyme and preformed redox polymer in a polymer matrix, such as poly(phenol) formed *in situ* at the electrode surface.[38] A very important advantage of electropolymerization in biosensor preparation is that film thickness and characteristics can be controlled by monitoring polymerization charge. It is also possible to functionalize the monomer before polymerization, or the electropolymerized film can be doped on the electrode surface.

Electrochemical polymerization can be effected by potentiostatic, galvanostatic, or potentiodynamic (i.e., cyclic voltammetric) methods. In the galvanostatic mode, an amount of current is passed through the solution containing the monomer, and the polymer film is formed on the working electrode. In the potentiostatic mode the working electrode is kept at a potential at which polymerization takes place on the electrode. In the cyclic voltammetric method, the potential of the working electrode is scanned within a range determined by the oxidation potential of the polymer film. Potentiostatic and cyclic voltammetric formations of polymer films have wider applications in biosensor fabrication than the galvanostatic method.[36]

2. AMPEROMETRIC BIOSENSORS

Electrochemical sensor fabrication has dominated the analytical application of polymers. In some sensors the polymer film acts as a membrane for the preconcentration of ions or elements before electrochemical detection. Polymers also serve as materials for electrode modification that lower the potential for detecting analytes. In addition, some polymer films function as electrocatalytic surfaces. Using a polymer in biosensors is a very rapidly developing area of electroanalytical chemistry. Polymeric matrix modifiers have been applied as diffusional barriers in constructing not only sensitive amperometric biosensors, but also electrochemical sensors that apply potentiometric, conductimetric, optical, and gas-sensing transducer systems. The principles, operations, and application of potentiometric, conductimetric, optical and gas sensors are described in Refs. 13, 39–41. In this chapter, we focus mainly on amperometric biosensors based on redox enzymes.

2.1. Redox Centers of Oxidoreductases

Redox enzymes, or oxidoreductases, are proteins with prosthetic groups (cofactors) that undergo oxidation and reduction reactions during their catalytic cycle. Cofactors may be large organic groups or organometallic moieties containing transition metals. Organic cofactors can be a quinone, such as pyrroloquinoline quinone (PQQ), topaquinone (TPQ), and tryptophanyltryptophanquinone (TTQ);[42–45] or a nucleotide, such as flavin mononucleotide (FMN), flavin adenine dinucleotide (FAD), and nicotinamide adenine dinucleotide (NAD^+). Transition metals found in the organometallic cofactors include Co, Cu, Fe, and Mo. The location of the redox center in the protein has serious implications for transferring electrons between the enzyme and the electrode surface, and hence it determines the configuration of the biosensor. This is because the protein interior is usually insulating, with the result that electrical communication between deep-seated, or protected, redox sites and the electrode surface is not possible.[19–24,36,46–48] Three classes of redox enzymes most commonly applied in biosensors are oxidases, dehydrogenases, and peroxidases. In a normal physiological milieu, dioxygen accepts either two hydrogens ($2H^+ + 2e^-$) or four hydrogens ($4H^+ + 4e^-$) from oxidases to form H_2O_2 and H_2O, respectively. Dehydrogenases transfer hydrogen to acceptors other than dioxygen, e.g., NAD^+. Hydrogen peroxide and other organic peroxides are the hydrogen acceptors for peroxidases. For most oxidases the redox site or cofactor is FAD (Fig. 11.1) strongly held within the protein interior. Some oxidases used in analytical systems and their redox centers are given in Table 11.1. Of these enzymes, the glucose oxidase electrode is the most widely applied biosensor in analytical studies.[6,36]

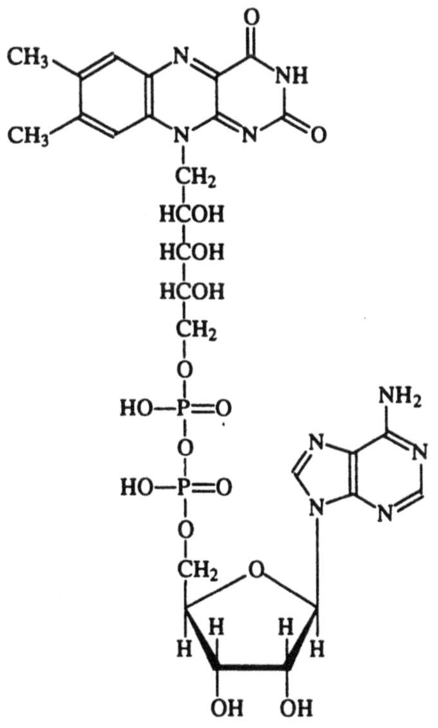

FIGURE 11.1. Structure of FAD.

TABLE 11.1
Examples of Analytically Relevant Oxidases and Their Redox Centers (Cofactors)

Enzyme	Redox Center
Alcohol oxidase	FAD
L-Amino acid oxidase	FAD
Cholesterol oxidase	FAD
Choline oxidase	FAD
Galactose oxidase	Cu
Glucose oxidase	FAD
Glycolate oxidase	FMN
Lactate oxidase	FMN
Sulphite oxidase	Haem, Mo
Urate oxidase	Cu
Xanthine oxidase	FAD, Fe, Mo

2.2. Glucose Oxidase Electrodes

Amperometric detection of glucose, which has dominated both analytical and fundamental studies of enzyme electrodes, is carried out mainly using glucose oxidase (GOx), a dimeric glycoprotein containing one FAD redox site per monomer.[47] Biosensors based on glucose dehydrogenase (GDH) containing the PQQ prosthetic group have also been applied in amperometric glucose sensing.[48] In the amperometric detection of glucose involving GOx, two electrically insulated FAD groups oxidize glucose to gluconolactone, and these are reduced to $FADH_2$. An electron transfer mediator is needed for indirect electrochemical oxidation of $FADH_2$ to continue the catalytic cycle. In biological systems, dioxygen serves this purpose and oxidizes $FADH_2$ to regenerate FAD and form the electroactive H_2O_2, which is electrochemically detected.[36,49–52]

$$\text{Glucose} + \text{GOx(FAD)} \rightarrow \text{Gluconolactone} + \text{GOx(FADH}_2\text{)} \qquad (1)$$

$$\text{GOx(FADH}_2\text{)} + O_2 \rightarrow \text{GOx(FAD)} + H_2O_2 \qquad (2)$$

$$H_2O_2 \rightarrow 2H^+ + O_2 + 2e^- \qquad (3)$$

Though the current arising from H_2O_2 oxidation is proportional to glucose concentration if measured under steady-state conditions, biosensors based on H_2O_2 detection on both modified and unmodified electrode surfaces present some serious drawbacks.[53] One of the problems is that some biological compounds in the serum along with glucose are electroactive at the working potential for H_2O_2 detection

(+0.7 V versus SCE). These compounds include ascorbic acid, uric acid, and acetaminophen. In addition to the high working potential required, the loss of sensor sensitivity due to high peroxide activity, and the sensitivity of the sensor to ambient oxygen pressure, are other drawbacks associated with sensors using dioxygen as a diffusional electron transfer mediator.[53–55] Nonphysiological electron transfer mediators, such as ferrocene and its derivatives, ferricyanide, Ru complexes, tetrathiafulvalene, quinones, etc., which are electroactive, can be employed to shuttle electrons between the $FAD/FADH_2$ redox couple and the electrode surface. Consequently the $GOx(FADH_2)$ formed in Eq. (1) reacts with the mediator to regenerate the oxidized form of the enzyme:

$$GOx(FADH_2) + 2Med_{ox} \rightarrow GOx(FAD) + 2Med_{red} + 2H^+ \quad (4)$$

$$2Med_{red} \rightarrow 2Med_{ox} + 2e^- \quad (5)$$

Incorporating electron transfer mediators into biosensors by using preformed and electropolymerized polymer films has led to a variety and variations of amperometric biosensors not only for glucose but for other analytes of biological and environmental interest, such as phenols, cholesterol, choline, amino acids, peroxides, and alcohols.[13,35–37] Recently, polymer films have been modified to detect H_2O_2 on nonconducting films with minimal interference on the glucose sensor.[56–60] Details of these and other applications of polymeric film are given in the following sections.

3. CONDUCTING POLYMERS

These polymers possess the conjugated π-electron systems, and characteristically they are easily reduced and oxidized reversibly. Charge transfer species (dopants) affect this oxidation or reduction process, with the result that insulating polymers are converted into those with near metallic conductivity in many cases. Since the synthesis of the first conducting polymer, i.e., poly(acetylene),[61] many conducting polymers have been electrochemically or chemically produced.

3.1. Electrochemical Conducting Polymers

Poly(pyrrole), known for over 50 years, is one of the most studied polymers.[35–37] First synthesized by electrochemical polymerization in 1968,[62] poly(pyrrole) is electrically conductive, stable, and insoluble when doped. The mechanism of its electrosynthesis was studied by Genies *et al.*[63] Thin films of electropolymerized poly(pyrrole) ($< 0.1\ \mu m$) are electroactive, and they can be switched between the neutral and the oxidized state at +0.1 V versus SCE. This particular property of polypyrrole can be applied in constructing conductivity biosensors.[13] Conducting polymers of aniline, phenol, pyrrole, and their derivatives have also been used in preparing enzyme electrodes.[64–68] A review of enzymes immobilized in electropolymerized films, particularly poly(pyrrole) and its derivatives, was carried out recently by Bartlett and Cooper.[36] Some applications of conducting polymers include their use in amperometric sensors based on detecting H_2O_2 or electron transfer mediators,[64] construction of reagentless sensors, enhancement of permselectivity and other biosensor characteristics, and the use of conducting polymers directly as mediators.

3.1.1. H_2O_2-Based Biosensors

For redox enzymes, such as GOx with insulated redox sites, a variety of electron mediator compounds were used to shuttle electrons between the enzyme redox couple and metal transducer. Dioxygen (the natural electron acceptor in glucose biosensor) is reduced to H_2O_2, which is detected at the electrode surface. The GOx electrodes have been used to detect H_2O_2 in poly(pyrrole),[54,55,65,66] poly(N-methylpyrrole),[67] and poly(aniline)[68] films. Horseradish peroxidase (HRP)-based biosensors that depend on H_2O_2 detection have also been prepared by galvanostatic growth of poly(pyrrole) film on SnO_2 at 0.1 $mA\,cm^{-2}$.[69] An L-lactate sensor, for which H_2O_2 is the electroactive product, was prepared by trapping lactate oxidase in a potentiostatically grown poly(*o*-phenylenediamine) film.[70] However, due to the problems associated with H_2O_2 detection described earlier and because of the need to lower operational potentials of the biosensor, diffusional electron transfer mediators, such as ferrocenes, quinones, ruthenium amines, and organic metals have been used in place of dioxygen. The success of these so called second-generation biosensors has been mainly due to the entrapment of both the enzyme and mediator in conducting polymers grown electrochemically. The result is the construction of reagentless sensors suitable for *in vivo* application. The weakness of this type of biosensor is the possibility of the mediator leaching into the surrounding solution, particularly in its soluble form. This may have some deleterious effect when applied *in vivo*.

3.1.2. Functionalized Conducting Polymers

Functionalized conducting polymers are very suitable for preparing reagentless biosensors. This is achieved by polymerizing mediator-functionalized monomers in the presence of the enzyme.[71,72] Functionalization of the polymer film can also be accomplished by incorporating the mediator as anionic counterions compensating the charge of the cationic polymer matrix.[73,74] Substituting surface-bound polymer counterions by electrostatically binding ion exchangers, such as Nafion, poly(lysine), and PVP, gives rise to functionalized polymers that are permselective. By electrostatic binding, amperometric biosensors can be prepared to possess exclusion properties based on size, shape, charge, or polarity of ana-

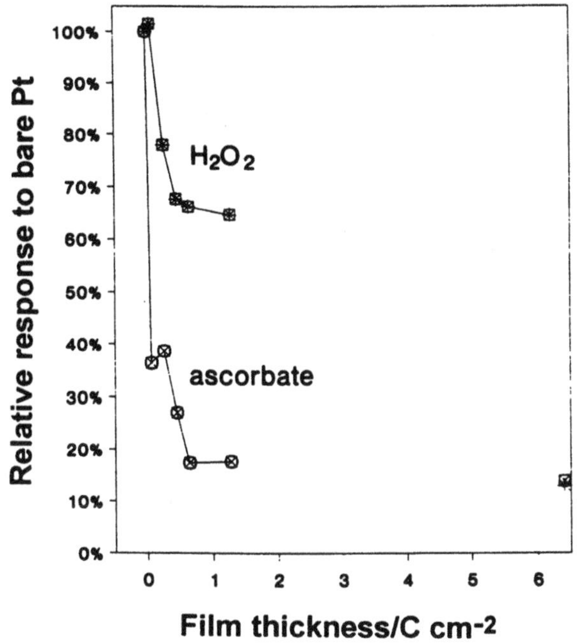

FIGURE 11.2. Dependence of the signal for H_2O_2 and ascorbic acid oxidation on the thickness of the polypyrrole film. Values are normalized by the response at the bare platinum electrode surface. (From Ref. 101).

lytes.[8,75–80] Electropolymerization is a very effective method for growing highly controllable size exclusion films.[8,80,81] There are differences in size exclusion properties of different conducting polymers. For example in the amperometric amino acid sensor based on L-amino-acid oxidase electrode, poly(pyrrole) was found to be more permeable to amino acids than poly(aniline).[80] Also, permeability of the electropolymerized film depends on polymer morphology and thickness, which can be controlled by the amount of charge passed during film formation;[8] e.g., poly(aniline) layers thicker than 100 mC cm^{-2} prevented direct oxidation of amino acids but permitted oxidation of analytically relevant H_2O_2.[80] Figure 11.2 shows that the thickness of poly(pyrrole) film can be used to decrease the interference of ascorbate during H_2O_2 detection.

3.1.3. Conducting Polymers As Mediators

Direct electrochemistry of fructose dehydrogenase trapped in conducting poly(pyrrole) films has been reported.[82] In an attempt to prepare biosensors that give rapid response to analytes and where mediators do not leach, ferrocenes were incorporated into monomers before polymerization. However, it was demonstrated that electron transfer mediators can be polymerized directly into the working electrode[53,83] in the presence of enzymes. In these electrode systems, the roles of electrode modification, enzyme entrapment, and shuttling electrons between the enzyme redox site and the transducer are played by the same polymer film. Poly(ferrocenylenemethylene) and methoxyphenyl-substituted polyferrocene were used as mediators with GOx at carbon electrodes.[84] An amperometric glucose sensor has also been prepared which was based on the immobilization of GOx in the conducting poly(ether amine quinone).[53] The CV of the response of the enzyme electrode is shown in Fig. 11.3. Polymers containing *p*- and *o*-quinone groups, and polysiloxanes with attached quinone groups have also been shown to be effective

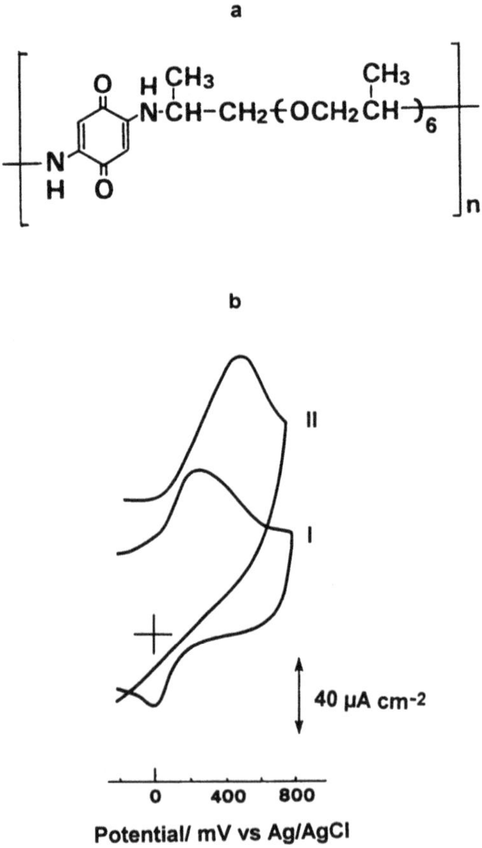

FIGURE 11.3. (a) Structure of poly(ether amine quinone). (b) Cyclic voltammogram for the cross-linked poly(ether amine quinone)/glucose oxidase/carbon paste electrode in phosphate buffer of pH 7.0; scan rate: 5 mV/s, added electrolyte: 0.1 M KCl; (I) glucose absent and (II) 0.1-M glucose present. (From Ref. 53).

mediators in GOx electrodes.[85–88] Poly(ether amine quinone), just like conventional quinone polymers, acts as an electron transfer relay system, and not as a diffusional mediator. This property of conducting polymerized mediators makes them suitable for producing biosensors for *in vivo* application.

3.2. Redox Polymers for Electrically Wiring Enzymes

Redox polymers, which have also been referred to as electron exchangers, redox exchangers, electron exchange polymers, oxidation-reduction exchangers, electron exchange polymers, oxidation-reduction polymers and redoxites,[92] are substances that can transfer or accept electrons from their surrounding environment. Redox polymers are characterized by chemical inertness and rapid electron self-exchange rates, which make them useful as electrocatalytic films. In addition, redox polymers exhibit excellent catalytic features, such as the ability to entrap or adsorb reactive species, accessibility of multiple oxidation states, and the ability to transfer both protons and electrons.[93–96] Simple, sensitive, and reagentless amperometric biosensors have been produced by immobilizing enzymes in redox polymer films without requiring a membrane to contain the enzyme or additional electron transfer mediators.[19–24,97–100]

Enzyme electrodes have been constructed by covalent immobilization of enzyme on functionalized conducting polymers, such as poly(3-aminopyrrole).[101–103] In some cases the polymer incorporates the mediator, producing a redox macromolecule, e.g., poly[(ferrocenyl)amidopropyl]pyrrole,[71] ferrocene-modified poly(ethylene oxide),[72,104] and highly flexible ferrocene-modified siloxane.[104] These redox polymers have short chains and low molecular weights (≤ 20 kDa); they function as a special kind of diffusional mediators[100] and do not alleviate the loss of a mediator to the bulk solution during amperometric detection.

Constructing amperometric glucose sensors that incorporate nondiffusional mediators was accomplished with GOx by chemical modifications of the enzyme that perturbs the carbohydrate shell.[71,89,105,106] Covalent binding of the enzyme to the electrode was also used as a method of overcoming the insulation of the redox-active FAD by the carbohydrate cover. An example of this approach is the use of ferrocene-modified n-type silicon in a glucose biosensor.[107] Nondiffusional mediation in GOx-based biosensor was also achieved by using conducting polymers that act as molecular relay redox couples[99,100,104] when chemically attached to the enzyme. However, chemical modification of enzymes used in biosensors may have an adverse effect on enzyme activity. Also for good sensor performance, modified enzyme must be contained in the vicinity of the electrode in a membrane rather than bound on the surface of the electrode.[100] This is because the modified enzyme must move freely to assume the orientation that gives the shortest distance between the redox site and the surface of the electrode. Using membranes makes this method expensive and leads to slow sensor response.

Direct unmediated electrical communication between enzyme redox centers and the electrode by molecular wiring has been attracting attention recently.[108–111] In this technique the natural ability of polyanionic oxidoreductases to form electrostatic complexes with redox proteins for effective electron transfer can be applied in biosensor technology.[108] However, in place of the redox protein, a polycationic redox macromolecule is used. The result is a three-dimensional network of an electrostatic complex of the enzyme and the redox polymer. In this system membranes are not needed, nor is unique orientation of the enzymes necessary. This is

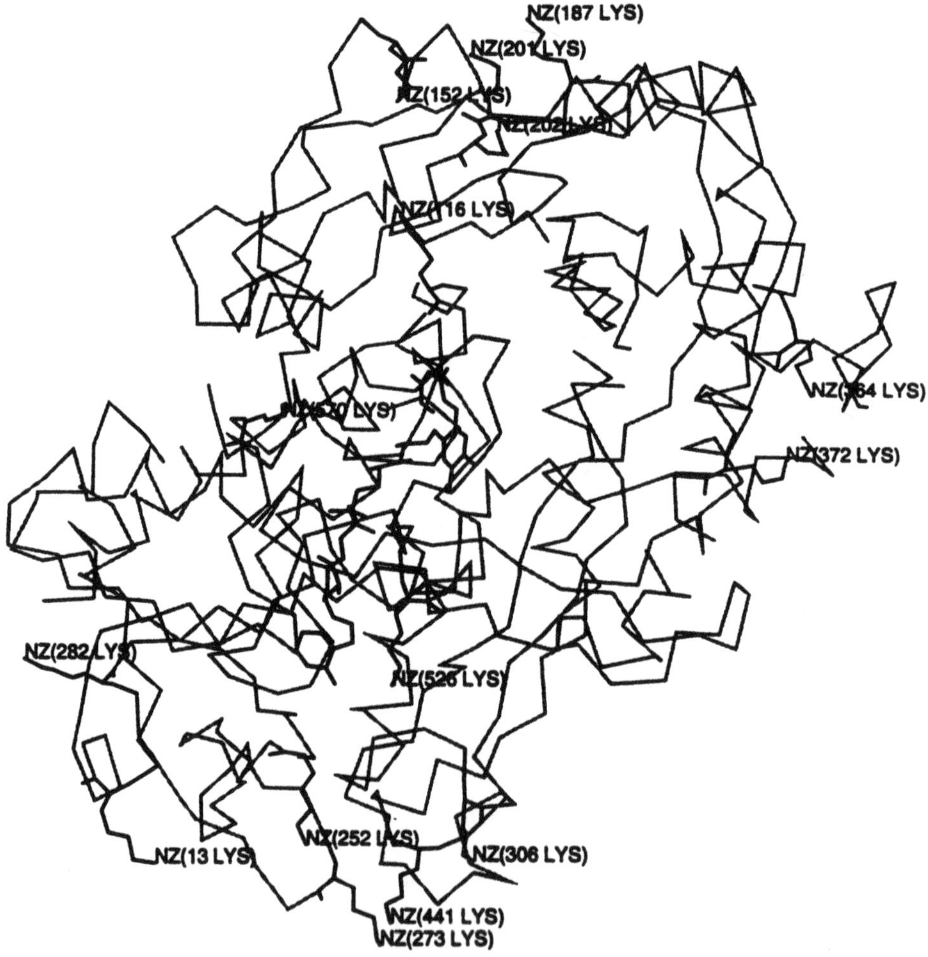

FIGURE 11.4. Computer graphics display of the C_α backbone of the GOx monomer, showing the location of the 15 lysine residues at the surface of glucose oxidase. (From Ref. 101).

because redox centers of the enzyme are electrically connected to the electrode and the enzyme–polymer complex is physically attached to the electrode surface.

Poly(pyrroles) and other redox macromolecules that adsorb too strongly on the electrode surface do not make good molecular wires. Their segments are also usually too rigid to fold along protein chains. Hydrogels (i.e., cross-linked network of polymers swollen with water)[(112)] of ferrocene-modified acrylamide have flexible and mobile side chains containing redox group at the ends. They have been used to wire the enzyme electrically to the electrode.[(112)] For a glucose biosensor, wiring can be achieved by complexing the high molecular weight polycationic GOx (MW 160,000, radius 86 Å, 15 lysine residues, as shown in Fig. 11.4)[(113–114)] with a long-chain polycationic redox macromolecule, such as siloxane polymer,[(22–24,104)] or $[Os(bpy)_2Cl]^{+/2+}$ poly(vinylpyridine) polymers.[(19,21,99–100,115–116)] Reviews of the application of siloxane and osmium polymers to molecularly wiring enzymes in sensors appear in Refs. 100 and 104.

Biosensors prepared with these redox macromolecules achieve fast direct electron transfer between GOx ($FAD/FADH_2$) and the electrode due to the presence of stable redox couples that permit rapid electron self-exchange. Ling Ye and coworkers reported preparing a wired quinoprotein glucose dehydrogenase (GDH) electrode that produced high current density.[(48)] The authors reported the formation of an O_2-insensitive glucose sensor reaching a 1.8 mA cm^{-2} current density. The sensor was prepared by complexing GDH and poly(ethylene glycol diglycidyl ether) (PEGDE) cross-linked with bromoethylamine quarternized Os-polymer (POs-EA).

Cross-linking an Os-polymer containing 20 PVP (Fig. 11.5) and GOx on a platinum or glassy carbon electrode (GCE) surface with glutaraldehyde, produces biosensors that are very sensitive and stable in organic phases.[(116)] The CV of the enzyme electrode shows that complexing the polymer with GOx does not change

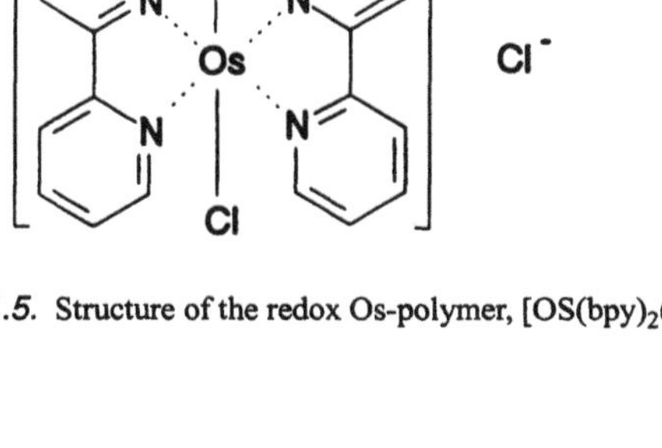

FIGURE 11.5. Structure of the redox Os-polymer, $[OS(bpy)_2(PVP)_{20}Cl]Cl$.

the reversible electrochemistry of mediator species and the electrocatalytic action of GOx is not impeded. This behavior of the sensor is shown in Fig. 11.6.

Gregg and Heller[117] reported a simple and gentle cross-linking of the lysine side chains on the surface of GOx and the pendant amines on the Os-containing redox polymer, POs-EA, by PEGDE. The authors observed that this epoxy-curing reaction on the electrode surface of a vitreous carbon disk electrode allows the simultaneous immobilization and electrical connection of several enzymes, including glucose oxidase, lactate oxidase, glycerol-3-phosphate oxidase, and D-amino acid oxidase. Cured enzyme/POs-EA/PEGDE films were described as redox-conducting epoxy cements.[110] They are reproducible, highly permeable, water-insoluble, strongly bound hydrophilic films in which the enzyme retains much of its activity and specificity.

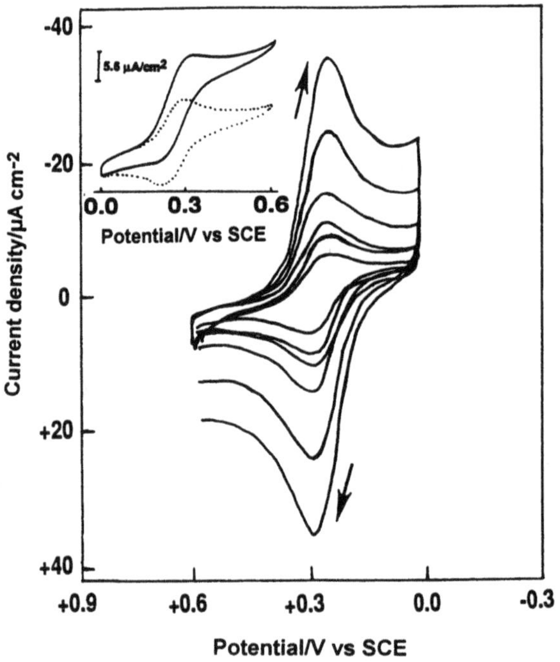

FIGURE 11.6. Cyclic voltammograms of an $Os^{2+/3+}$ couple in a GOx/[$Os(bpy)_2(PVP)_{20}Cl$]Cl electrostatic complex immobilized on a 3-mm glassy carbon disk electrode. The CVs (*outward*) are for potential sweep rates 2, 5, 10, 20, 50, and 100 mV/s. The $E_{p,a}$, $E_{p,c}$, and $E^{o\prime}$ calculated from these CVs are 0.295 V, 0.250 V, and 0.273 V versus SCE, respectively. The $\Gamma_{Os2+/3+}$ 5.85×10^{-8} mol cm^{-2}; GOx concentration = 8.405×10^{-9} mol cm^{-2}. Background solution: CH_3CN-H_2O (90 + 10% v/v), 0.05-M tetraethylammonium p-toluene sulphonate. Inset: Cyclic voltammetric response of the glucose biosensor at 10 m V/s potential sweep rate in absence (...) and presence (—) of 0.025-M glucose.

4. NONCONDUCTING POLYMERS

Advances in the design and preparation of reagentless amperometric glucose biosensors notwithstanding, systems that operate by detecting electroactive H_2O_2 are still very popular.[54–60,65–69,118,119] This is more so for sensors required for clinical applications, where the natural electron transfer mediator for glucose oxidation, dioxygen, and its reduction product H_2O_2 do not create toxicity problems.[60,120] The reason is that the enzyme catalase in the human body decomposes H_2O_2 into H_2O and dioxygen.[100] However, direct faradaic interference by endogeneous electroactive species, such as ascorbic acid, uric acid, acetaminophen, and cysteine, pose serious problems, so these must be contained. Nonconducting polymer films have been employed in biosensor preparations not only to block electroactive interferents during H_2O_2 oxidation but also to protect electrode surfaces from fouling species (proteins) and surfactants (lipids).

4.1. Electropolymerized Nonconducting Films

Nonconducting polymers used for enzyme immobilization can be classified into three groups: electropolymerized, preformed, and gamma irradiation-induced films. Electropolymerization gives rise to self-regulating nonconducting films with uniform thickness that completely cover the electrode surface. Uniformity of the film is achieved because it grows only thick enough to become an insulator.[121,122] Polymerization continues until the polymerization current reaches a minimum when the whole electrode surface is completely covered and the monomer can no longer penetrate the film.

Yacynch and coworkers[121] reported the application of electropolymerized diaminobenzene (DAB) and dihydroxybenzene (DHB) isomers in preparing glucose sensors based on H_2O_2 detection. Various electropolymerized films on reticulated vitreous carbon (RVC) or platinized RVC were grown potentiodynamically by scanning the electrode potential from 0.00 to +0.80 V versus SCE. The biosensor prepared with poly(1,3-DAB/resorcinol) films formed over 24 hours was stable over a period of 5 months and effective in protecting the electrode from interferences and fouling in complex serum matrices (Fig. 11.7). In the study the enzyme was immobilized on the Pt-RVC by cross-linking with glutaraldehyde, and the polymer coating was grown electrochemically. The enzyme electrode produced by this procedure gave no response to physiological levels of acetoaminophen, urate, and ascorbate. Other authors[57,123–125] reported the effectiveness of electropolymerized poly(1,2-DAB) films in preventing interferences and fouling in human serum glucose determination. Though this film stabilized the enzyme and selectively allowed free diffusion of glucose and H_2O_2, no comments were made by these workers on the severe effects of acetoaminophen.

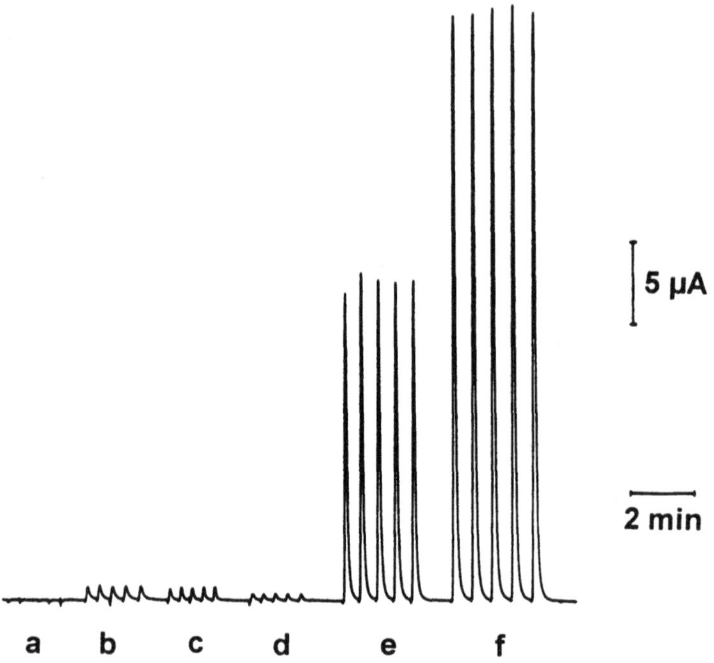

FIGURE 11.7. Response of an RVC/poly(1,3-DAB/resorcinol)/GOx glucose biosensor after 5 weeks to: (a) buffer, (b) 1-mM acetoaminophen, (c) 1-mM ascorbic acid, (d) 1- mM uric acid, (e) 2.5-mM glucose, (f) 5-mM glucose. (From Ref. 121).

Poly(*o*-phenylenediamine) (PPD) films grown electrochemically have been applied to biosensor construction because of their permselectivity, fast response, and stability in flow systems.[(56–60,127–129)] Malitesta *et al.*[(56)] have described a rapid-response GOx/PPD-based glucose sensor in which the PPD film was grown potentiostatically at 0.65 V versus SCE on a Pt electrode; H_2O_2 was detected at +0.8 V versus SCE. The biosensor exhibited high sensitivity, good storage stability, and it blocked interferences when applied to flow injection analysis (FIA) of undiluted human serum.[(129)] Working in batch analysis regime, Lowry and O'Neill[(58)] observed a depression of response by ascorbic acid in the GOx/PPD glucose sensor. This they attributed to a competing homogeneous reaction between H_2O_2 and ascorbic acid that depletes the electroactive H_2O_2 and produces dehydroascorbic acid (DAA). However, in another study,[(59)] Palmisano and Zambonin attributed the negative bias introduced by ascorbic acid in the batch analysis to sensor modifications caused by ascorbic acid electrooxidation. These authors argue that the homogeneous reaction between H_2O_2 and ascorbic acid is too slow to affect the fast response of the Pt/GOx/PDD sensor. They ascribe the observed signal depression phenomenon to fouling of the Pt surface by DAA, its dimers, and hydrated species formed by electrooxidation of ascorbic acid. In experiments monitoring brain glucose levels *in vivo* with systemic injection of ascorbic acid, Lowry *et al.*[60] reported the absence of the homogeneous reaction characteristic of *in vitro* batch analysis. This finding by the authors, and previous studies of the Pt/GOx/PPD electrode,[130] show that PPD films can be reliably applied to *in vivo* monitoring of glucose.

Electrochemical PPD films were applied by Wang *et al.*[127] in the construction of ultrathin enzyme probes. The GOx, alcohol oxidase, and tyrosinase electrodes were formed on 35–50 nm diameter carbon films. The authors reported rapid

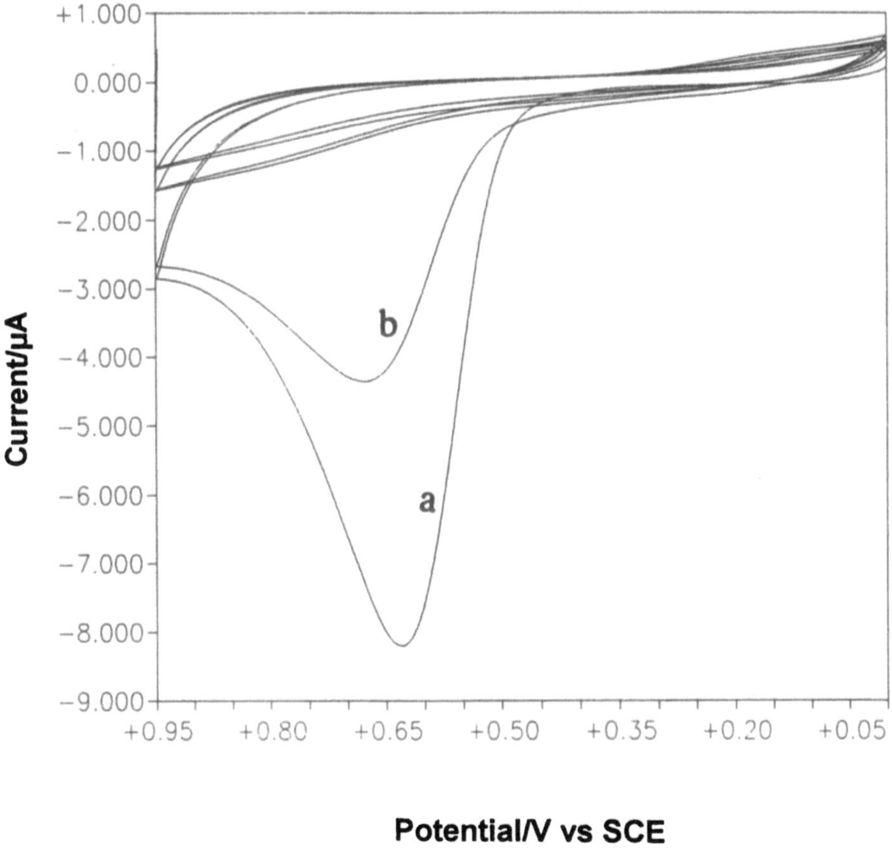

FIGURE 11.8. Electrochemical immobilization of GOx in poly(phenol) films on a 3–mm diameter platinum electrode. Experimental conditions: 0.1-M phosphate buffer, 7 mg/ml GOx; added electrolyte: 0.15-M tetraethylammonium tetrafluoroborate; potential sweep rate: 50 mV s^{-1}. Polymerization solution also contains (a) 50-mM phenol; (b) 50 μl/ml 1% w/w in methanol of $[Os(byp)_2(PVP)_{10}Cl]Cl$ and 50-mM phenol.

measurement of millimolar substrate concentrations. The PPD film was demonstrated as a means of immobilizing lactate oxidase for L-lactate monitoring via H_2O_2 detection at the electrode. The sensor characteristics include very fast response time, reproducible quantification of L-lactate in both batch and flow injection methods, high sensitivity, and blocking of normal interfering compounds in serum.

Insulating poly(phenol) films can be grown electrochemically by the potentiostatic or potentiodynamic mode from aqueous-buffered solutions at neutral pH. The films are usually thin (< 100 nm) because the polymerization process is self-controlled and stops as soon as the electrode surface is completely insulated by the polymer coating, as shown in Fig. 11.8. The unlabeled wave forms in Fig. 11.8 show the decrease in polymerization current to a minimum after three scans in a buffered solution containing the monomer at pH 7.0. Mechanisms for the electrochemical formation of poly(phenol) films have been described by Bartlett and coworkers.[36,68,131] The polymer is believed to be a mixture of para-linked and meta-linked units. Figure 11.9 compares the catalytic efficiency of biosensors based on a poly(phenol) film entrapping GOx and that containing GOx and Os-polymer.

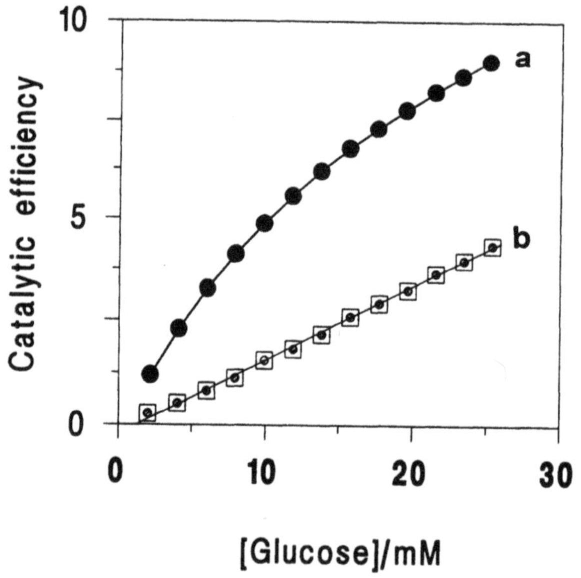

FIGURE 11.9. Catalytic efficiency (I_k/I_d) of (a) GOx-[Os(bpy)$_2$(PVP)$_{10}$Cl]Cl electrostatic complex and (b) GOx alone, entrapped in poly(phenol) films on a 3-mm glassy carbon electrode. The I_k and I_d are the oxidation currents at 0.4 V versus SCE from CVs in the presence and absence of glucose, respectively. The $E_i = 0.0$ V and $E_\lambda = 0.6$ V; potential sweep rate: 5 mV/s. Other conditions are as in Fig. 11.8. Ferricyanide ion in solution was used as the electron transfer mediator in (b).

Some workers indicate the suitability of the polymer as permselective films in enzyme electrodes.(132–134) However, studies in our laboratory with Pt/GOx/Poly(phenol) biosensors show strong interference by ascorbic acid in both ferricyanide and Os-polymer-mediated glucose detection.(38) A potentiostatically grown nonconducting poly(indole) film was also used to entrap GOx from the aprotic acetonitrile.(135) Glucose detection with the enzyme electrode was carried out at 1.56 V versus Ag/AgCl. This potential is too high for the glucose sensor to be analytically relevant in real matrices because interferents are electroactive at this potential. However, the stability of the sensor in CH_3CN suggests that the sensor may be useful in organic phase monitoring of glucose.

4.2. Insulating Overoxidized Poly(pyrrole)

Most studies on enzyme immobilization in electropolymerized films are based on pyrrole and its derivatives.(35–37) Conductivity of the polymer film is known to be destroyed by H_2O_2.(136) The GOx-based glucose sensors produce H_2O_2 even when a nonphysiological electron transfer mediator is used, provided the reaction medium is not completely oxygen free.(36) Incidentally, insulating poly(pyrrole) films do not affect H_2O_2 detection even in the presence of an enzyme.(118) Zambonin and coworkers(118,129) demonstrated the interference-free detection of H_2O_2 with an amperometric glucose sensor based on the electrochemical immobilization of GOx in overoxidized nonconducting poly(pyrrole) films. The Pt/GOx/Poly(pyrrole) electrode was prepared potentiostatically at +0.7 V versus Ag/AgCl and overoxidized overnight by polarizing the electrode at the same potential. These authors reported a rapid response time, high sensitivity, and the lowest recorded bias introduced by any electroactive species.(118) In particular, this group claimed complete elimination of exogeneous electroactive acetoaminophen, which is known to be the most difficult interferent to exclude.(126) The polymer film was shown to be very efficient in protecting the electrode from fouling effects. The authors demonstrated the application of the sensor to glucose determination in undiluted serum using an FIA regime and *in vivo* continuous monitoring of glucose via a hyphenated microdialysis sampling technique. The current time response of the enzyme electrode is shown in Fig. 11.10. This sensor appears to be the simplest and most efficient permselective glucose sensor ever reported.

4.3. Preformed Nonconducting Polymer: Poly(vinyl Chloride) Membrane

Poly(vinyl chloride) (PVC) is an excellent selective membrane barrier widely used in ion-selective electrodes.(137–140) The ion-selective properties of the membrane depend on incorporating a platicizer, ion exchanger, or neutral carrier molecule. Christie *et al.*(137) applied PVC membranes (with no incorporated ion

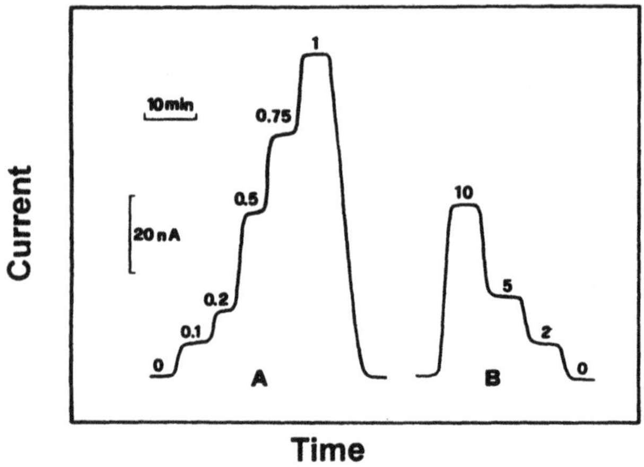

FIGURE 11.10. (A) Current time response obtained at different glucose concentrations without and (B) with Pt/GOx/poly(pyrrole) microdialysis probe. Numbers indicate a millimolar concentration of glucose. An experimental setup is given in Ref. 118.

exchanger) as permeability barriers for conferring selectivity to amperometric sensors. These authors report virtual exclusion of interferents found in serum, particularly ascorbate, urate, and NADH, which are ionic. According to them the PVC permselective membrane is suitable for H_2O_2 detection in GOx-based sensor, but it does not exclude paracetamol and 4-aminophenol, which are uncharged at neutral pH.

4.4. Gamma Irradiation-Induced Polymer Films

Biosensor technology based on enzyme entrapment in polymeric films grown by gamma irradiation-induced polymerization has received only modest attention.[119,138–146] Two techniques have generally been adopted in preparing such enzyme electrodes. One method involves preparing radiation-induced neutral, partially water-soluble, and biocompatible polymers, such as poly(vinyl alcohol) (PVA), that are applied as support for enzyme immobilization.[143–145] The enzyme is cross-linked with the polymer on a platinized graphite electrode by exposure to a suitable gamma radiation source.[143,145] The other method involves the *in situ* polymerization and simultaneous entrapment of an enzyme by irradiating a solution containing an enzyme and monomer with an adequate dose of low-temperature gamma rays[138,139,144] in a deaerated solution. Both methods give rise to hydrophilic, biocompatible, and inert polymers.

A variety of polymeric films were produced by this method; these include poly(2-hydroxyethyl methacrylate) that have been applied in GOx-based sensor,[144] PVA in glucose oxidase, and lactate oxidase immobilization,[145] and poly(N-vinylpyrrolidone) in GOx immobilization.[146] Glass-forming monomers[138–142] have also been polymerized from aqueous solutions containing α-amylase and glucose amylase.[138]

5. ION-EXCHANGE POLYMERS

5.1. Perfluorinated Sulphonic Acid Ion-Exchange Polymer (Nafion)

DuPont Nafion perfluorinated-sulphonated ion-exchange polymer and polyester sulphonated Eastman AQ polymer are the two most important ionomers applied in preparing biosensors. Nafion (see Fig. 11.11) is a permselective polymer known for its ability to incorporate positively charged ions and reject anions.[147] The use of Nafion in biosensors stems from its biocompatibility attributes: it is both hydrophilic and hydrophobic, chemically inert, and it has low adsorption potential for species in solution.[28] Nafion has had a wide range of applications in electroanalytical chemistry. These can be classified into applications based on its ion-exchange properties, membrane barrier behavior, and suitability as an immobilization matrix for biocomponents. Its ion-exchange characteristics have been applied in the preconcentration of cationic species[148] and in its use as a modifier matrix for cationic redox species and electrocatalysts.[30] The polymer was also used as a membrane barrier that excludes counterions in electroanalysis.[149,150] It was employed in a glucose biosensor designed as a dialysis membrane cover for GOx in a bovine serum albumin-glutaldehyde matrix.[28,151] This Nafion coating yields electrodes with a large linear range ($\leq$ 28 mM) and response times superior to those of cellulose as a dialysis membrane.[28] As a dialysis membrane coating on a GOx electrode for glucose determination in whole blood, Nafion selectively allowed the determination of the neutral glucose. Harrison *et al.*[28] reported reproducibility in the dialyzing action of the Nafion-coated electrode comparable to those of cellulose acetates and polyurethane films. Other groups used Nafion as a protective selective coating on electrodes in intracellular fluids in nerve and brain tissues to minimize ascorbic acid interference in determining various neurotransmitters.[149,152]

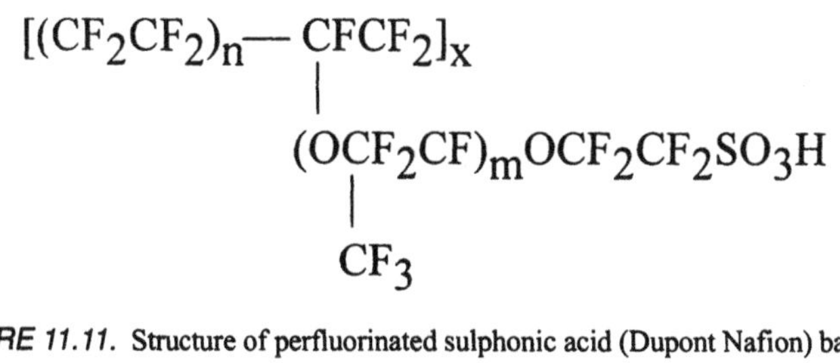

FIGURE 11.11. Structure of perfluorinated sulphonic acid (Dupont Nafion) backbone.

Nafion is also used as a suitable matrix for enzyme immobilization. In their study Fortier *et al.*[30] reported a one-step preparation of GOx/Nafion film on a Pt electrode. At its optimal performance conditions, the Pt/GOx/Nafion electrode also served as a membrane barrier to oxidizable anionic species. However GOx decreased the ability of the Nafion film to exclude anionic interferents.

5.2. Poly(ester Sulphonic Acid) Ion-Exchange Polymer (Eastman AQ Polymer)

Similar to the behavior of Nafion, poly(ester sulphonic acid) ionomers selectively exclude anionic species and large particles, but cations and neutral molecules are permeable.[153–155] Of the three Eastman Kodak AQ 29, AQ 38, and AQ 55 polymers,[26] AQ 55 (see the polymer backbone in Fig. 11.12) is the most studied and most applied in biosensor preparation.[153–165] Wang and coworkers[153–156] described have features of this polymer as an electrode material, including strong affinity toward hydrophobic counterions, prevention of electrode fouling from proteins, stability of the polymer film on the electrode, ability to preconcentrate catalysts in the film, and lowering the overpotential of many species difficult to oxidize or reduce. Several workers also showed that this poly(ester sulphonic acid) polymer is very stable as an organic phase electrode material.[159–165]

Eastman AQ 55 polymer was applied in trace analysis,[153] electrostatic binding of cations,[154] and as a coating on GOx-ferrocene/polysiloxane electrode to improve selectivity.[22] This latter application is based on the fact that unlike Nafion, which dissolves in methanol (and thus affects enzyme activity), the Eastman polymer can be dispersed easily in water. The permselective properties of the polyester ionomeric films make them valuable for *in vivo* electroanalysis. Wang and Lin[155] described a graphite epoxy microelectrode coated with polyester ionomeric films that was applied in sensing cationic neurotransmitters, such as dopamine in the presence of anionic ascorbic acid, without the latter interfering. Several enzymes were applied with AQ polymer films in biosensors. These include glucose oxidase, galactose oxidase, choline oxidase, L-amino acid oxidase, horseradish peroxidase, and tyrosinase.[104,157–165]

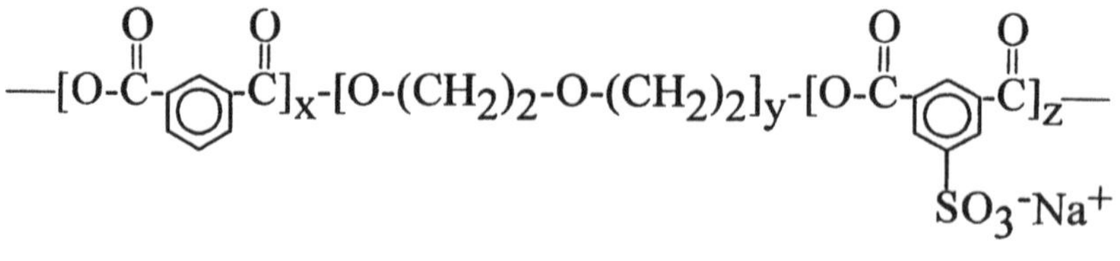

FIGURE 11.12. Structure of an Eastman AQ 55 poly(ester sodium sulphonate) ionomer backbone.

5.3. Organic-Phase Application of the Eastman AQ 55 Polymer

Starting with the work of Gennett and Purdy,[26] which indicated that Eastman AQ 55 polymer maintains its structural integrity and ion-exchange characteristics in acetonitrile, its application in organic-phase biosensors has received increasing attention. Phenol and peroxide detections in the organic phase were carried out with tyrosinase and HRP electrodes, respectively, in which the enzyme was entrapped in poly(ester sulphonic acid) film.[27,158] The upsurge in research involving the use of the AQ polymer in organic phases actually stemmed from the finding that enzyme electrodes are stable for organic-phase sensing. This was demonstrated for several enzymes, including cholesterol oxidase, peroxidases, polyphenol oxidase, tyrosi-

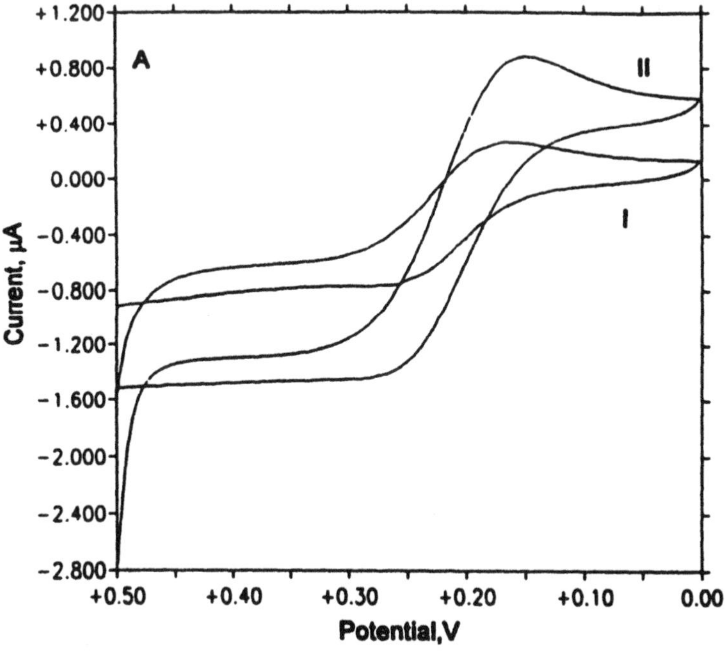

FIGURE 11.13. Cyclic voltammogram of 4-mM DMFc on a horseradish peroxidase (HRP)/Eastman AQ 55-modified glassy carbon electrode (GCE) in acetonitrile containing 1.5% v/v water at 20° C. (A) Response of the biosensor in the absence (I) and presence (II) of 1 mM butanone peroxide (BTP). (B) Response of the biosensor to 1 mM BTP (I) and on addition of 1 mM of the inhibitor, thiourea (II). A 20 μl drop of mixed polymer–enzyme solution containing 1.4% Eastman AQ55 polymer and 0.05 mg HRP was evaporated on a 3 mm diameter GCE; scan rate: 5 mV s^{-1}; support electrolyte: 0.1 M tetraethylammonium p-toluenesulphonate.

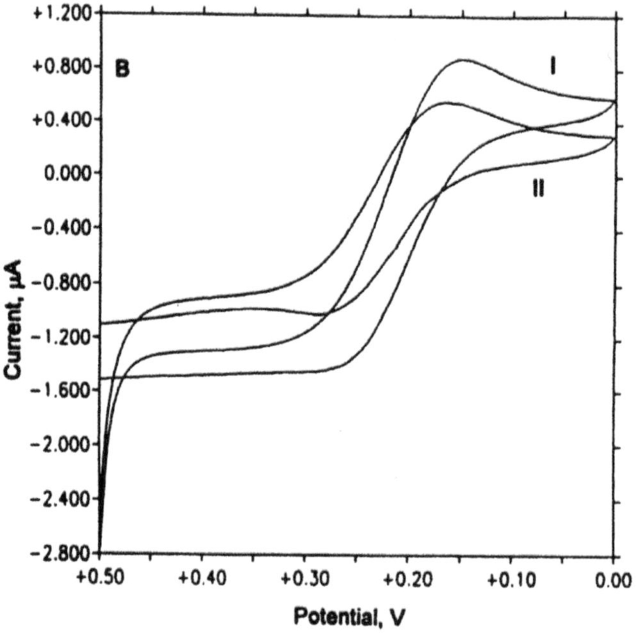

FIGURE 11.13. (Continued)

nase, and glucose oxidase.[27,116,166–171] Recently, in our laboratory, we applied biosensors prepared with the Eastman AQ 55 polymer to detect pesticide compounds, such as thiourea, ethylenethiourea, and mecaptoethanol.[162–165] Figure 11.13(a) shows the detection of butanone peroxide in acetonotrile using HRP immobilized in AQ 55. The inhibitory effects of thiourea on the biosensor is shown in Fig. 11.13(b) by the decrease in the butanone peroxide reduction current.

6. COMPOSITE ENZYME ELECTRODES

The search for enzyme electrodes whose modification matrices exhibit better mechanical and chemical properties, such as cohesion, adhesion, air stability, flexibility, selectivity, and sensitivity, has led to the design and application of composite enzyme films. Composite polymer films used in biosensors have two principal profiles, irrespective of whether the film formation is exclusively electrochemical or achieved by other physical or chemical techniques. These are multilayer and single-layer configurations.

6.1. Multilayer Composite Films

In this electrode type, layers of different polymeric materials are used to prepare the biosensor, with each layer contributing to overall sensor performance. For example, in preparing a glucose sensor, Rohde *et al.*(21) cross-linked Os-polymer with GOx on a platinum surface. This layer was then covered with an electrochemically grown conducting poly(pyrrole) layer containing GOx. The additional step, the authors argued, brought about a high loading of enzyme and improved the turnover of the analyte and stability of the electrode. It is also possible to form film layers by cross-linking enzymes and redox polymers on an electrode surface and then coating with Nafion or Eastman AQ polymer film.(104,107) The top layer behaves as a permselective phase, regulating and determining materials that reach the underlying phase where catalytic reaction takes place.

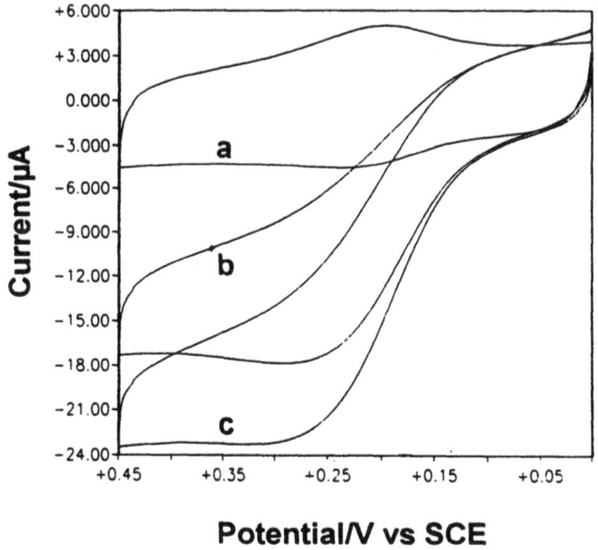

FIGURE 11.14. Cyclic voltammograms of a composite biosensor consisting of 125-mg graphite paste, 100-μl silicon oil, 250 μl of 1% methanolic solution of $[Os(bpy)_2(PVP)_{10}Cl]Cl$, and 125-mg GOx. Experimental conditions: 0.1-M phosphate buffer pH 7.2, potential sweep rate: 5 m V/s, E_i and E_λ: 0 and 450 mV, respectively. The CVs were obtained (a) in the absence, presence of (b) 10 mM and (c) 20 mM of glucose. (From Ref. 176).

6.2. Single-Layer Composite Films

Biosensors with composite films can be prepared by electropolymerization from a solution containing a preformed redox or ion-exchange polymer, enzymes and monomers of the encapsulating polymer. The procedure yields a single homogeneous layer of polymer film that contains all the biosensor reagents. Sung and Huang[172] have reported an interference-free sensor consisting of a Nafion-poly(aniline) composite film electrochemically deposited. Recently,[38] we prepared a glucose sensor by electrochemical formation of a composite film of Os-polymer/GOx/poly(phenol). The Os-polymer forming an electrostatic complex with GOx,[116] is entrapped in the poly(phenol) film during electropolymerization. The CV of the formation of the composite electrochemical film is shown in Fig. 11.8. Polymerization current for this film is lower than that obtained when only GOx is entrapped in the poly(phenol) film. Fast sensor response and storage stability are important features of the biosensor.

Biocomposite films are also formed by redox polymers and enzymes with carbon paste or graphite epoxy resin matrices.[104, 173–177] Hale *et al.*[104,173] have described this type of reagentless sensor based on the biocomposite containing

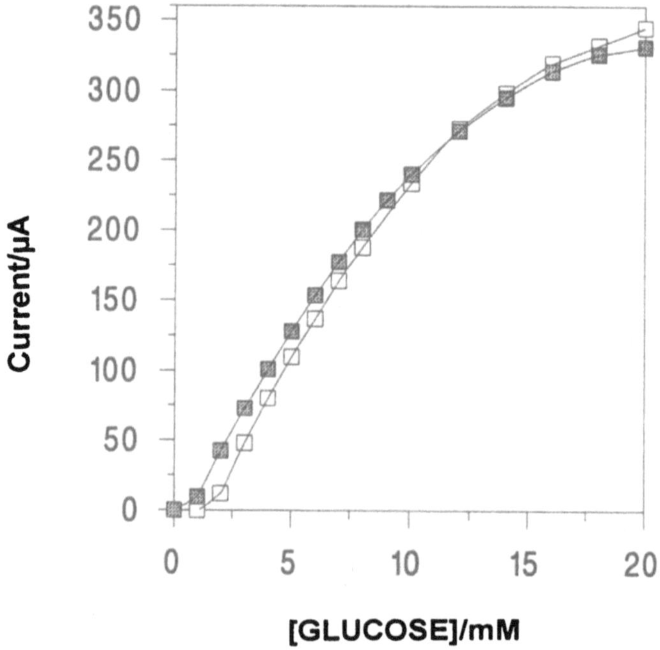

FIGURE 11.15. Glucose calibration curves for the composite biosensor described in Fig. 11.14. The electrode was polarized at + 400 mV versus Ag/AgCl in nitrogen-■ and air-saturated □ solutions.

GOx, carbon paste, and ferrocene-modified poly(ethylene oxide) or ferrocenyl siloxane polymers. The GOx–Os-polymer electrostatic complex was also incorporated into a carbon paste matrix (Figs. 11.14 and 11.15).(176) Electrodes based on tyrosinase in carbon paste were also reported.(177) One advantage of this kind of sensor is that the surface is renewable by polishing. Closely related to composites containing redox polymers in solid paste matrices is the incorporation of an enzyme into a graphite epoxy resin matrix.(175,178–182) These authors report mechanical and storage stability, and fast sensor response as some of the advantages of the enzyme electrode. The epoxy resin composite electrode was applied in biosensors for phenol,(175) glucose,(182) and alcohol(180) determinations.

7. CONCLUSION

Chemical modification of electrodes with polymeric films has expanded the scope and application of electroanalytical techniques. The availability of many polymeric materials and procedures for constructing polymer-based biosensors have provided a lot of options for use in analytical systems. For example, species that are electrochemically unstable at polymerization potentials may be entrapped in polymer films grown by gamma irradiation or immobilized in preformed polymer matrices. Electropolymerization techniques for biosensor preparation are particularly attractive because they offer easy, versatile, and controllable means for immobilizing the bio-components and redox components of sensing devices. Through this approach film thickness can be determined from the polymerization charge, and the composition and morphology of the modifying film can be tailor-made for detecting a particular analyte in a specific matrix.

Recent applications of nonconducting polymers, such as PPD and overoxidized poly(pyrrole), as permselective and biocompatible membranes hold great promise for the future of biosensors used for *in vivo* monitoring. Also the suitability of polymeric films (e.g., Eastman AQ 55) for organic-phase biosensors has led to a new opportunity for amperometric detection of analytes in real nonaqueous matrices. Since enzymes are stable in nonaqueous media, many analytes can be detected amperometrically with organic-phase biosensors.

Despite advances in the research on polymer-based biosensors, the need for sensor systems that are not only very sensitive but also interference free still remains a challenge. Coupled to this problem is the issue of storage stability of biosensors, which is a prerequisite for mass producing amperometric sensors. Another challenge for biosensor technology is that the mechanism for entrapping biocomponents in polymeric films grown *in situ* is still not well understood. Solving this problem may lead to better control of the distribution and activity of enzyme within the polymer matrix and long-term stability of biosensors.

REFERENCES

1. S. J. Updike and C. P. Hicks, *Anal. Chem.* **38**, 726 (1966).
2. R. F. Lane and A. T. Hubbard, *J. Phys. Chem.* **77**, 1401 (1973).
3. R. F. Lane and A. T. Hubbard, *J. Phys. Chem.* **77**, 1411 (1973).
4. E. Wang, H. Ji, and W. Hou, *Electroanalysis* **3**, 1 (1991).
5. J. Labuda, *Sel. Electr. Rev.* **14**, 33 (1992).
6. L. Gorton, E. Csöregi, E. Domínguez, J. Emnéus, G. Jönsson-Petterson, G. Marko-Varga, and B. Persson, *Anal. Chim. Acta* **250**, 203 (1991).
7. S. A. Wring, and J. P. Hart, *Analyst* **117**, 1215 (1992).
8. J. Wang, *Electroanalysis* **3**, 255 (1991).
9. H-L. Schmidt, W. Schuhmann, F. W. Scheller, and F. Schubert, in *Sensors: A Comprehensive Survey*, vol. 3, *Chemical and Biological Sensors, Part II* (W. Gopel, J. Hesse, J. N. Zemel, W. Gopel, T. A. Jones, M. Kleitz, I. Lundtromand, T. Seiyama, eds.) (VCH, Weinheim, 1992), pp. 717–817.
10. J. Kulys, in *Advances in Biosensors*, vol. 1 (A. P. F. Turner, ed.) (JAI Press Ltd., London, 1991), pp. 107–24.
11. A. M. Yacynch, in *Advances in Biosensors*, vol. 2 (A. P. F. Turner, ed.) (JAI Press Ltd., London, 1992), pp. 1–52.
12. F. Scheller, F. Schubert, R. Renneberg, H-G. Mueller, M. Jaenchen, and H. Weise, *Biosensors* **1**, 135 (1985).
13. R. S. Sethi, *Biosen. Bioelectron.* **9**, 234 (1994).
14. R. W. Murray, in *Electroanalytical Chemistry*, vol. 13 (A. J. Bard, ed.) (Marcel Dekker, New York, 1984), pp. 191.
15. J. Q. Chambers, *J. Electroanal. Chem. Interfacial. Electrochem.* **130**, 381 (1981).
16. N. Oyama and F. C. Anson, *J. Am. Chem. Soc.* **101**, 739 (1979).
17. G. J. Samuels and T. J. Meyer, *J. Am. Chem. Soc.* **103**, 307 (1981).
18. K. D. Gleria and H. A. O. Hill, in *Advances in Biosensors*, vol. 2 (A. P. F. Turner, ed.) (JAI Press Ltd., London, 1992), pp. 53–78.
19. B. A. Gregg and A. Heller, *Anal. Chem.* **62**, 258 (1990).
20. A. Aoki and A. Heller, *J. Phys. Chem.* **97**, 11014 (1993).
21. E. Rohde, E. Dempsey, M. R. Smyth, J. G. Vos, and H. Emons, *Anal. Chim. Acta* **278**, 5 (1993).
22. L. Gorton, H. I. Karan, P. D. Hale, T. Inagaki, Y. Okamato, and T. A. Skothiem, *Anal. Chim. Acta* **228**, 23 (1990).
23. P. D. Hale, L. I. Boguslavsky, T. Inagaki, H. I. Karan, H. S. Lee, T. A. Skotheim, and Y. Okamoto, *Anal. Chem.* **63**, 677 (1991).
24. P. D. Hale, H. S. Lee, and Y. Okamoto, *Anal. Lett.* **26**, 1 (1993).
25. J. Wang and T. Golden, *Anal. Chem.* **61**, 1397 (1989).
26. T. Gennett and W. C. Purdy, *Anal. Chem.* **62**, 2155 (1990).
27. J. Wang, Y. H. Lin, and Q. Chen, *Electroanalysis* **5**, 23 (1993).
28. D. J. Harrison, R. F. B. Turner, and H. P. Bates, *Anal. Chem.* **60**, 2002 (1988).
29. H. Gunasingham and C. B. Tan, *Analyst* **114**, 695 (1989).
30. G. Fortier, M. Vaillancourt, and D. Bélanger, *Electroanalysis* **4**, 275 (1992).
31. P. J. Peerce and A. J. Bard, *J. Electroanal. Chem. Interfacial. Electrochem.* **108**, 121 (1980).
32. C. D. Ellis, W. R. Murphy, Jr., and T. J. Meyer, *J. Am. Chem. Soc.* **103**, 7480 (1981).
33. P. Denisevch, K. W. William, and R. W. Murray, *J. Am. Chem. Soc.* **103**, 4727 (1981).
34. T. Ikeda, C. R. Leidner, and R. W. Murray, *J. Am. Chem. Soc.* **103**, 7422 (1981).
35. A. Ivaska, *Electroanalysis* **3**, 247 (1991).
36. P. N. Bartlett and J. M. Cooper, *J. Electroanal. Chem.* **362**, 1 (1993).
37. M. D. Imisides, R. John, P. J. Riley, and G. G. Wallace, *Electroanalysis* **3**, 879 (1991).

38. M. Pravda, C. M. Jungar, E. I. Iwuoha, M. R. Smyth, K. Vytras, and A. Ivaska, *Anal. Chim. Acta* **304**, 127 (1995).
39. G. G. Gilbault and R. D. Schmid, in *Advances in Biosensors*, vol. 1 (A. P. F. Turner, ed.) (JAI Press Ltd., London, 1991), pp. 257–89.
40. P. R. Coulet, in *Advances in Biosensors* vol. 1 (A. P. F. Turner, ed.) (JAI Press, Ltd., London, 1991), pp. 151–80.
41. G. G. Guilbault and J. Jordan, *CRC Reviews* **19**, 28 (1988).
42. J. A. Duine, J. Frank, and J. A. Jongejan, *Adv. Enzymol.* **59**, 169 (1987).
43. J. A. Duine, *Eur. J. Biochem.* **200**, 271 (1991).
44. J. A. Duine, J. Frank, and J. K. Van Zeeland, *FEBS Lett.* **108**, 443 (1979).
45. K. Matsushita, E. Shinagawa, O. Adachi, and M. Ameyam, *FEMS Microbiol. Lett.* **55**, 53 (1988).
46. P. D. Hale, T. Inagaki, H. S. Lee, H. I. Karan, Y. Okamoto, and T. A. Skotheim, *Anal. Chim. Acta* **228**, 31 (1990).
47. R. Wilson and A.P. F. Turner, *Biosen. Bioelectron* **7**, 165 (1992).
48. L. Ye, M. Hämmerle, A. J. J. Olsthoorn, W. Schuhmann, H-L. Schmidt, J. A. Duine, and A. Heller, *Anal. Chem.* **65**, 238 (1993).
49. L. C. Clark, Jr. and C. Lyons, *Ann. N.Y. Acad. Sci.* **102**, 29 (1962).
50. G. Joensson and L. Gorton, *Anal Lett.* **20**, 839 (1987).
51. B. F. Y. Yon-Hin, R. S. Sethi, and C. R. Lowe, *Sensors and Actuators B* **1**, 550 (1990).
52. K. Narasimham and L. B. Wingard, *Anal Chem.* **58**, 2989 (1986).
53. T. Kaku, H. I. Karan, and Y. Okamoto, *Anal. Chem.* **66**, 1231 (1994).
54. D. Bėlanger, J. Nadreu, and G. Fortier, *J. Electroanal. Chem.* **274**, 143 (1989).
55. M. Umana and J. Waller, *Anal. Chem.* **58**, 2979 (1986).
56. C. Malitesta, F. Palmisano, L. Torsi, and P. G. Zambonin, *Anal. Chem.* **62**, 2735 (1990).
57. S. V. Sasso, R. J. Pierce, R. Walla, and A. M. Yacynch, *Anal. Chem.* **62**, 1111 (1990).
58. J. P. Lowry and R. D. O'Neill, *Anal. Chem.* **64**, 453 (1992).
59. F. Palmisano and P. G. Zambonin, *Anal. Chem.* **65**, 2690 (1993).
60. J. P. Lowry, K. McAteer, S. S. El Atrash, A. Duff, and R. D. O'Neill, *Anal. Chem.* **66**, 1754 (1994).
61. T. Ito, H. Shirakawa, and S. lkeda, *J. Polym. Sci. Polym. Chem. Ed.* **12**, 11 (1974).
62. A. Dall'Olio, G. Dascola, V. Bocchi, and V. Varacca, *Compt. Rendu. Aca. Sci. Paris Ser. C* **267**, 443 (1968).
63. E. M. Genies, G. Bidan, and A. F. Diaz, *J. Electroanal. Chem.* **149**, 101 (1983).
64. E. A. H. Hall, *Int. J. Biochem.* **20**, 357 (1988).
65. N. C. Fould and C. R. Lowe, *J. Chem. Soc. Faraday Trans. I* **82**, 1259 (1986).
66. J. M. Dicks, S. Hattori, I. Karube, A. P. F. Turner, and T. Tokozawa, *Ann. Biol. Clin.* **47**, 607 (1989).
67. P. N. Bartlett and R. G. Whitaker, *J. Electroanal. Chem.* **224** 37 (1987).
68. P. N. Bartlett and R. G. Whitaker, *Biosensors* **3**, 359 (1987/88).
69. T. Taksuma, M. Gordaira, and T. Watanabe, *Anal. Chem.* **64**, 1183 (1992).
70. E. Dempsey, J. Wang, and M. R. Smyth, *Talanta* **40**, 445 (1993).
71. N. C. Foulds and C. R. Lowe, *Anal. Chem.* **60**, 2473 (1988).
72. P. D. Hale, T. Inagaki, H. S. Lee, H. I. Karan, Y. Okamoto, and T. A. Skotheim, *Anal. Chim. Acta* **251**, 31 (1991).
73. Y. Kajiya, H. Sugai, C. Iwakura, and H. Yoneyama, *Anal. Chem.* **63**, 49 (1991).
74. S. Yabuki, H. Shinohara, Y. Ikariyama, and M. Aizawa, *J. Electroanal. Chem.* **277**, 179 (1990).
75. J. Wang, S-P. Chen, and M. S. Lin, *J. Electroanal. Chem.* **273**, 231 (1989).
76. T. Schalkhammer, E. Mann-Buxbaun, G. Urban, and F. Pittner, *J. Chromatogr.* **510**, 355 (1990).
77. T. Schalkhammer, E. Mann-Buxbaun, F. Pittner, and G. Urban, *Sensors and Actuators B* **4**, 273 (1991).
78. W. Schuhmann, T. J. Ohara, H-L. Schmidt, and A. Heller, *J. Am. Chem. Soc.* **113**, 1394 (1991).
79. M. Hämmerle, W. Schuhmann, and H-L. Schmidt, *Sensors and Actuators B* **6**, 106 (1992).

80. J. C. Cooper, M. Hämmerle, W. Schuhmann, H.-L. Schmidt, *Biosen. Bioelectron.* **8**, 65 (1993).
81. J. Wang, S. P. Chen, and M. S. Lin, *J. Electroanal. Chem.* **64**, 231 (1989).
82. G. F. Khan, E. Kobalake, H. Sinohara, Y. Ikariyana, and M. Aizawa, *Anal. Chem.* **273**, 1254 (1992).
83. A. M. Yachynch, in *Advances in Biosensors*, vol. 2 (A. P. F. Turner, ed.) (JAI Press Ltd., London, 1992), pp. 1–52.
84. A. P. F. Turner, S. P. Hendry, M. F. Cardosi, and E. W. Neuse, *Diabetic Med.* **2**, 5 (1988).
85. T. Inagaki, H. S. Lee, P. D. Hale, T. A. Skotheim, and Y. Okamoto, *Macromolecules* **22**, 4641 (1989).
86. H. I. Karan, P. D. Hale, H. L. Lan, H. S. Lee, L. F. Liu, T. A. Skotheim, and Y. Okamoto, *Polym. Adv. Technol.* **2**, 229 (1991).
87. N. K. Cenas, A. K. Poius and J. J. Kulys, *Bioelectrochem. Bioenerg.* **12**, 583 (1984).
88. J. J. Kulys, *Biosensors* **2**, 3 (1986).
89. M. Umana and J. Waller, *Anal. Chem.* **38**, 2979 (1986).
90. P. N. Bartlett, Z. Ali, and V. Eastwick-Field, *J. Chem. Soc. Faraday Trans.* **88**, 2677 (1992).
91. P. N. Bartlett and D. J. Carana, *Analyst* **117**, 1287 (1992).
92. M. Kaneko and D. Wohrl, *Adv. Poly. Sci.* **84**, 143 (1988).
93. W. J. Albery and P. N. Bartlett, *J. Electroanal. Chem.* **131**, 137 (1982).
94. D. A. Buttry and F. C. Anson, *J. Am. Chem. Soc.* **106**, 59 (1984).
95. F. C. Anson, C-L. Ni, and J-M. Saveant, *J. Am. Chem. Soc.* **107**, 3442 (1985).
96. M. E. G. Lyons, D. E. McCormack, and P. N. Bartlett, *J. Electroanal. Chem.* **261**, 51 (1989).
97. P. D. Hale, T. Inagaki, H. I. Karan, Y. Okamoto, and T. A. Skotheim, *J. Am. Chem. Soc.* **111**, 3482 (1989).
98. M. W. Pishko, I. Katakis, S-L. Lindquist, L. Ye, B. A. Gregg, and A. Heller, *Agnew. Chem. Int. Ed. Engl.* **29**, 82 (1990).
99. A. Heller, *Acc. Chem. Res.* **23**, 128 (1990).
100. A. Heller, *J. Phys. Chem.* **96**, 3579 (1992).
101. W. Schuhmann, *Sensors and Actuators B* **4**, 41 (1991).
102. W. Schuhmann, R. Lammert, B. Uhe, and H-L. Schmidt, *Sensors and Actuators B* **1**, 537 (1990).
103. W. Schuhmann and H-L. Schmidt, in *Advances in Biosensors*, vol. 2 (A. P. F. Turner, ed.) (JAI Press Ltd., London, 1992), pp. 79–130.
104. P. D. Hale, L. I. Boguslavsky, T. A. Skotheim, L. F. Liu, H. S. Lee, H. I. Karan, H. L. Lan, and Y. Okamoto, in *Biosensors and Chemical Sensors: Optimizing Performance through Polymeric Materials*, vol. 487 (P. G. Edelman and J. Wang, eds.) (ACS Symposium Series, ACS, New York 1992), chap. 9, pp. 111–24.
105. A. E. G. Cass, G. D. Davis, G. D. Frances, H. A. O. Hill, W. J. Aston, I. J. Higgins, E. V. Plotkin, L. D. L. Scott and A. P. F. Turner, *Anal. Chem.* **56**, 667 (1984).
106. A. Badia, R. Carlini, A. Fernandez F. Battaglini, S. R. Mikkelsen, and A. M. English, *J. Am. Chem. Soc.* **115**, 7053 (1993).
107. J. M. Dicks, M. F. Cardosi, A. P. F. Turner, and I. Karube, *Electroanalysis* **5**, 1 (1993).
108. A. H. Heller, *J. Am. Chem. Soc.* **111**, 2357 (1989).
109. B. A. Gregg and A. Heller, *Anal. Chem.* **62**, 258 (1990).
110. B. A. Gregg and A. Heller, *J. Physical Chem.* **95**, 5970 (1991).
111. M. V. Pishko, A. C. Michael, and A. Heller, *Anal. Chem.* **63**, 2268 (1991).
112. E. J. Calvo, C. Danilowicz, and L. Diaz, *J. Chem. Soc. Faraday Trans.* **89**, 377 (1993).
113. R. Wilson and A. P. F. Turner, *Biosen. Bioelectron.* **7**, 165 (1992).
114. K. R. Frederick, J. Tung, R. S. Emeric, F. R. Masiarz, S. H. Chamberlin, A. Vasavada, S. Rosenberg, S. Chakraborty, L. M. Schopter, and V. Massey, *J. Biol. Chem.* **265**, 3793 (1990).
115. R. J. Foster and J. G. Vos, *Macromolecules* **59**, 4372 (1990).
116. E. I. Iwuoha, M. R. Smyth, and J. G. Vos, *Electroanalysis* **6**, 932 (1994).
117. B. A. Gregg and A. J. Heller, *Phys. Chem.* **95**, 5976 (1991).

118. F. Palmisano, D. Centonze, A. Guerrieri, and P. G. Zambonin, *Biosen. Bioelectron.* **8**, 393 (1993).
119. L. Doretti, D. Ferrara, and S. Lora, *Biosen. Bioelectron.* **8**, 443 (1993).
120. A. Heller and Y. Degani, in *Redox Chemistry and Interfacial Behavior of Biological Molecules*, (G. Dryhurst, and K. Niki, eds.) (Plenum Press, New York, 1988), pp. 151–71.
121. R. J. Geise, J. M. Adams, N. J. Barone, and M. Yacynch, *Biosen. Bioelectron* **6**, 151 (1991).
122. P. W. Stoecker and A. M. Yacynch, *Sel. Electr. Rev.* **12**, 137 (1990).
123. R. J. Geise and A. M. Yacynch, in *Chemical Sensors and Microinstrumentation*, vol. 43 (R. W. Murray, R. E. Dessy, W. R. Heineman, J. Janata, and W. R. Seitz, eds.) (ACS Symposium Series ACS, New York, 1989), pp. 65–77.
124. G. M. Heider, S. V. Sasso, K-M. Huang, A. M. Yacynch, and H. J. Wieck, *Anal. Chem.* **62**, 1106 (1990).
125. A. M. Yacynch, S. V. Sasso, E. R. Reynolds, and R. J. Geise, in *Proceedings of Biosensors International Workshop*, vol. 10, (R. D. Schmid, ed.) (GBF Monographs, VCH Publishers, New York, 1987) pp. 69–74.
126. G. Palleschi, M. A. Nabirahni, G. J. Lubrano, J. N. Ngwainbi, and G. C. Guilbault, *Anal. Biochem.* **159**, 114 (1986).
127. J. Wang, N. Naser, and C. L. Renschler, *Anal. Lett.* **26**, 1333 (1993).
128. E. Dempsey, J. Wang, and M. R. Smyth, *Talanta* **40**, 445 (1993).
129. D. Centonze, A. Guerrieri, C. Malitesta, F. Palmisano, and P. G. Zambonin, *Fresenius J. Anal. Chem.* **342**, 729 (1992).
130. T. W. Sohn, P. W. Stoecker, W. Carp, and A. M. Yacynch, *Electroanalysis* **3**, 763 (1991).
131. P. N. Bartlett, P. Tebbutt, and C. Tyrrell, *Anal. Chem.* **64**, 138 (1992).
132. Y. Ohnuki, T. Oshaka, T. Matsuda, and N. Oyama, *J. Electroanal. Chem.* **158**, 55 (1984).
133. T. Oshaka, T. Hirokawa, H. Miyamoto, and N. Oyama, *Anal. Chem.* **59**, 1758 (1984).
134. J. Wang, S. P. Chen, and M. S. Lin, *J. Electroanal. Chem.* **273**, 231 (1989).
135. P. C. Pandey, *J. Chem. Soc. Faraday Trans. I* **84**, 2259 (1988).
136. P. N. Bartlett, Z. Ali, and V. Eastwick-Field. *J. Chem. Soc. Faraday Trans.* **88**, 2677 (1992).
137. I. M. Christie, P. H. Treloar, and P. Vadgama, *Anal. Chim. Acta* **269**, 65, (1992).
138. M. Yoshida, M. Kumakura, and I. Kaetsu, *Polymer* **20**, 9 (1979).
139. G. J. Moody, R. B. Oke, and J. D. R. Thomas, *Analyst* **95**, 910 (1970).
140. W. E. Morf and W. Simon, in *Ion-Selective Electrodes in Analytical Chemistry* vol. 1, (H. Freiser, ed.) (Plenum Press, New York, 1978), p. 236.
141. I. Kaetsu, M. Kumakura, T. Fujimura, M. Yoshida, M. Asano, N. Kasai, and M. Tamada, *Rad. Phys. Chem.* **27**, 245 (1986).
142. M. Carenza, S. Lora, G. Palma, E. Boccù, R. Largaiolli, and F. M. Veronese, *Rad. Phys. Chem.* **31**, 657 (1988).
143. C. Galiatsatos, Y. Ikariyama, J. E. Mark, and W. R. Heineman, *Biosen. Bioelectron.* **5**, 47 (1990).
144. K. Hajizadeh, H. B. Halsall, and W. R. Heineman, *Talanta* **38**, 37 (1991).
145. K. Hajizadeh, H. B. Halsall, and W. R. Heineman, *Anal. Chim. Acta* **243**, 23 (1991).
146. I. Gürsel and V. N. Hasirci, *Biomaterials* **13**, 150 (1992).
147. I. Rubinstein, *J. Electroanal. Chem.* **188**, 227 (1985).
148. L. C. Whiteley and C. R. Martin, *Anal. Chem.* **59**, 1746 (1987).
149. G. Nagy, G. A. Gerhardt, A. F. Oke, M. E. Rice, and R. N. Adams, *J. Electroanal. Chem.* **188**, 85 (1985).
150. E. W. Kristensen, G. K. Werner, and R. M. Wrightman, *Anal. Chem.* **59**, 1752 (1987).
151. D. S. Bindra and G. S. Wilson, *Anal. Chem.* **59**, 2566 (1987).
152. J. Wang and P. Tuzhi, *Anal. Chem.* **58**, 3257 (1986).
153. J. Wang and Z. Lu, *J. Electroanal. Chem.* **266**, 287 (1989).
154. J. Wang and T. Golden, *Anal. Chem.* **61**, 1397 (1989).
155. J. Wang and M. S. Lin, *Electroanalysis* **2**, 253 (1990).

156. J. Wang, D. Leech, M. Ozsoz, S. Martinez, and M. R. Smyth, *Anal. Chim. Acta* **245**, 139 (1991).
157. G. Fortier, R. Béliveau, E. Leblond, and D. Bélanger, *Anal. Lett.* **23**, 1607 (1990).
158. G. Bremle, B. Persson, and L. Gorton, *Electroanalysis* **3**, 77 (1991).
159. J. Wang, in *Uses of Immobilized Biological Compounds*, (G. G. Guilbault and M. Mascini, eds.) (Kluwer Academic Publishers, Dordrecht, 1993), pp. 255–62.
160. J. Wang, Y. Lin, and L. Chen, *Analyst* **118**, 277 (1993).
161. J. Wang, E. Dempsey, A. Eremenko, and M. R. Smyth, *Anal. Chim. Acta* **279**, 203 (1993).
162. O. Adeyoju, E. I. Iwuoha, and M. R. Smyth, *Anal. Lett.* **27**, 11 (1994).
163. O. Adeyoju, E. I. Iwuoha, and M. R. Smyth, *Talanta* **41**, 1603 (1994).
164. E. I. Iwuoha, O. Adeyoju, E. Dempsey, M. R. Smyth, J. Liu, and J. Wang, *Biosen. Bioelectron.* **10**, 661 (1995).
165. O. Adeyoju, E. I. Iwuoha, and M. R. Smyth, *Anal. Chim. Acta* **305**, 57 (1995).
166. S. Saini, G. F. Hall, M. E. A. Downs, and A. P. F. Turner, *Anal. Chim. Acta* **249**, 1 (1991).
167. G. F. Hall, D. J. Best, and A. P. F. Turner, *Anal. Chim. Acta* **213**, 113 (1988).
168. J. Wang and Y. Lin, *Anal. Chim. Acta* **271**, 53 (1993).
169. E. I. Iwuoha and M. R. Smyth, *Analyst* **119**, 265 (1994).
170. O. Adeyoju, E. I. Iwuoha, and M. R. Smyth, *Anal. Proc.* **31**, 19 (1994).
171. E. I. Iwuoha and M. R. Smyth, in *Uses of Immobilized Biological Compounds* (G. G. Guilbault and M. Mascini, eds.) (Kluwer Academic Publishers, Dordrecht, 1993), pp. 245–54.
172. J. Y. Sung and H. J. Huang, *Anal. Chim. Acta* **246**, 275 (1991).
173. P. D. Hale, H. L. Lan, L. I. Boguslavsky, H. I. Karan, Y. Okamoto, and T. A. Skotheim, *Anal. Chim. Acta* **251**, 121 (1991).
174. L. Boguslavsky, P. D. Hale, L. Geng, T. A. Skotheim, and H. S. Lee, *Solid State Ionics* **60**, 189 (1993).
175. J. Wang, L. Fang, and D. Lopez, *Analyst* **119**, 455 (1994).
176. M. Pravda, O. Adeyoju, E. I. Iwuoha, J. G. Vos, M. R. Smyth, and K. Vytras, *Electroanalysis* **7**, 619 (1995).
177. M. Bonakdar, J. Vilchez, and H. Mottola, *J. Electroanal. Chem.* **266**, 47 (1989).
178. S. Wring, J. Hart, and B. Birch, *Anal. Chim. Acta* **229**, 63 (1990).
179. J. Wang and K. Varughese, *Anal. Chem.* **62**, 318 (1990).
180. J. Wang, E. Gonzalez-Romero, and M. Ozsoz, *Electroanalysis* **4**, 539 (1992).
181. B. Shaw and K. Creasy, *Anal. Chem.* **60**, 1241 (1988).
182. F. Cespedes, E. Martinez-Fabregas, J. Bartoli, and S. Algeret, *Anal. Chim. Acta* **273**, 409 (1993).

Index